P. N. Tung
PO Box 1800
Vacaville, CA 95696-1800

P9-BYC-136

THE BOOK OF LIFE

THE BOOK OF LIFE

GENERAL EDITOR

Stephen Jay Gould

CONTRIBUTING SCIENTISTS

AND ILLUSTRATORS

Peter Andrews,

John Barber, Michael Benton,

Marianne Collins, Christine Janis, Ely Kish,

Akio Morishima, J. John Sepkoski Jr,

Christopher Stringer,

Jean-Paul Tibbles

WITH

Steve Cox

W. W. Norton & Company

New York • London

TITLE PAGE
In the late Cretaceous in Mongolia,
Tarbosaurus *chases* Saurolophus
into a lake.

New introduction © Stephen Jay Gould 2001
Preface © Stephen Jay Gould 1993
Chapter 1 © J. John Sepkoski, Jr 1993
Introduction and Chapters 2, 3, 4 © Michael Benton 1993
Chapter 5 © Christine Janis 1993
Chapter 6 © Peter Andrews and Christopher Stringer 1993
Paleontological reconstructions © Marianne Collins 1993, © Ely Kish 1993,
© John Barber 1993, © Akio Morishima 1993, © Jean-Paul Tibbles 1993

For information about permission to reproduce selections from this book, write to Permissions,
W. W. Norton & Company, Inc., 500 Fifth Avenue, New York, NY 10110

The text and display of this book are composed in Palatino
Designed by Paul Welti with Toucan Books Ltd.
Manufacturing by Colorprint Offset

Library of Congress Catalog Card Number: 93-028409

ISBN 0-393-05003-3 hc
ISBN 0-393-32156-8 pbk

W. W. Norton & Company, Inc., 500 Fifth Avenue, New York, N.Y. 10110
www.wwnorton.com

W. W. Norton & Company Ltd., 10 Coptic Street, London, WC1A 1PU

1 2 3 4 5 6 7 8 9 0

CONTENTS

A New Introduction

A FLAWED WORK IN PROGRESS

Stephen Jay Gould

The Book of Life made its first appearance way back in the last millennium, less than ten years ago. Both the book and its subject may be described by the title of this introduction—flawed works in progress. And, as we hope for continuity and perseverance for the macrocosm of the subject, so too might we designate the microcosm of this book as worthy of survival (at least for a moment longer, geologically speaking). Life itself needs no explicit rationale, but the reissue of a book does require some overt validation, a task perhaps best fulfilled by recalling Kant's famous aphorism that "percepts without concepts are blind, but concepts without percepts are empty."

Only a desiccated and lazy scholar could feel entirely at ease with a proposal to reissue, without a thorough overhaul and revision of content, a scientific work in a field that has advanced so much in such a short interval of time—a marvelous sign of vitality and health for the discipline of paleontology, but also a guarantee of swift obsolescence for products that embodied the height of excellence and modernity of thought just a few years ago in 1993.

Moreover, and now speaking metaphorically, this editor must be especially wary since this reissue emerges on the other side of a millennial transition, and the classic text for millennial prophecy (Revelation, Chapter 20) tells us that, following the completion of the blessed millennium (the 1,000-year future reign of Christ, not the secular stretch that we have just entered in the year 2000), the "dead shall be raised incorruptible"—to appear before God's throne in a Last Judgment that shall allocate the good guys to heaven and relegate the less worthy to that fiery inferno where, in compensation, they may at least enjoy the eternal company of most of history's truly interesting characters. God, we are further told, will make his judgment by reading our deeds from an opened book that just happens to bear the same title as this work—so I feel especially wary about the reissue of this volume. Indeed, I'd better watch out, and neither shout nor pout, for you never know who might be coming to town. We read in Revelation (20:12, 15):

And another book was opened, which is *the book of life:* and the dead were judged out of those things which were written in the books, according to their works . . . and whosoever was not found written in the book of life was cast into the lake of fire.

Well, we haven't written everyone into this *Book of Life* because science can only operate as a work in progress without perfect knowledge, and we must therefore leave a great deal out from ignorance—especially in a historical field like paleontology, where we must work with the strictly limited evidence of a very imperfect fossil record. So I don't worry about things left out of this *Book of Life* because no one knows about them yet. I *do* have concern for things discovered since 1993 but still left out because we have reissued, rather than extensively revised and updated, this particular *Book of Life.*

These "things left out" reside in Kant's two inclusive categories of percepts and concepts, as mentioned in my first paragraph—and many important new items have accumulated since 1993. The editor of a reissued book like this has an obvious and primary duty to identify the major items in his introduction to any reprinted edition. But I am caught in the happy irony of woefully insufficient space—happy because the weight of novelty spells such robust health for my discipline, but woeful because the full weight cannot be adequately encompassed herein, making this book an even more flawed work in progress.

Nonetheless, I feel quite content—in fact, strongly positive—about reissuing *The Book of Life* in its original form. The factual and conceptual novelties that I shall outline below represent additions and sophistications, not overturns and replacements that relegate the 1993 contents to serious and substantial error. In other words, the 1993 text is incomplete by the standards of 2000 but not wrong. (The few flat-out errors of fact have been corrected.) In short, *The Book of Life* should not yet become extinct because, with all its flaws, nothing better has ever been produced in the important and popular genre of well-illustrated narratives for the history of life on Earth.

No other book has ever succeeded even this far in integrating the sequential tales of actual organisms with the history of their environments, their interactions with other organisms in ecological communities and habitats, and their utility in illustrating the principles of evolutionary theory (and, reciprocally, in providing key evidence for changes in the theory since its Darwinian foundation in 1859). Moreover, no previous book has featured as many scrupulously accurate pictures (within the limits of inference from incomplete fossil information) of extinct organisms across the full taxonomic range of life (and not only the dinosaurs that pique our interest, or the putative human ancestors that nurture our genealogical fascination)—all done with the aesthetic sensitivity of the best modern artists in the small, but vibrant, speciality of reclothing and reanimating the bones and shells of our forebears. This incomplete *Book of Life* stands proud as a work in progress. Better an informed child of promise than a fully fleshed incarnation of outdated ideas and images.

On the subject of primarily factual discoveries (Kant's percepts) unknown in 1993 and sufficiently important in the conceptual structure of our understanding to secure space in a short preface (in other words, more than mere tidbits of novel information without much conceptual weight, however attractive or intriguing), the cascade of so much new material in so few years has been nurtured by the coincidence of powerful new techniques and discoveries in three cognate fields: paleontology for direct data of ancient organisms, molecular genetics for reconstruction of evolutionary trees based on interrelationships among living organisms, and developmental biology for a revolution in our understanding of how genetic programs and organic materials become translated into complex adult organisms replete with signs of their evolutionary history.

Some important advances have been based almost entirely upon the discovery of novel fossils that either chronicle the existence of previously unknown (and often utterly unsuspected) forms or else extend the ages or habitats of known forms into previously unsuspected times or places. To cite four examples in chronological order, each with important theoretical implications of a distinctive sort:

1. *Precambrian embryos of bilaterian animals of modern design.* In discussing the Earth's first assemblage of multicellular animals—the Ediacara fauna that arose about 600 million years ago and may bear little relation to modern groups—*The Book of Life* states that animals of modern design probably lived in these faunas as well, even though extensive evidence for such creatures does not appear in the fossil record until the "Cambrian explosion," beginning about 530 million years ago. In support of this assertion, the 1993 text tells

us that organisms of modern design "left their mark, though not their bodies, at various Ediacaran locations" in the form of "trace fossils"—trails and burrows that could not have been made by Ediacaran creatures and must therefore represent the activity of coeval organisms of modern design. Now, in 2000, although we still have not recovered body fossils of adult bilaterian animals (mobile creatures of modern design and bilateral symmetry) from Ediacaran rocks, the most sensational paleontological find of the late 1990s revealed delicate embryos, showing preserved patterns of cell division known only in bilaterian animals of modern design. Moreover, these fossils lie deep in Ediacaran times—in beds some 570 million years old, thus proving the contemporaneity of Ediacaran creatures with bilaterian ancestors of modern phyla.

2. *Greatly expanded evidence of the Cambrian explosion.* Nearly all phyla of modern animals make their first appearance in the fossil record during a short span of only 5 to 10 million years, called the Cambrian explosion, and extending from about 530 to 520 million years ago. In 1993, virtually all our important evidence for the full range of results from this central episode in the history of life—a range that can only be recorded by rare preservation of soft anatomies and cannot be inferred from the usual fossil evidence of hard parts alone—came from the celebrated Burgess Shale of Canada, a deposit that postdates the explosion by some 10 million years. The 1993 text also mentions that these Burgess results have been "reinforced by finds in China, Greenland, and elsewhere," but it gives no details about the additional material because virtually none had been published. But now, less than ten years later, Chinese and Western scientists have published several books of descriptions and interpretations for the Chengjiang fauna—the Chinese equivalent of the Burgess Shale, but perhaps even more informative because soft anatomies are equally well preserved, and the Chengjiang fauna is several million years older than the Burgess, and probably lived right at the end of the explosion itself.

3. *New discoveries about the origin and early history of vertebrates.* The 1993 text identifies the Burgess shale *Pikaia* as the only known Cambrian member of the phylum that includes all vertebrates (even though *Pikaia* belongs to one of the phylum's "cousin" lineages as a maximally close relative, but not a true vertebrate itself). The text then identifies the jawless fish *Astraspis*, from Ordovician rocks of the next geological period (and some 30 million years younger), as the first true vertebrate—with nothing in between but some

ambiguous late Cambrian fragments regarded as parts of fishes by some paleontologists. Now, in 2000, the fossil record of early vertebrates has improved dramatically, and for two distinct reasons. First, new discoveries, particularly from the Chinese Chengjiang fauna, have revealed several other nonvertebrate members of our chordate phylum and, in the interpretation of most experts, even some genuine (if rudimentary) vertebrates as well. Second, the enigmatic conodonts, an abundant group of fossils (but of previously unresolved zoological affinity), have now been interpreted, with fair confidence, not only as chordates, but probably as true vertebrates as well. The fossil record of probable conodonts extends well back into the Cambrian period.

4. *The full sequence of intermediary forms to trace the evolution of whales.* The 1993 text uses information from both genetics and paleontology to site whales taxonomically among artiodactyl ungulates (large, hoofed, terrestrial mammals, including deer, cattle, and hippos), probably most closely related to (and perhaps directly descended from) an extinct group (carnivorous in lifestyle, but ungulate by genealogy) known as the mesonychids. The text also includes a lovely picture of the Eocene *Basilosaurus,* labeled as a "prototype" whale. But *Basilosaurus,* while retaining tiny and vestigial hind legs, is already a fully marine creature, and cannot be designated as a true structural intermediate between fully terrestrial mesonychid ancestors and our modern behemoths of the oceans.

In fact, the lack of fossil intermediates had often been cited by creationists as a supposedly prime example for their contention that intermediary forms not only haven't been found in the fossil record but can't even be conceived, because, or so they argued, how could a creature function in the nether-nether, in-between world of transition between a terrestrial ancestor and a marine descendant—that is, as half adapted for land, half for sea, and obviously, therefore, unfit for either? (I never understood the force of this argument because several groups of seals live perfectly well in this intermediary state today.) In any case, in one of the great triumphs of modern paleontological exploration, a lovely series of intermediary steps have now been found in rocks of the right age (50–55 million years old) and environments (shallow waters near shorelines) in Pakistan. This elegant series, giving the lie to creationist claims, includes the almost perfectly intermediate *Ambulocetus* (literally, the walking whale), a form with substantial rear legs to complement the front legs already known from many fossil whales, and

clearly well adapted both for swimming and for adequate, if limited, movement on land.

Other factual advances in the last decade have integrated new paleontological data with the molecular and developmental biology of modern organisms. On two ends of the taxonomic spectrum, for example, the family tree of procaryotic unicells (the simplest living creatures with no internal cellular structure, that is, no nucleus, chromosomes, mitochondria, etc.) has been revolutionized by the recognition that bacterial species of very distant genealogical relationship can nonetheless move genes from one to the other—a phenomenon called lateral transfer. (This process does not occur extensively in multicellular organisms—that is, nature does not make a new species of mammals by mixing half a dolphin's genome with half a bat's and evolving a creature that can both swim and fly.) Lateral transfer represents a strikingly different mode of evolution that violates the basic topology of life's treelike structure (with its implication of continual separation in branching) by permitting distantly related lineages to share their genetic heritage.

Our ability to read the genetic sequences of organisms has also led to a probable resolution of a major issue that, in 1993, I thought would not be answered in our lifetime, if ever—the genealogical relationships among the major phyla of animals. The first fossils of these creatures already show full separation at their initial appearance in the Cambrian explosion—so I expected no paleontological resolution. Moreover, I thought, the potential rapidity of separation for these lineages from their last common ancestor probably lies beyond the resolving power of evolutionary trees inferred from genetic differences among modern organisms.

But the power and sophistication of genetic techniques has already reached a point where this "holy grail" among taxonomic questions—the interrelationships among the animal phyla—may already have yielded its basic solution. The cnidarians (corals and their allies) do seem to be the sister group (as long suspected) of all triploblast phyla (modern bilaterally symmetrical and mobile organisms). Among triploblasts, the protostomes (mollusks, arthropods, and nearly all other "invertebrate" phyla) are indeed separate from the deuterostomes (our chordate lineage and the echinoderm phylum among "invertebrates"), as long inferred from anatomical and paleontological evidence. Moreover, among the protostome phyla—the issue that I did not expect to be resolved—a basic division into two great groups now seems well established: an "ocdvsozoan" line of phyla that molt their external skeletons (mostly arthropods and related groups) and a "lophotrochozoan" line (mollusks, annelids, and related groups) with distinctive larval forms and feeding structures. Surprisingly, the annelids (segmented worms) rank with the mollusks, and not with the arthropods—thus disproving the so-called great group of Articulata (annelids plus arthropods) that had been a staple of taxonomy since Cuvier's proposal in the early nineteenth century.

Moving from the reading of genetic sequences themselves to the translation of these sequences into adult organisms during growth, the field of developmental biology has contributed the most startling and novel information to evolutionary studies during the past decade. As the most general statement of the import of these discoveries, we now know that distantly related phyla—arthropods and vertebrates for example, to cite the most striking case of evident, if parochial, concern to us—share (by conserved descent from their common ancestry in Precambrian times) far more genetic instructions for building basic features of anatomy than conventional views had thought possible (for example, establishment of fundamental topologies of front and back, top and bottom; segmentation; and even more specific features like the underlying developmental pathways for the generation of eyes).

Standard Darwinian arguments had long regarded such differently constructed and anciently separated lineages as genealogically independent for too long, and too extensively adapted to different habitats and functions, to share any conserved genetic heritage for the construction of basic anatomy. But, quite to the contrary, certain aspects of body segmentation now seem to be homologous (retained from common ancestry) in vertebrates and insects. Even the classic textbook cases of "convergence" (completely independent evolution of similar adaptations in entirely separate lineages) turn out to illustrate not so much the wondrously detailed craftsmanship of natural selection in building such similar forms from several independent starting points (the old interpretation) but more the power of retained history to channel future developments. The overt anatomy of the squid eye and the human eye, for example, must be separately evolved (since the strikingly similar structures arise from different tissue layers); but nature's capacity to craft such independent similarity not only expresses the power of external natural selection, it also records the influence of internally retained history—for we have learned, since 1993, that squids and vertebrates (and insects as well, for a third case) hold important genetic pathways in common for the development of eyes. In fact, either the human or the squid gene can, when expressed in flies, "turn on"

the appropriate "downstream" fly genes to produce a normal fly eye.

In the second Kantian domain of concepts that give "sight" and meaning to percepts, I have already discussed the important ideas that emerge from the factual discoveries previously highlighted—the radical concepts of lateral transfer in bacterial evolution and the expanded role for retained history in developmental homologies between insects and vertebrates; as well as the confirmatory concepts of adaptive continuity in major evolutionary transitions between strikingly different habitats, now documented in the evolution of whales.

But if I wished to present an honest criticism of concepts underlying this book's design and basic execution, and if I wished to anticipate the truly thorough revision that time and courage may allow in the near future, I would make the following suggestion (already advanced, albeit in undeveloped and tentative fashion, in my 1993 introductory essay): This book broke new ground in many ways and will surely rank as more innovatory than hidebound in the judgment of future generations. But, on one of the most crucial of all issues, we punted and bowed to a convention so entrenched that we ourselves never thought to mount a challenge. That is, we followed the two great traditions of all narrative accounts of the history of life. First, we proceeded chronologically. Second, our order of chapters followed the seemingly invariable sequence of invertebrates first, followed by discussions of fishes, reptiles (including dinosaurs), and mammals, and ending with a final treatment of human evolution. This sequence inevitably supports the oldest of all incorrect and biased constructions of life's history—the idea that evolution's central and defining thrust yields a history of progressively increasing complexity toward a predictable human pinnacle.

I will continue to defend the ordering of chapters by chronology, for time must remain the fundamental matrix for any narration, and the history of life is—fundamentally, and as its last two syllables assert—*a story*. Stories unfold in time, and this defining framework must be honored. But what kind of a story does life's history tell? On this basic question, I think that our conventional accounts—from monad to man, or from amoeba to America, to include some chauvinistic biases in our linguistic rendering as well—have not only proven woefully inadequate as accounts of life's grandeur and richness, but have also fed our oldest and most comforting biases in validating inherent human superiority and domination. The primary task of intellectual work, I believe, must lie in a responsibility to challenge such conventions—and in this sense, *The Book of Life* fails.

Yes, humans evolved late—so our particular story must appear at or near the end of a chronological account. But the convention of dropping invertebrates from the text once fishes evolve, then dropping fishes once terrestrial vertebrates appear, then dropping reptiles as soon as mammals replace dinosaurs, and then ending with a chapter exclusively devoted to a single species (albeit a fascinating creature, and the only animal capable of presenting such an account in any case) can only support the false impression that evolution progresses toward us, and this sequence represents a privileged and definitional pathway among the billions of lineages represented by branches on the grand tree of life.

Invertebrates never "went away" or stopped evolving, even though our narrative texts largely omit them after the first appearance of fishes. Nearly 90 percent of all living animal species belong to the arthropod phylum. Forty thousand species of vertebrates represent a small contribution to the totality of animal life. Moreover, more than 20,000 of these vertebrate species live in the sea as fishes, and also get erased from our narrative books after the origin of terrestrial vertebrates. Finally, and most important, the bulk of the organic world is now, and always has been, bacterial—the true titans and rulers of the planet by any legitimately evolutionary and geological criterion of length of dominion, prospects for survival, tenacity, sheer abundance, genetic and biochemical diversity, or range of current habitats (from nearly boiling waters in Yellowstone Park to pore spaces within rocks lying two miles beneath the Earth's surface).

How could we write a *Book of Life* that would honor (or even just properly depict) the true diversity of life's expansions and contractions within all groups through time? As a first requirement, we would surely have to abandon the conventional sequence of chapters that suggests a ladder rising predictably toward an exalted human state as the controlling and overarching theme of our narrative. We would need, as the first principle of iconographic reform, to substitute a luxuriantly (and largely unpredictably) branching tree for an inevitably rising line. Admittedly, standard books must be linear, and themes of multiple branching fare less well than tales of unilinear direction within such a format. But other professions have solved the problem of presenting a fundamentally branching story in the linear format of a book—most notably, the *Oxford English Dictionary,* which presents a strictly chronological account for each meaning of a word but traces the numerous and multiple branching definitions of single words by a complex, yet easily comprehensible, system of naming for branches and subbranches (admittedly in a necessarily linear sequence).

I don't know how we should write the *Book of Life* in a notably different format, based upon a substantially different iconography (branchings rather than ladders). For starters, I would certainly add a Chapter 7 to follow the current and ultimate Chapter 6 on human evolution—a new finale devoted to modern bacteria (with subsections on the less important plants, fungi, and invertebrate animals). But this largely cosmetic and merely additive stopgap cannot suffice. Trees must certainly replace ladders as a central concept—if only so that rodents, bats, and antelopes, the true success stories of mammalian evolution, can finally achieve their due by wresting primary notice from our absurd tendency to honor depleted lineages of single surviving twigs, with humans and horses as the standard textbook examples, merely because our prejudices lead us to confuse the single twigs of desiccated and dying lineages with acmes of linear trends.

But even conventional trees nurture our biases, for we so easily conflate upward (meaning only geologically more recent) with better—thus relegating such magnificently successful (but now extinct) groups as dinosaurs to metaphorical union with failed businesses and outmoded thought. Some historians of life have experimented with branching systems that radiate in a circle from a common point, with time now represented as a set of concentric circles rather than a single upward direction. But these schemes entail further problems, particularly in the suggestion (as concentric circles must increase in circumference) that temporally later must imply richer in diversity—another, if less noxious, incarnation of progressivistic bias.

I do not know the best way for telling this most wonderful and intricate of all stories (although I hope that we may spotlight the conventions that should be avoided, or at least be labeled as traditions rather than facts, as we explore a range of alternatives and innovations). But I do know that the history of earthly life is the greatest story ever told (or at least the greatest story that we know how to explore with a rich storehouse of factual information). *The Book of Life* therefore deserves and demands the best and most imaginative format that a unique (at least so far) and astonishing evolutionary invention on one tiny, accidental, and late-arising branch of a copious tree can devise to tell the full story with maximal respect for the entire glorious structure.

RECONSTRUCTING (AND DECONSTRUCTING) THE PAST

Stephen Jay Gould

Framed on the bias

The Great Exhibition of 1851 did wonders for the morale of two central figures in Victoria's Britain – for her husband Prince Albert, who directed this magnificent show of might and industry at the Crystal Palace, and who thereby won respect and accolades from his previously suspicious subjects; and for Charles Darwin, a frequent visitor, who viewed this vast yet transparent edifice as a sign that his previously fragile nation had now become a stable and fertile ground for an intellectual revolution that he had been guarding in silence since the late 1830s.

When the Exhibition closed at its original home in Hyde Park, workmen dismantled the innovative modular building of steel and glass and re-erected the Crystal Palace in the London suburb of Sydenham. Among the varied attractions commissioned for the grounds of the Crystal Palace's new home, none was as spectacular, as innovative, as fecund, and as enduring as the set of life-sized models of prehistoric beasts built by the London sculptor Waterhouse Hawkins (1807–89) with the close collaboration of England's greatest anatomist, Richard Owen (1804–92), inventor of the term "dinosaur."

A Victorian view of the model dinosaurs at Crystal Palace.

The Crystal Palace was destroyed by fire in 1936, but Hawkins's models are still in Sydenham (recently brightened, in fact, by a welcome restoration and a fresh coat of paint). They still stand in their naturalistic settings, in and around an artificial pond (which, in Hawkins's original plan, once ebbed and flowed to mimic an oceanic tide, thus exposing and submerging the ichthyosaurs and plesiosaurs). Their present setting may even surpass the original situation, for surrounding trees, many planted for the site, have flourished and now envelop the *Iguanodon* and *Megalosaurus* in a veil of mystery.

I have twice made my pilgrimage to Crystal Palace by commuter train from London. And I have been both awed and amused by the current display – awed that such an extensive and impressive project was realized just ten years after Owen coined the term "dinosaur," but amused at the inevitable errors. The ichthyosaurs and plesiosaurs are reconstructed as denizens of the shoreline, half in and half out of the shallow waters – though they lived in the open seas. (The later discovery of ichthyosaur dorsal and tail fins proved the elegance of their hydrodynamic design for powerful swimming.) *Iguanodon* still bears the infamous spike on its nose (actually a structure of the organism's hand, but misplaced atop from the first discovery), and walks on four legs (though we now recognize the animal as a biped).

These errors are just the usual stuff of imperfect knowledge. *Iguanodon's* spike was an isolated fragment, so who knew where it went? We could not identify ichthyosaur fins (as they lack bony supports) until the magnificent fossils of the Holzmaden, preserving outlines of soft body parts, were discovered later in the nineteenth century.

But another error at Sydenham strikes me as far more instructive, for it records,

concretely, the kind of interaction between science and human life that makes the subject of fossil iconography so interesting (and also sets the primary theme of this introductory essay). Note Hawkins's reconstruction of *Labyrinthodon*, an early amphibian. We now know that this animal was elongate, with four roughly equal legs. But Hawkins, who had little more than a skull to guide his work, reconstructed the animal as the canonical amphibian of our own times – as a frog, with powerful jumping hind limbs and a shortened body. The point is obvious, but profound – we reconstruct according to our own prejudices and our standard images.

The chronicle of changing restorations for fossil beasts therefore becomes a fascinating epitome of our social and intellectual history as well. The interplay of these two factors – the externally empirical and the internally social – captures the central dynamic of change in the history of science.

On the one hand, we do gain objectively ascertainable knowledge as we learn more and more about the fossil record. *Iguanodon* did not have a spike on its nose, and *Ichthyosaurus* did sport a dorsal fin. When we learn these facts, we improve our restorations and thus record a genuine gain in knowledge. A common mythology, much promoted by professionals for self-serving benefit, argues that science always proceeds in this way – intrinsically and uniquely. The history of changing views should therefore record a simple progress to greater knowledge, mediated by our application of that infallible guide to empirical truth: the scientific method.

Yet, on the other hand, science must proceed in a social context and must be done by human beings enmeshed in the constraints of their culture, the throes of surrounding politics, and the hopes and dreams of their social and psychological construction. We scientists tend to be minimally aware of these human influences because the mythology of our profession proclaims that changing views are driven by universal reasoning applied to an accumulating arsenal of observations. But all scientific change is a complex and inseparable mixture of increasing knowledge and altered social circumstances.

Moreover, we must not conclude that, of these two spheres of influence, factual accumulation is a pure boon and social

context a pure impediment. Data are often misread (and never come to us as self-evident and unambiguous things-in-themselves) – while changing social views can shake old prejudices and free our minds for fruitful novelty. Charles Darwin derived natural selection more by wondering how he might transfer the *laissez-faire* principle of Adam Smith's economics into nature than by observing tortoises on the Galapagos Islands. The widespread abandonment of scientific racism after World War II owed as much to our contemplation of what Hitler had done with such doctrines as to any increment of genetic knowledge.

Iconography, in my view, provides the best domain for grasping this interplay of social and intellectual factors in the growth of knowledge – and the iconography of ancient beasts opens a revealing window upon ourselves, while providing a series of images for creatures of the distant past. Peer inside a *Brontosaurus*, and Pan or Puck peeps out.

Pictorial imagery catches us unawares because, as intellectuals, we are trained to analyze text and to treat drawings or photographs as trifling adjuncts. Thus, while we may pore over our words and examine them closely for biases and hidden meanings, we often view our pictures as frills and afterthoughts, simple illustrations of a natural reality or crutches for those who need a visual guide. We are most revealed in what we do not scrutinize.

The social construction of fossil iconography lies best exposed in the conventions that create such an enormous departure between scenes as sketched and any conceivable counterpart in nature. (The

A photograph by Stephen Jay Gould of the Ichthyosaur at Crystal Palace.

A study for the Yale mural, "The Age of Reason," by Rudolph Zallinger.

point is obvious and easily grasped once presented, but I have often been surprised to discover how many people do not recognize the difference – and the need for such disparity – between nature and our conventions for painting her. Many of us have been looking at such paintings all our lives and have come to accept them as accurate snapshots of the natural world.) All artistic genres follow social conventions, but few others also grapple with the assumption that finished products represent a natural reality. Consider just three conventions, all defendable, that distinguish painted fossil scenes from inferred actualities.

1. *Number*. At most natural moments in most places, nothing much is happening to rather few organisms. But illustrations of such reality would be empty and boring. Moreover, museums and books grant little space for painted scenes, so artists must make the most of limited opportunities. If I am allowed but one picture of a Mesozoic landscape, I will therefore try to wedge everything in – predator and prey, pond-dweller and mountain climber. Consider the famous mural at Yale University by Rudolph Zallinger. We accept the necessary pedagogical theme, and rarely consider that such scenes represent artistic convention rather than natural landscape.

2. *Activity*. There is an old quip that the life

of a soldier features long periods of boredom interspersed with rare moments of terror. We paint animals during their few incidents of interesting behavior, and our concept of "interesting" changes with time. The Victorians loved Tennyson's description of "nature red in tooth and claw," and, by social convention, shied away from scenes of mating. Their paintings of nature almost invariably feature predation as a centerpiece (usually sanitized in showing little blood and gore). Iconographies of the last twenty years, especially if done largely for children, tend to focus on themes that are more "politically correct" – maternal behavior, herding and helping.

The prototype for "scenes from deep time" (to use Martin Rudwick's felicitous phrase from the title of his recent book) is Henry de la Beche's *Duria antiquior* ("An Older Dorset"), first lithographed in 1830, but reproduced endlessly (in both legitimate and pirated editions), and also serving as a model, often shamelessly copied, for almost all later artists. De la Beche, English to the core despite his francophonic name, was the first director of the British Geological Survey. To be sure, he designed this figure partly with humorous intent (and as a charitable act to benefit, through sales, the impoverished collector Mary Anning who had provided so much aid to British paleontologists). But de la Beche's

image became the canonical figure of ancient life at the inception of this genre. Note how he follows both conventions of unnatural crowding and pervasive predation. Almost every creature is either a feaster or a meal, and the centerpiece of an ichthyosaur biting into the neck of a plesiosaur became the image *par excellence* of early-nineteenth-century reconstruction. (We must also note de la Beche's unconventionalities, particularly his depiction of feces descending from several of the larger beasts – a feature that most of his later plagiarists eliminated.)

3. *Emphasis*. We switch now from necessary artistic conventions (though still distorting reality and promoting false impressions) to the primary social influence (for sales and acceptability) that makes these tableaux such a biased representation of the fossil world. Consider life's full history, at least from the beginning of modern multicellular animals (already a biased emphasis) more than 500 million years ago. Taxonomists have described more than a million species (most of them insects), divided into more than 20 phyla. Of this plethora, vertebrates represent only part of one phylum, and a mere 40,000 species or so. We are but one branch of life's more copious tree (albeit an unusually successful limb that has spawned the largest organisms).

I don't object to a primary emphasis on vertebrates, for we have a legitimately parochial interest in ourselves and our immediate ancestry. But consider two ways in which the conventional depiction of life's history as a parade of scenes from invertebrates to humans distorts the major pattern of our fossil record.

First, more recent geological periods may add new kinds of vertebrates, but the invertebrates (and the early vertebrates) don't go away; they persist and continue to dominate in most habitats. Thus, the conventional tableau of the Cambrian is a sea-bottom filled with trilobites and brachiopods, while the standard snapshot of the much later Tertiary is a landscape studded with mammals. But oceans never disappeared. They still dominate our planet and cover some 70 percent of the Earth's surface; they still teem with invertebrate life, different in fascinating ways from the Cambrian faunas. Yet no conventional set of tableaux for life's history ever includes an invertebrate marine scene from any time following the rise of terrestrial vertebrates.

A defender of the traditional faith might respond that everyone understands this convention. The usual account only presents the history of vertebrates as a particularly interesting sample of totality. Not so. The titles of these pageants lay claim to inclusivity. Consider the three most

Henry de la Beche's cartoon of life in "a more ancient Dorset" (1830).

Henry de la Beche's cartoon of life in "a more ancient Dorset" (1830).

influential presentations of our century – Charles R. Knight's *Before the Dawn of History* (1935), his later *Parade of Life through the Ages* (1942), and the 1956 volume by J. Augusta and Z. Burian on *Prehistoric Animals*. None draws a single invertebrate animal from any period following the origin of vertebrates. Even the present book, though a great improvement in extended coverage of invertebrates, does not violate this tradition of tableaux parading toward *Homo sapiens*. Yet we name this work, comprehensively, *The Book of Life*.

Even when people realize that invertebrates and "lower" vertebrates persisted, this biased iconographic tradition encourages a belief that such "primitive" forms stopped at their early plateau (and can therefore be subsequently ignored), as the torch of novelty passed to higher vertebrates (who must therefore be chronicled). No such thing. All major forms of life continue to diversify and adapt, continue their fascinating roles in life's endless ebb and flow of extinction and origination. We foster a seriously skewed account when we abandon the later history of animals that arose early, and pretend that the twig of vertebrates can act as a surrogate for all later history. Moreover, the bias thus introduced is the worst and most harmful of all our conventional mistakes about the history of our planet – the arrogant notion that evolution has a predictable direction leading toward human life.

Second, even when artists deign to give some early space to invertebrates, the amount allotted is never commensurate with true importance or time elapsed. Most of life's history gets scrunched into an introductory picture or two. Augusta and Burian devote the first three of their sixty plates to Paleozoic invertebrates. Knight (1942) grants them two of twenty-four paintings – one on animals of the Burgess Shale, the other on eurypterids, largest and most spectacular of all invertebrates. Knight's earlier work (1935) is marginally more generous to the pre-vertebrate Earth, with the first four of forty-four illustrations entitled "The world before life," "Blue-green algal pools," "Ordovician seashore," and "Silurian coral reef – Chicago area." But these four figures display another traditional feature that demotes invertebrate life to a periphery – this time harking back to the category of artistic convention rather than social bias. None of these four scenes shows invertebrates in their natural habitat – that is, under water. All follow an old artistic convention, dating at least to the seventeenth century and thoroughly documented by Rudwick (1992), of drawing invertebrates as dead creatures cast up upon the shoreline. We can scarcely get a good idea about life in the sea if we only encounter such animals as their exuviae, or desiccating bodies, tossed up into an alien environment.

By 1942, Knight had improved his treatment and presented his two invertebrate scenes as living creatures underwater. Rudwick notes that Western artists may have had great difficulty even conceptualizing a subaqueous scene before the great aquarium craze of the mid-nineteenth century made such vistas familiar to all. In this sense, de la Beche's *Duria antiquior* of 1830 is a pathbreaking icon, though infrequent repetition (beyond simple copying of de la Beche's figure), and Knight's continued allegiance to the old convention a century later, demonstrate that a fairer deal for invertebrates did not come easily.

To those who persist in viewing iconography as peripheral or subsidiary to text, I can only respond with a primal fact of our evolutionary biology. Primates are quintessentially visual animals, and have been so endowed since the first tree-dwellers of earliest Tertiary reconstructions had to move nimbly among the branches, or fall to their deaths away from further scrutiny by natural selection. Humans, as legatees of this heritage, learn by seeing and visualizing. Confucius was not dispensing an oracular item of arcane Eastern wisdom, but epitomizing a central truth of primate evolution, when he proclaimed that one good picture is worth ten thousand words.

In this context, I have never understood why large-format volumes of illustrations are often contemptuously dismissed by academics and intellectuals (though usually by the posturers rather than the folks of substance) as "coffee-table books." I do not despise my coffee table as a low form of furniture (except in the strictly literal sense), and I regard beautiful and informative books of pictures as among the most sublime products of the publishing industry.

For all these reasons – the marginal reputation of illustrated books, combined

Apatosaurus, a painting by Charles R. Knight.

with the opposite theme of our deep propensity to be moved and influenced by images; the maximal display of social and cultural biases in a medium that we do not scrutinize for their presence and pervasiveness – iconography is a subject of enormous importance to scholars and historians of ideas. It is, in fact, my personal favorite of all disregarded and understudied themes in the chronicles of human thought (see Gould, 1987, 1993).

Iconography comes upon us like a thief in the night – powerful and remarkably efficacious, yet often so silent that we do not detect the influence. If I ask who was the man most responsible for setting our conventional concept (until recent years) of dinosaurs as grand but cumbersome, most respondents will search for the name of a leading scientist who defended this notion in words. But the question has an undeniable and unambiguous answer – Charles R. Knight (though many have never heard of him). Knight (1874–1953) was the greatest iconographer of dinosaurs at

a time when his superb work formed a one-man show without credible competition anywhere in the world. He painted all the great murals done before World War II in museums in the United States – New York, Chicago, Los Angeles. His elegant, anatomically accurate, ecologically detailed, and visually exciting paintings filled books and magazines. In the absence of alternative imagery, Knight created the canonical picture of dinosaurs for professionals and the public alike. I cannot think of a stronger influence ever wielded by a single man in such a broad domain of paleontology.

Similarly, the most telling sign of our changing concept comes from the new generation of dinosaur artists who are finally superseding these grand conventions and providing alternative images for an astonishing range of products from kiddie books to cereal boxes to postage stamps. Just consider the contrast between Knight's classic *Brontosaurus*, buoyed up in a swamp because even such elephantine legs could not support

"Crossing the flat," a restoration of Mamenchisaurus hochuanensis *by Mark Hallett.*

such a bulky body on land, and Mark Hallett's corresponding sauropods, nimbly marching forward with head and tail outstretched. Was Confucius right, or do you want another 20,000 words from me to explain the conceptual shift? "A word fitly spoken is like apples of gold in pictures of silver" (Proverbs 25: 11).

Some waystations in a history of fossil iconography

Verbal descriptions of fossils can be found in ancient Greek texts. Woodcut illustrations of fossils are featured in the founding documents of modern natural history – the sixteenth-century treatises of Konrad Gesner and Ulysse Aldrovandi. The first complete reconstruction of a fossil vertebrate from scattered bones is conventionally (but, I think, dubiously) attributed to the German physicist

and inventor of the air pump, Otto von Guericke (1602–86), for a preposterous unicorn strung together from a heap of disparate material, including some mammoth bones. The first systematically accurate reconstructions of fossil vertebrates were not achieved until Georges Cuvier, armed with a museum-full of modern skeletons for comparison, founded the field of vertebrate paleontology in the last years of the eighteenth century.

But we are interested, in this book, in a different tradition – flesh and blood, interaction and ecology. Our genre is the fusion of artistic *and* scientific imaginations to produce encompassing images of the past – scenes from deep time, in the apt words of Rudwick's title. This vital subject has hardly been studied by scholars, though Rudwick (1992) has presented a fine first attempt. I cannot therefore provide an accepted or firmly documented history of the development of prehistoric iconography. Instead I offer some suggestions in the form of

From The Copper Bible *of Johann Jakob Scheuchzer.*

four sequential waystations in a largely unexplored chronology.

1. *The Copper Bible (1731–3) of J. J. Scheuchzer.* If adequate reconstructions of individual fossil vertebrates do not antedate Cuvier's time, then properly executed artistic scenes of prehistoric life cannot exist before the nineteenth century. But the genre of "scenes from deep time" contains several formative threads, and some can be traced into a world with little direct knowledge of fossils. If an accurate account of fossils provides one thread, a second, and equally important, derives from other cultural traditions committed to presenting pictorial and chronological accounts of the pageantry of history.

We may now draw our histories from geology and other texts of natural science, but our forebears inferred their chronologies from another and more literal kind of text – the words of Scripture, deemed sacred and infallible. We may now reject the short time scale, the literality of interpretation, the invocation of miracle as an agent of change, but such biblical history shares many elements with our later genre of scenes from deep time; it is comprehensive (starting with the origin of the Earth), sequential, progressive (in the six "days" of Creation), and, above all, a damned-fine, intellectually and morally informative story. Rudwick has therefore argued (correctly, in my view, and I follow him here by choosing the same beginning) that traditions of pictorial Bible history form the basis for our genre of prehistoric art as chronological pageantry.

Our surprise at such a claim (and the desire that many of us feel to reject it initially) should teach us something useful and corrective about our prejudices. We have been trained to consider science and religion as institutions intrinsically at odds, but this theme applies only (and in a restricted manner) to a limited domain where theology once overextended a taxonomic wing into regions since properly claimed for empirical inquiry. In other areas, science and religion offer different and incommensurable (but equally necessary) insight for a full life – spheres of knowledge and ethics which, in their union, constitute wisdom. In still other realms, science and religion may share common interests – as in their joint tendency

From The Copper Bible *of Johann Jakob Scheuchzer.*

to render pathways of history as pageants.

The great Swiss savant Johann Jakob Scheuchzer (1672–1733) lived in an age that recognized no sharp boundary among disciplines later separated into different faculties of our universities. He was a physician and professor of mathematics in Zurich, but he also published a topographic map of Switzerland and wrote a 29-volume work on the history of that country. His important paleontological publications, particularly on fossil plants, rank him among the founders of our science.

Late in his life, Scheuchzer conceived, and managed to execute just before his death, one of the great triumphs in the history of publishing. We can only comprehend its scope today if we make a proper analogy to the comparable phenomenon in contemporary culture – the lavishly funded, multiple-hour, no-holds-or-expenses-barred, camera-crews-around-the-world, TV special series with Scheuchzer as Carl Sagan or Bill Moyers, and all his collaborators as the team of hundreds (and corporate support in millions) behind every such enterprise.

The *Physica sacra* (or "Sacred Physics," construing the second term broadly to include all that we would now define as science) is a series of large folio volumes, embellished with 745 full-page, copperplate engravings presenting biblical history in chronological order from the beginning of time, and commenting *in extenso* upon every conceivable link between a biblical passage and any scientific or cultural art. When Jesus says to the Pharisees and Sadducees, "O generation of vipers," Scheuchzer responds with a beautiful set of plates presenting a full taxonomy of known snakes. The building of Solomon's Temple calls forth a magnificent series of plates on the norms and practices of public architecture. The Flood inspires several versions of the Ark, all drawn to scale and accompanied with a discussion of how the animals might fit.

Scheuchzer's plates were first drawn by his friend J. M. Fueslin and then engraved by a team of nineteen artists working under the direction of Johann Andreas Pfeffel, the imperial engraver of the Holy Roman Empire. Some of these artists worked only on the elaborate baroque frames surrounding the scenes; one man did all the titles in both Latin and German gothic script. The *Physica sacra* appeared in three editions, one in Scheuchzer's native German, the others in Latin and French, the older and the newer international languages of science and scholarship. The German edition featured the engravings in its title – *Die Kupfer-Bibel* ("The Copper Bible"). Many of the later conventions in our genre of prehistoric art, obeyed even today, can be traced to this grand inception.

Plate 125, on the plague of frogs visited by Moses upon the Egyptians, shows a graphic scene within the frame, as frogs scale the city wall and scamper over a human corpse. As a scientific commentary, the surrounding frame presents a complete life cycle of frogs, from egg, through tadpole to the large creatures in the lower left that unite the two parts of the plate by jumping from the frame into the scene.

Plate 19, *Opus quintae diei* ("Work of the fifth day"), presents the creation of mollusks, but uses modern species rather than fossils, and follows the iconographic convention of depicting shells draped upon the land, rather than living in the sea. Noah's sons build the Ark in plate 34, and the frame presents a pictorial discussion on the taxonomy of conifers that might have been used in the construction. Scheuchzer follows his account of the Flood with fourteen plates of fossils entitled *Restes du déluge* ("Relics of the Flood"). Plate 49 is by no means the best in this beautiful series (that honor goes to an engraving of ammonites), but it is the most telling in presenting the famous fossil that Scheuchzer himself had discovered and described in 1726 as *Homo diluvii testis*, "the man who witnessed the flood." Great effort always carries the danger of great error, and no blame should be attached to mistakes in this mode: nothing ventured, nothing gained. But Scheuchzer's *Homo diluvii testis* did turn out to be a salamander, and the correction was made by none other than Cuvier himself. (The specimen may still be seen at the museum in Haarlem in the Netherlands.)

2. *Hawkins and Figuier stabilize the genre.* Rudwick (1992) has collected and presented nearly all the scenes of prehistoric life composed between the first successful reconstructions of vertebrate fossils about 1800 and the establishment of the genre by mid-century. They are few and far between, and often presented with abject apologies by

scientists quite leery about the speculation implied in fleshing out such fragmentary information. Some major works on popular geology and paleontology, though filled with plates of actual fossils, presented no artistic reconstructions at all (most notably, William Buckland's Bridgwater Treatise of 1836, *Geology and Mineralogy Considered with Reference to Natural Theology*). Moreover, most scenes were single efforts, not series – and they may have fostered, while not so intending, the false impression that one might contrast the present world with a single tableau of ancient life representing an unordered and undifferentiated antediluvian time. Since the essence of history is chronological *sequence*, with antecedent states as causes of (or at least major constraints upon) subsequent scenes, such a two-valued view of life (past vs. present) cannot rank as a genuine science of ancient time.

The establishment of a time scale, and the working out of a consistent and worldwide sequence of changes in fossils through the stratigraphic record, represents the major triumph of the developing science of geology during the first half of the nineteenth century. In 1800, the fact of extinction had barely been established, and only a few fossil organisms had been properly reconstructed. By 1850, geology had developed a coherent global chronology based on life's history. This discovery and construction of history itself must rank as the greatest contribution ever made – indeed, I would argue, ever makable – by geology to human understanding. Such a spectacular success had to be accompanied by a maturing iconography committed to representing the past as a series of successive stages rather than a single undifferentiated past contrasted with our present world.

This new convention – merging the traditions of Scheuchzer's chronological perspective with new knowledge of entire faunas of fossil organisms – had become established by the 1860s, only a few decades after the first crude attempts, and therefore a tribute to the enormous energy and speed of development in geological science. I would single out two major efforts as most symbolic of this first credible work in comprehensive iconography of the past.

By building full-scale, three-dimensional models of an entire reptilian fauna (aerial

From The Copper Bible *of Johann Jakob Scheuchzer.*

From The Copper Bible *of Johann Jakob Scheuchzer.*

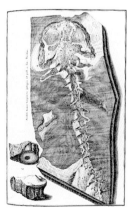

pterosaurs, terrestrial dinosaurs, and marine ichthyosaurs and plesiosaurs), Waterhouse Hawkins so upped the ante on what could be accomplished that artistic reconstruction of the past became a central activity, even accompanied by substantial prospects for commercial success, rather than a frill on a periphery. By collaborating so closely with Richard Owen, Hawkins also established a style followed ever since (Knight with Osborn, Burian with Augusta), and committed to the proposition that scientific accuracy and artistic excellence can march hand in hand. Moreover, Hawkins's Crystal Palace extravaganza added so many legends to the history of iconography that his full-scale models grew larger than life. Most famous, of course, is the dinner that Hawkins and Owen held *within* the body of his half-completed *Iguanodon* model, with Professor Owen seated appropriately at and in the head, on 31 December 1853. At midnight, according to legend, the assembled group of twenty-one guests sang a song composed by geologist Edward Forbes for the occasion:

> The jolly old beast
> Is not deceased.
> There's life in him again.

What a way to ring in the new!

But a full establishment of the genre required a complete set of chronological scenes, arranged in order of the geological time scale, just as Scheuchzer had followed the biblical sequence. Rudwick (1992) reports that the first such effort can be found in F.-X. Unger's 1851 work, *Die Urwelt in ihren verschiedenen Bildungsperioden* – "The Primitive World in its Different Periods of Formation." This set of fourteen lithographed plates by Josef Kuwasseg represented a good first try, but did not set the genre for popular culture. Unger's work appeared as a large and expensive folio atlas, printed in rather few copies in French and German editions. Moreover, Unger was a botanist, and most of his scenes include few if any animals – and these usually in the distant background. (I realize, of course, that zoocentricity is just another lamentable bias, but we are animals after all, and canonical iconography will have to feature our kingdom.)

The breakthrough occurred in 1863 when the famous French popularizer of science Louis Figuier (1819–94) commissioned the Parisian landscape painter and illustrator Edouard Riou (1833–1900) to produce some two dozen full-page engravings of scenes from deep time, arranged in sequence from the prebiotic world to the "creation of man." The collaboration of Figuier and Riou, who would also work for Jules Verne in the different but not so disparate domain of science fiction, produced one of the great successes in the history of popular science

"Marine life from the Carboniferous period," from Louis Figuier's Earth before the Deluge.

publishing – *La terre avant le déluge* ("The Earth before the Deluge"). This inexpensive octavo work received maximal circulation in numerous printings and translations into several languages (my own copy is a leatherbound, fourth English edition printed in New York in 1867, just four years after the original publication in French).

Figuier and Riou followed the conventions of the time. The early plates of invertebrates show shells cast up upon the shoreline, not animals living in the ocean. They also fudged the contentious issue of human origins and opted for an Edenic scene of a Caucasian family carrying no weapons and surrounded by modern animals. But later editions showed more spunk and better perspective. Riou added one Paleozoic marine scene of fishes and invertebrates in their natural setting, and Figuier then bowed to increasing evidence for the coexistence of early humans with extinct Ice Age mammals, thus making human origins part of the natural sequence, rather than a late imposition from on high. My edition removes the former Edenic scene from the textual sequence and substitutes a clan with axes confronting an ancient fauna including mammoths and Irish elks (though note the chasm, meant symbolically as well as literally no doubt, between humans and beasts). The people are still Caucasian and anatomically modern, but only so many conventions can be challenged at once –

particularly in works intended for big sales to mass audiences. Figuier and Riou continued to play both ends, for they reinserted the old Edenic scene into an even more conspicuous spot – as a frontispiece for the entire book.

3. *The genre matures in the canonical figure of C. R. Knight.* Unger, Figuier, Hawkins, and others originated a genre based on sequences of comprehensive scenes. But nothing so establishes an art-form as its own Picasso – a person so talented, and so obviously superior to all competition, that his figures become canonical (both a blessing in accuracy and excellence, and a danger in potential rigidification). The iconography of fossils did not achieve this canonical status until Charles R. Knight began his work and finally managed to render extinct animals as vibrant, vital, believable, and exciting creatures.

Knight was born and raised in New York City and became a commercial artist employed by the church-decorating firm of J. & R. Lamb in 1890. But Knight had always loved natural history most of all, and he soon won the job of drawing all plants and animals for the firm's stained-glass windows. Knight slowly built his skills by spending several mornings each week sketching animals at the Central Park Zoo. He eventually decided to devote his career to zoological art, and he began to build a reputation both for fossil reconstructions and for paintings of living organisms. The superiority of his prehistoric

"Appearance of man," *from Louis Figuier's* Earth before the Deluge.

"Appearance of man," revised version, from Louis Figuier's Earth before the Deluge.

paintings rested squarely upon his unparalleled expertise in the musculature and motions of modern organisms. No previous artist of ancient life had ever understood the universal principles of animal design so intimately.

When Knight finally teamed up with the brilliant and politically powerful Henry Fairfield Osborn, leading vertebrate paleontologist and president of the American Museum of Natural History, his future as the "official" painter of ancient life seemed assured. Osborn stated, for once without

hype: "Charles R. Knight is the greatest genius in the line of prehistoric restoration of human and animal life that the science of paleontology has ever known. His work in the American Museum will endure for all time." When I was a child, I used to visit the museum monthly. I always stopped before the statue of Osborn and I learned the inscription on its pedestal by heart. I can still recite the words, but I now think that they apply best to Charles R. Knight: "For him the dry bones came to life, and giant forms of ages past rejoined the pageant of the living."

When we look at Knight's dinosaurs today, we feel a whiff of archaism because changing opinion during the last two decades has substituted sleek, motile, highly efficient, and reasonably intelligent (even potentially warm-blooded) beasts for the slow, lumbering, dimwitted "primitives" often depicted by Knight. But, in drawing these heavy and encumbered beasts, Knight was only translating the ideas of leading paleontologists, not expressing an artistic limitation. He could paint animals as active and as sleek as any depicted by the most radical revisionists of our current schools – as in his insufficiently known but brilliant painting, completed in 1897, of the small carnivorous dinosaur *Dryptosaurus*, or his

Dryptosaurus, a painting by Charles R. Knight.

A painting of the 30 ft-(9 m) long mosasaur Tylosaurus, *by Charles R. Knight.*

more famous image of a mosasaur pursuing Cretaceous fish.

Yet for all the improvements in representing ancient animals as competent creatures rather than primitive heaps of inefficiency, authors of the new iconographies did not succeed in breaking free from the strongest constraint of progressivist imagery – the notion of life as a pageant of improvement, perhaps externally directed on its upward course, but certainly and predictably culminating in human intelligence, even if propelled by purely natural forces. The ghost of Scheuchzer's biblical chronology haunts us still. Knight, for example, wrote in 1935:

> Those of us whose minds are imbued with a proper amount of religious conviction will detect in this apparent selection [for increased human intelligence] the intervention and assistance of a power higher than ourselves – a certain definite purpose, divine or otherwise, whose control has shaped our destiny, and whose ultimate goal is a perfecting of all our faculties … Others, more scientifically inclined, will dispute this attitude, and will ascribe to a purely logical and physiological

development all our mental and physical improvement from a being of lowly mien and mentality to the more or less perfectly constructed modern man, physically impressive and, by virtue of his superior cerebral equipment, monarch of all he surveys.

With a few isolated exceptions (like the Zallinger mural at Yale University), the first seventy years of our century belong to Knight as master of fossil iconography. The only successful and widely popular competition arrived just after Knight's death in a wonderful book of sixty plates painted by the Czech artist Zdenek Burian, with a text by Joseph Augusta, Professor of Paleontology at Prague. These plates were also produced as large charts, and they adorn many museums and academic departments in this form. Yet even this text, with its overlay of Marxist and materialist rhetoric from an origin in Communist Czechoslovakia, continues to preach the conventional claims of pageantry. Scheuchzer's transmogrified piety reigns even here:

> From the very beginning of the history of life on Earth we see how life constantly develops and progresses, how

it is constantly being enriched by new, ever higher and more complex forms, how even man, the culmination of all living things on Earth, is tied to it by his life.

4. *Postmodern iconography*. We are now, as this book proves, in the midst of the greatest vigor and upheaval in fossil iconography since the genre emerged. The reasons are many, and partly due to mammon – a morally ambiguous reality in our commercial world, but glitzy models and paintings now command big bucks in our era of theme parks and cereal boxes, where moving and roaring plastic models bring more people into museums than magnificent skeletons of real bone.

But better reasons inhere in the rush of new ideas that have reinvigorated the science of paleontology and forced us to recast old primitives as efficient and worthy creatures in their own terms. We no longer condemn prehistoric beasts to ineptitude for the irrelevant reason that they lived a long time ago, and we have finally granted them respect with the recognition that extinction is no shame in our largely random world – and that a creature, namely us, who has lived for so short a time can construct no rationale for casting aspersions upon animals, like

dinosaurs, who dominated the Earth for a hundred million years. Above all, interest in ancient life is now so widespread, and consumer demands for its iconography so great, that the field will not soon lapse again into the general desuetude that limited nearly all available work to one man – even to so great an artist as Charles R. Knight.

I shall not pursue the analysis of current iconography here, for the field is in too rapid a transition, and the best fruits of modernity can be enjoyed in the pages of this book – so why look to preaching from me? But I am intrigued to note how closely the trends in prehistoric iconography match the winds of change labeled "postmodernism" in so many other fields from literature to architecture – so we are once again taking part in a general social movement, not merely following the local norms of science by responding to improvements in factual knowledge.

If postmodernism is diverse, non-hierarchical, playful, personal, pluralistic, iconoclastic, and multifarious in its points of view – whereas modernism sought a simplified and rulebound canonical consensus – then the maddeningly varied modern art of fossils certainly qualifies for the label, with its garish dinosaurs and its literally new dimensions and perspectives (often down from a pterosaur's eye view, or up from the vantage point of a baby dinosaur just hatched). As just one example, consider Gregory Paul's "What happens when *Apatosaurus ajax* seeks aquatic refuge from *Allosaurus fragilis*." Even the title is a satire, and a tweak of the old modernist consensus. You need to know the history of previous argument to grasp the playfulness and the "gotcha" of Paul's sardonic picture. For an old chestnut of dinosaurology proclaimed that sauropods retreated to the water to avoid theropods, who would not venture after them therein. But no one ever really asked why not, and Paul shows us that a herd of allosaurs might well have pursued and nabbed its quarry after all.

I only wish that this iconoclastic attitude toward canonical views of individual creatures were matched by an equally critical reappraisal of the most pervasive and constraining convention of all – the tradition of depicting life's history as a pageant leading from invertebrates up the vertebrate ladder to humans (a scheme still generally followed in this book, despite all its innovations). We often imagine that Darwin and evolution represent the great watershed that forever changed everything in biology. But many themes sailed right through this greatest of barriers, emerging relatively unscathed on the other side – reclothed in an evolutionary explanation, but unaltered in basic content. The idea of an "ascent to man" (to use the old gender-biased language) ranks most prominently among these unaltered, but woefully flawed, Western certainties. Pre-Darwinian paleontologists attributed such a progressive pattern to God's scheme of successive creations; but post-Darwinian evolutionists (like Charles R. Knight) told the same story, with natural selection substituting for God.

We are still awaiting the real revolution – the recognition that all lineages owe the details of their history largely to contingent good fortune rather than to predictable development. Darwin's revolutionary worldview actually implies such a pattern of sensible explanation after the fact, but no prediction beforehand – but we have resisted this implication because, to say it one last time, the ghost of Scheuchzer still hovers over us in our unwillingness to abandon human centrality as the ordering principle of life's

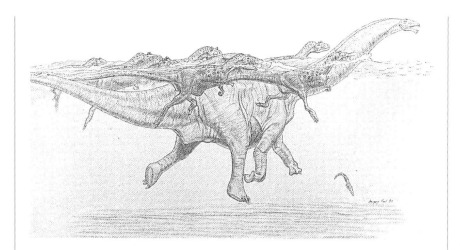

history. Evolutionists do understand how historical constraint affects paleontological lineages, but we are oblivious to the same theme when it clamps a conceptual lock upon our own mental schemes.

We do not even know how to conceptualize, much less to draw, the worldview that would place *Homo sapiens* into proper relationship with the history of life. We know the iconography of directional pageantry, for we have been drawing history according to this scheme for centuries, but what is the iconography of contingency? I love this book of life, but it continues to bring us the past as through a glass darkly. Someday, perhaps, we shall meet our ancestors face to face.

"What happens when Apatosaurus ajax *seeks aquatic refuge from* Allosaurus fragilis" *by Gregory Paul.*

BIBLIOGRAPHY

Augusta, J. and Z. Burian, 1956. *Prehistoric Animals.* London: Spring Books.
Buckland, William, 1836. *Geology and Mineralogy Considered with Reference to Natural Theology.* 2 volumes. London: William Pickering.
Figuier, Louis, 1863. *La terre avant le déluge: ouvrage contenant 24 vues idéales de paysages de l'ancien monde dessinées par Riou.* Paris: Hachette.
Figuier, Louis, 1867. *The World before the Deluge: A New Edition, the Geological Portion Carefully Revised, and Much Original Matter Added, by Henry W. Bristow, FRS…* London: Chapman & Hall.
Gould, Stephen Jay, 1987. *Time's Arrow, Time's Cycle.* Cambridge, MA: Harvard University Press, 222 pp.
Gould, Stephen Jay, 1993. *Eight Little Piggies.* New York: W.W. Norton, 479 pp.
Knight, Charles R., 1935. *Before the Dawn of History.* New York: McGraw-Hill.
Knight, Charles R., 1942. "Parade of Life through the Ages." *National Geographic Magazine,* Vol. 81, No. 2, pp. 141–84.
Rudwick, Martin J. S., 1992. *Scenes from Deep Time.* Chicago: University of Chicago Press, 280 pp.
Scheuchzer, Johann Jakob, 1731–5. *Physica sacra Johannis Jacobi Scheuchzeri … iconibus aeneis illustrata procurante & sumtus suppeditante Johanne Andrea Pfeffel …* Augsburg and Ulm.
Unger, Franz-Xaver, 1851. *Die Urwelt in ihren verschiedenen Bildungsperioden: 14 landschaftliche Darstellungen mit erlauternden Text; Le monde primitif à ses différentes époques de formation: 14 paysages avec texte explicatif.* Vienna: Beck.

Introduction

LIFE AND TIME

Michael Benton

The story of life is long, and growing longer as science pushes its beginnings ever further back, deep into time. As late as the eighteenth century, it was commonly believed that the universe itself had been created less than 6,000 years ago, and life a few days later. By the nineteenth century, the new science of geology had shown that the Earth must be hundreds of millions of years old, and we know now that its age is closer to 4.5 billion years. In the time of Charles Darwin, the fossil record seemed to start with the remains of shelly animals found in rocks of the Cambrian period, which may date back as long as 530 million years ago. In the twentieth century we are finding traces of microscopic life in ancient rock formations upward of 3.5 billion years old.

So *The Book of Life* tells a story that starts sometime between the formation of the Earth and the first recorded signs of living things. The story has a science of its own, which is paleontology, the study of ancient life, but life is an expansive, exploratory, inquisitive force, and few sciences are immune to its call. As well as biology and geology (the study of modern life, and the history and structure of the Earth), this book has enlisted sciences as various as astrophysics, chemistry, ecology, meteorology, and oceanography.

This is a story so vast that it could easily collapse under the weight of the facts that make it up, yet so vital that it does not belong only to the sciences that have pieced it together. On the other hand, when popular science sets out to address a non-specialist readership it is easy to fall into one of two main traps. One is the walk on the wild side, the Evolution Spectacular: giant sharks, dinosaurs, saber-tooth cats. The other is the Royal Road, the rounded and perfect narrative that runs inexorably from the primeval sludge to modern humans, as if pre-planned. The authors of *The Book of Life* have set out to preserve the sense of wonder that their subject demands by heading for the open space between oversimplification and academic obscurity. That means presenting not only the latest evidence and the most exciting new research, but also the key disagreements and the hottest controversies, so as to offer readers a hands-on feel for a dynamic subject.

Paleontology has taken huge strides during the past few decades, both in terms of new discoveries and by developing a stream of new theories and ideas. Much of this story could not have been told even as recently as twenty or thirty years ago, because it had not been unearthed. But there is far more left buried in the ground, or as yet unseen in our growing mass of evidence. There are areas where we have only the echoes of the story, and where a few handfuls of facts have to speak for millions of years. Here the direct evidence will only provide a shadowy picture of what was really happening, and closely argued guesswork sheds what light there is. The purpose of this Introduction is to review some of the kinds of data we rely on, and to examine a few of the ideas that string these facts together.

Geological time

Human history deals in days and centuries. Its longest unit is a thousand years, which is not much more than the blink of an eye for geologists and paleontologists, whose common currency is millions and billions of years. As animals go, human are long-lived, but surely it takes a limitless imagination to grasp the sheer extent of geological time. A standard comparison is to imagine the age of the Earth as a single cosmological day; then dinosaurs appear quite late in the evening, at 10.42 p.m., humans at one and a half minutes before midnight, and civilization with less than one second to go.

The geological time scale breaks down these deserts of eternity into slightly more manageable units, established by international agreement among the Earth's scientists since the mid-nineteenth century. It is a calendar that divides the Earth's history into eons, eras, periods, and finer subdivisions, based on the evidence of rocks whose age can be measured as far back as 4 billion years. Visualized as layers lying one on top of the other in the order of their formation, they would make a single "geological column" that contained all of the history, both chemical and biological, that our planet can bequeath to us. In fact, the Earth's crust has a record of geological violence which has melted, eroded, or otherwise removed or transformed much of the rock forced upward or laid down as sediment since the beginning. Yet somewhere on the surface geologists have located remnants of past ages, and worked out their chronological order.

By the early nineteenth century it had become obvious to most students of landscape and of nature that if the same rate of change had operated in the past as they observed in the present, it must have taken hundreds of millions of years to produce such huge thicknesses of rock and such depths of erosion. They saw that the sedimentary rocks – sandstones, mudstones, limestones – were laid down in layers or strata in recognizable sequences, and they found that any of these layers might contain a set (in paleospeak, an assemblage) of fossils that were distinct from the set found in the layers above or below. Each assemblage of fossils was unique to its own age, and could be used to identify rocks of similar age in another part of the country, or the world.

The early geologists worked out the overall order of the sedimentary rocks by matching sequences first in Wales, and then all over the world. Then they identified the basic periods recorded in fossil-bearing rocks, and between 1799 and 1879 they devised the names that we still use for these clear stages in the history of the world. Some are called after regions where typical sequences survive – the Jurassic after the Jura mountains of France and Switzerland, the Cambrian after the Latin name for Wales. Other names may describe the typical rocks of their period – Cretaceous means chalky, Carboniferous means coal-bearing. The boundaries between the periods were marked by major changes in the kinds of rocks laid down, and most crucially in the fossil plants or animals they included.

These periods are grouped into Paleozoic ("ancient life"), Mesozoic ("middle life"), and Cenozoic ("recent life") eras. The names were given by John Phillips in 1841, because of the key transitions found at the boundaries between these eras – as he saw them, from no life to life at the start of the Paleozoic, a

One of the wonders of paleontology is that new worlds are revealed. Fossils tell us about animals that are weirder than any invention of a science-fiction writer, and striking new discoveries are being made every year. For example, in the early 1970s, giant bones of a flying reptile, a pterosaur, were found in Texas. These were named Quetzalcoatlus, after an Aztec god, and they belong to the largest flying animal of all time. Quetzalcoatlus had wings 40 ft (12 m) across, and was the size of a small airplane. This is an impossible animal; the calculations show that no flying animals could be so big. And yet, it existed, and it flew.

turnover of more than half the living groups at the start of the Mesozoic, and the loss of many major groups at the Mesozoic/Cenozoic boundary.

The fossil-bearing ages are called the Phanerozoic ("abundant life"), and the previous vast span of rocks and time is the Precambrian. But that is a period of some 4 billion years, and as we learn more about its history, and the story of Precambrian life, we have begun to break it up into more manageable sections: the Hadean, ending about 3.9 billion years ago, when the oldest surviving rocks were formed; the Archean, ending 2.5 billion years ago, and the Proterozoic ("earlier life"), which ends where the Cambrian begins.

The science of stratigraphy maps and studies the *relative* ages of the various rock divisions. Exact ages are a question only soluble with the technology of the nuclear age, which has developed methods of radiometric dating based on the fact that the Earth's crust contains several elements that may exist in unstable forms (isotopes) that "decay" by emitting atomic particles. As soon

as the particle is emitted, an atom of potassium 40 turns into argon 40, rubidium 87 into strontium 87, uranium 238 into lead 206, carbon 14 into nitrogen 14, and so on. With the exception of carbon, these elements are found mainly in igneous rocks formed at high temperatures either deep in the crust (like granites) or on the surface (like lavas), when the rock crystallizes out from a cooling fluid into a solid state.

The new rock may contain any of several isotopes, which begin to decay from an unstable "parent" into a stable "daughter" form. Each isotope has a different rate of decay which is defined by its "half-life," the time it takes for half of the parent to decay into the daughter form. By measuring the ratio of the parent to the daughter element, we can estimate the time when the process began. The longer the half-life, the longer the span of measurable time, but the less accurate the calculation. Uranium 238 has a half-life of 4.5 billion years. Carbon 14 has a half-life of only 5,730 years, so it will not measure millions of years, but is accurate for thousands.

GEOLOGICAL TIME

The geological time scale is an internationally used reference point. The basics are shown here, but geologists have developed a much more detailed version which divides the history of the Earth into dozens of precisely defined periods. This is particularly true for the time represented by the Paleozoic, Mesozoic, and Cenozoic eras, when fossils were abundant.

The time scale is founded on the relative dating of rocks by fossils. Since life has evolved, no particular species lasts forever. Species come and go, and it is therefore possible to recognize any particular short span of time in the past by assemblages of fossil species. This is the basis for matching rocks in different parts of the world.

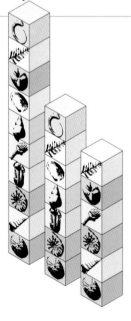

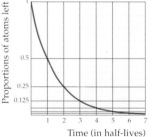

Radiometric dating is based on the fact that certain chemical elements, such as uranium, argon, potassium, and carbon, occur in two or more forms, or isotopes. One of the isotopes may be more unstable than the other, and it will tend to break down, over time, into the other isotope, or even into another element

altogether, giving off natural radioactivity as it does so. These rates of breakdown are predictable, and the age of a rock can be estimated if the original proportions of two isotopes are known and can be compared with the current isotopic proportions. The isotopes break down, or decay, at a steady geometric rate within any particular rock sample. The rate is proportional to the sample size, and hence the decay pattern follows a curve that never reaches zero. Hence, the "half-life" is measured, which is the time it takes for half of the sample to decay.

Formation of the Sun

Formation of Earth and Moon

Origin of life

Oldest known rocks

Oldest known fossils

Archean

4,800   4,000   3,800   2,500

Geological forces

What forces shape the surface of the globe? Geologists assume that energies active today are likely to have worked in the past. They can examine the effects of modern volcanoes, and use their research to interpret the impact of ancient volcanism 200 million years ago. They can measure the rate at which rivers erode the landscape, or sediments form on river bends or seabeds, and assume that similar rates applied 500 million years ago.

Charles Lyell, a key figure in geology in the early nineteenth century, believed that gradual changes would account for everything. The great French paleontologist Georges Cuvier and the geologist William Buckland believed in the power of sudden exceptional events, great cataclysms that ushered in new ages. These supposedly opposing beliefs became known as "uniformitarianism" and "catastrophism." The first theory does explain a lot, but that does not exclude the possibilities raised by the second. The Deccan Traps of central India, for example, cover an area of about 200,000 square miles (52 million ha), all that remains of huge lava beds caused by a volcanic episode that happened about 65 million years ago. Nothing on that scale has happened since. Some of the boundaries between geological periods seem to represent mass extinction events whose causes have become a major target of research. Asteroid impacts are a strong possibility, and other suggestions implicate changes in climate, sea levels, and evolutionary rates.

Since the 1960s, a revolution has transformed our understanding of a driving force that builds up mountains, opens and closes seas, assembles and disassembles continents, and ferries them slowly but inexorably over the surface of the globe. Alfred Wegener (1880–1930) is the best-known early champion of the idea of continental drift. Like others before him, he looked at the map of the world and noticed how well the continents would fit together if South America were moved eastward to nestle against Africa, and North America and Greenland against northern Africa and Europe. Geologists noticed that rock

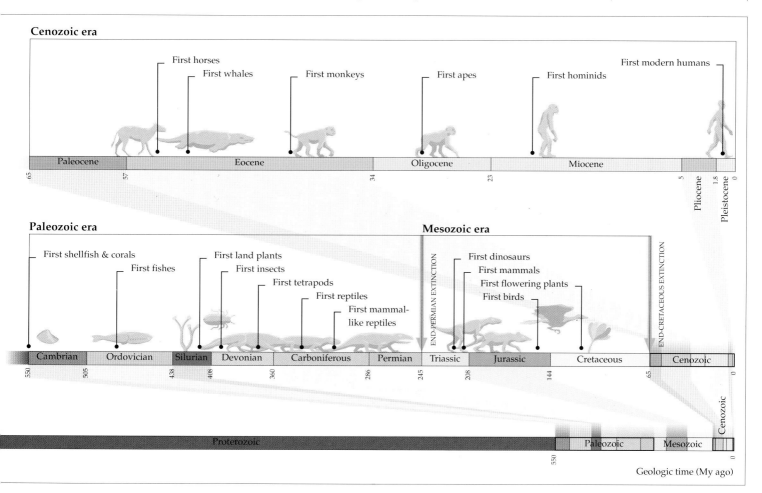

Geologic time (My ago)

formations and mineral deposits thousands of miles apart mirrored each other with a precision that would be all the more uncanny if it was only accidental. Paleontologists knew of hundreds of fossil species too similar not to be closely related (if not identical), but located on different continents. The plant *Glossopteris* and the reptile *Lystrosaurus* appeared in South America, southern Africa, India, and Australia. How had they got there? Probably not by sea, and even where some lengthy overland route was possible in theory, why had the species left no fossils anywhere along it? Coal deposits in the Antarctic were the fossil remains of tropical forests. Rocks in the Sahara desert were scored by the movement of ancient glaciers. Torrid poles and frozen tropics: could climates have flipped so completely?

Wegener's simple answer was that the continents had moved. In Paleozoic times they had been united in one supercontinent, which he called Pangaea. Later it had divided into a northern and a southern landmass, which he called Laurasia and Gondwana. But the continents of Earth were known to be made of comparatively light material lying on a denser base. Paper does not cut scissors. No one could think of a mechanism that would move a softer object through a harder one. Wegener's theory was rejected.

Yet the dossier of contradictions kept growing, and finally burst. In the 1950s, the new science of paleomagnetism ("ancient magnetism") discovered that when new rocks are created their magnetism aligns with the Earth's, and they keep their original polarity except when heat transforms them. The strength of their magnetic field depends on the quantity of iron compounds they contain, but the field will point toward the north and south magnetic poles as they were when the field was first impressed. The magnetism locked into ancient rocks in Europe and North America pointed nowhere near the present position of the poles, so either the poles had moved, or the land had. If you could date the rocks in any region, they would yield a record of the region's north–south alignment, and you could estimate where on the globe that region must have lain, in order for those northward- and southward-pointing magnetic arrows to have intersected at the North or South Pole. Only if Europe and North America were lying next to each other 200 million years ago could they have produced the alignment recorded in the rocks.

The long-established vision of Earth as a basically stable planet with continents rooted in their place had become a scientific "paradigm," a concept so powerful that it shut out not only alternative ideas but inconvenient facts. Overthrow the paradigm, and its prisoners are free to examine new concepts and to feed discarded data into new research. Almost as soon as science was compelled to recognize that continents did move, new theories began to explain what happens on the Earth's thin crust as the product of forces deep beneath it.

The Earth is a spherical heat engine, primed by the primordial heat generated during and shortly after the planet's formation, and stoked by the decay of long-lived radioactive elements located in the interior. Wrapped around a core kept solid by gravitational pressure lie a lower and an upper mantle. The topmost layer of the upper mantle consists of a more or less solid shell, the lithosphere, about 45 miles (70 km) thick. Embedded in this shell is a layer of oceanic crust, an average of 2 miles (5 km) deep, and the thicker crust of the continents, about 4–20 miles (10–50 km) deep. Beneath it is the hot and partially molten layer of the asthenosphere, reaching over 100 miles (250 km) down beneath the surface. Heat convection cells inside the asthenosphere force magma upward, and it emerges mainly in mid-oceanic ridges that run from north to south in both the Pacific and the Atlantic. The magma makes new crust, and forces outward the crust already formed.

This is the machinery that moves the continents by moving the crust on which they sit, which is split up into eight major plates and dozens of minor ones. They grind and jostle on the surface, moving and changing the continents they carry. In some regions, oceanic crust forced outward from the mid-oceanic ridges, and growing cooler and denser as it moves, dives under the thicker but lighter continental crust, back into the asthenosphere, where it melts once again. This process is called "subduction," and the zones where it happens are volcanically unstable, and may give rise to coastal mountain ranges, as in the case of the Sierras and the Andes of North and South America.

The movements of the continents through time have been reconstructed on the basis of a variety of geological and paleontological evidence. As one looks further back in time, the reconstructions become more and more contentious. Some very strong evidence comes from the distribution of particular rocks and fossils (see bottom left). Quantitative evidence comes from studies of paleomagnetism. When iron-rich rocks are laid down or crystallized, they lock into their structure a clear north–south magnetic orientation. In ancient rocks, most of these directions do not correspond to the modern north–south orientation at all. Hence the continents containing the rocks may be rotated back to their original alignments and the earlier continental positions deduced.

CONTINENTAL DRIFT

Over 130 years ago, geographers noticed the striking similarity between the eastern coast of South America and the western coast of Africa; these two continents even fitted together like two pieces of a jigsaw.

One problem had been to explain the distribution of the small aquatic reptile *Mesosaurus*, known from Permian rocks of eastern Brazil and western Africa. Then there was the Southern Hemisphere *Glossopteris* flora. *Glossopteris* was a seed fern, known from late Permian rocks of South America, southern Africa, India, and Australia. How could these rather odd distributions be explained when these parts of the world are separated by great oceans?

Another problem was the late Carboniferous Southern Hemisphere glaciation. Some 300 million years ago, in all the *Glossopteris* flora continents, there was clear evidence for freezing conditions, and the advance of a great ice sheet from the South Pole across the equator to encompass India.

Rearranging the continents as shown below solved all of these problems.

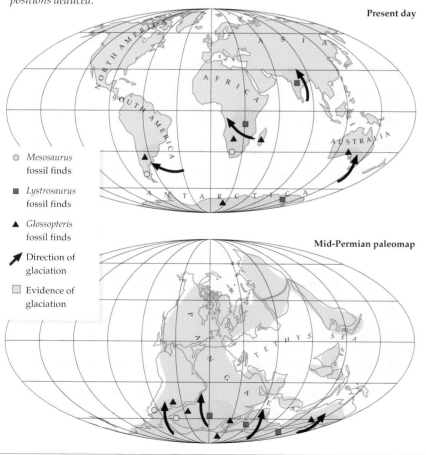

Present day

○ *Mesosaurus* fossil finds

■ *Lystrosaurus* fossil finds

▲ *Glossopteris* fossil finds

↗ Direction of glaciation

▨ Evidence of glaciation

Mid-Permian paleomap

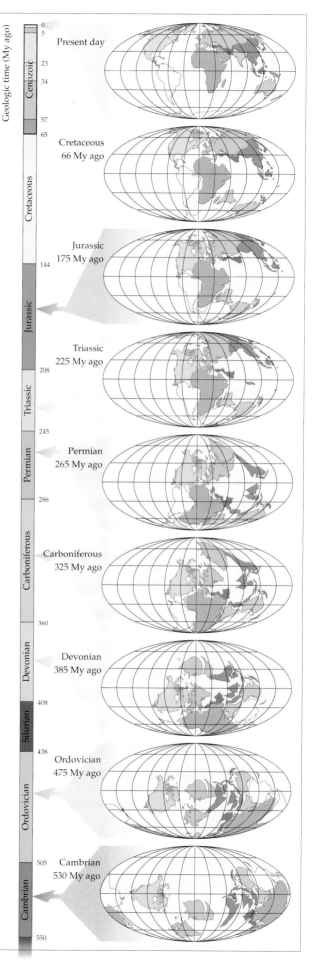

Geologic time (My ago)

Cenozoic — 0, 5, 23, 34, 57, 65

Present day

Cretaceous — 66 My ago

144

Jurassic — 175 My ago

208

Triassic — 225 My ago

245

Permian — 265 My ago

286

Carboniferous — 325 My ago

360

Devonian — 385 My ago

408

Silurian

438

Ordovician — 475 My ago

505

Cambrian — 530 My ago

550

Where two plates collide, and drive one continent against another, the land is forced upward, caught between colossal bulldozers. This happened when the Indian subcontinent plate left its original moorings next to Africa, Australia, and Antarctica, and slid northward over the asthenosphere to collide with the main Asian plate. The Himalayas are young mountains, and still growing.

The study of all these processes is known as "plate tectonics." There is evidence that they have been operating for at least 2 billion years, and for some of that time we have enough information to draw approximate maps of the Earth's surface by combining tectonic data with the paleomagnetic record and with fossil evidence. None of these sources offers a simple message, and conflicting interpretations are rife, and often furiously argued. It is never possible to say for certain what the Earth looked like in the deeps of time, only to draw what seem to be the most logical conclusions from the available facts.

All the same, it is clear that a mobile Earth provides no permanent conditions for any of its life forms. In the long run, not only can continents drift from one climate to another, but their combination or separation can cool, warm, or annihilate the sea, and transform the Earth's climate. Plains can be drowned, or folded and uplifted to become mountains. A species brilliantly successful in its time and place may turn out to be so closely interlocked with its conditions that even a minor change is fatal. Another may thrive because its design was less perfectly adapted, and so more versatile.

Evolution

Life has changed through time. No other explanation will account for the sequence and variety of the life forms preserved as fossils, or the history recorded since humankind began to draw, paint, and carve, about 30,000 years ago. Nothing else will explain the relationships among living species revealed by the techniques of modern biology. All life is interrelated, in a pattern reaching back to the first life forms that appeared on Earth some 4 billion years ago. Every plant and every animal, all of the great kingdoms of life, both living and extinct, could be displayed – if

we could trace them all – as a single system of pathways, roads, and many, many cul-de-sacs, which lead back to a single path, a species that was the common ancestor of us all: slime molds, bacteria, butterflies and buddleia, oysters and oyster-catchers, plankton and people.

Charles Darwin saw a way for evolution to have worked, but he was not the first to recognize that life had changed, and was still changing. Fossil bones too huge to belong to any living species were plowed up or mined in the pre-Darwinian past, and identified as belonging to monsters, giants, dragons. Scholars explained that life must have started with a full cargo of such creatures, and maybe they still lurked at the frontiers of the known world, but they had simply died out. The brilliant French naturalist Jean Baptiste Lamarck (1744–1829), who introduced key words such as "biology" and "invertebrate," was one of many thinkers in the eighteenth century who believed in the *scala naturae*, the ladder of nature. He visualized a chain, or chains, of being, "possibly one for animals and one for plants," with each stage occupied by creatures from the simplest to the most complex, each of them capable of being transformed into the next along this line, or into one of nature's numerous side alleys.

The man who overthrew the paradigm of the chain of being was Charles Darwin. He had spent five years (1831–6) working as ship's naturalist on board HMS *Beagle*, assigned by the Admiralty to survey the coasts of South America. The fossils he collected there, and the living animals he collected and observed on the Galapagos Islands, started him doubting that species could not change. In Argentina and Chile he came across remains of mammals recently extinct, and he asked himself why they were so similar to living forms if they were not somehow ancestral.

In 1859, Darwin published *On the Origin of Species by Means of Natural Selection*. His book both marshaled the evidence for evolution and also gave the framework on which science could organize a mass of information previously too shapeless to have meaning. It was Darwin who provided a structural principle, the first convincing mechanism of long-term change. Natural selection was based on a set of simple propositions:

1 Organisms produce more offspring than can survive and reproduce.

2 The organisms that survive tend to be better adapted to local environments.

3 The characters of the parent appear in the offspring.

4 So generation by generation, hundreds of thousands of times over, the better-adapted lines will survive to pass on the features that give them advantage in local environments.

The strengths of plants and animals are their adaptations, the features that equip them to make the best use of their environment and to pass on that ability. Do they run faster, digest more efficiently, resist cold, or heat, absorb more sunlight?

It was not Darwin but his supporter Herbert Spencer who coined the phrase "the survival of the fittest," and the phrase has caused trouble ever since. Some Victorians decided that the "fittest" must contain some inborn superiority preserved throughout the history of life, and saw evolution as a tree on which humans – European humans – occupied the topmost branch, rather than as a process in which they happened to be enjoying provisional success in conditions that were liable to change.

Darwin saw evolution as a series of gradual transformations, generally much too slow to be observable in present time. In his day, the fossil record was not known well enough to tell the large-scale story nowadays called "macro-evolution." But he had faith in the record yet to be unearthed, and fully expected that new discoveries would eventually be so well scattered over time that evolutionary pathways could be recognized essentially by "joining up the dots." This has not happened. Fossil preservation is too chancy, and even when it happens, the history of the Earth's crust has been far too violent to preserve much more than a random sample, mainly of organisms tough enough to fossilize.

So paleontology has to squeeze more information out of the fossils that do survive, unpacking the meaning compressed into "natural selection," and finding more broad patterns, both inside and outside Darwin's thinking. Since about 1970, a number of dramatic evolutionary models have emerged, based on both fossil and living evidence.

One approach has been to examine the way in which life has diversified. If the whole of our living world arose from a single ancestor some 4 billion years ago, how did that ancestor diversify into our 10 million or more species, not to mention the billions more that have died along the way? It now seems clear that the pattern has been episodic. Long phases of little change in overall global diversity have alternated with bursts of rampant growth or sudden loss. The booms may relate to the conquest of richer or safer new habitats – the shift on to land, or into the trees, or the air, or ever-deeper burrows – and the origin of radical new sets of adaptations, such as the ability to metabolize oxygen, or to fly, and the development of hard skeletons, or warm blood.

Around these topics other questions cluster. What are the main factors that spur evolutionary changes: is it competition with other plants or animals, or is it changes in the physical environment? Could there have been changes in the very ability to change? Push Darwinian natural selection hard, and does it mean that only the perfectly adapted survive, or are most organisms more like *Homo sapiens*, by no means ideally adapted to their modes of life? (Flat feet, arthritic hips, back problems, and hernias would not afflict the ideal biped. Why suffer agonies with only two sets of teeth, when fishes, amphibians, and reptiles can keep replacing theirs?) What is evolutionary success anyway? Is a modern horse "better" than a horse that lived 50 million years ago?

The curve of life's diversity has climbed through time, but a number of steep drops interrupt it, and some of these must stand for mass extinctions. Two powerful ideas of recent times have been those of "nuclear winter" and of the "greenhouse effect," both of them representing threats of severe global changes caused by human behavior. Studies of mass extinctions that have been made since the early 1980s have asked key questions about the speed of these events, their causes and long-term effects, and what factors selected for survival or extinction. Chapter Five examines a few of the more influential theories fostered when human self-interest and scientific curiosity coincide.

There are enough well-dated rock

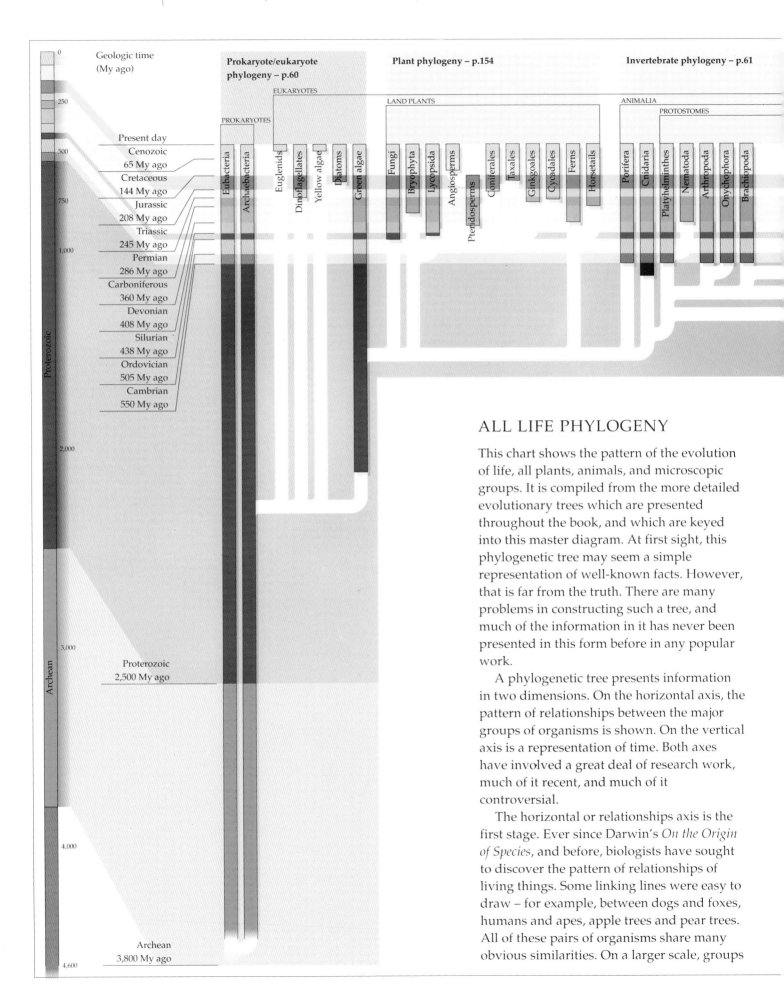

Geologic time
(My ago)

Prokaryote/eukaryote phylogeny – p.60

Plant phylogeny – p.154

Invertebrate phylogeny – p.61

EUKARYOTES

PROKARYOTES

LAND PLANTS

ANIMALIA

PROTOSTOMES

0
250
500
750
1,000
2,000
3,000
4,000
4,600

Present day
Cenozoic
65 My ago
Cretaceous
144 My ago
Jurassic
208 My ago
Triassic
245 My ago
Permian
286 My ago
Carboniferous
360 My ago
Devonian
408 My ago
Silurian
438 My ago
Ordovician
505 My ago
Cambrian
550 My ago

Proterozoic
2,500 My ago

Archean
3,800 My ago

Proterozoic

Archean

Eubacteria
Archaebacteria
Euglenids
Dinoflagellates
Yellow algae
Diatoms
Green algae
Fungi
Bryophyta
Lycopsida
Angiosperms
Pteridosperms
Coniferales
Taxales
Ginkgoales
Cycadales
Ferns
Horsetails
Porifera
Cnidaria
Platyhelminthes
Nematoda
Arthropoda
Onychophora
Brachiopoda

ALL LIFE PHYLOGENY

This chart shows the pattern of the evolution of life, all plants, animals, and microscopic groups. It is compiled from the more detailed evolutionary trees which are presented throughout the book, and which are keyed into this master diagram. At first sight, this phylogenetic tree may seem a simple representation of well-known facts. However, that is far from the truth. There are many problems in constructing such a tree, and much of the information in it has never been presented in this form before in any popular work.

A phylogenetic tree presents information in two dimensions. On the horizontal axis, the pattern of relationships between the major groups of organisms is shown. On the vertical axis is a representation of time. Both axes have involved a great deal of research work, much of it recent, and much of it controversial.

The horizontal or relationships axis is the first stage. Ever since Darwin's *On the Origin of Species*, and before, biologists have sought to discover the pattern of relationships of living things. Some linking lines were easy to draw – for example, between dogs and foxes, humans and apes, apple trees and pear trees. All of these pairs of organisms share many obvious similarities. On a larger scale, groups

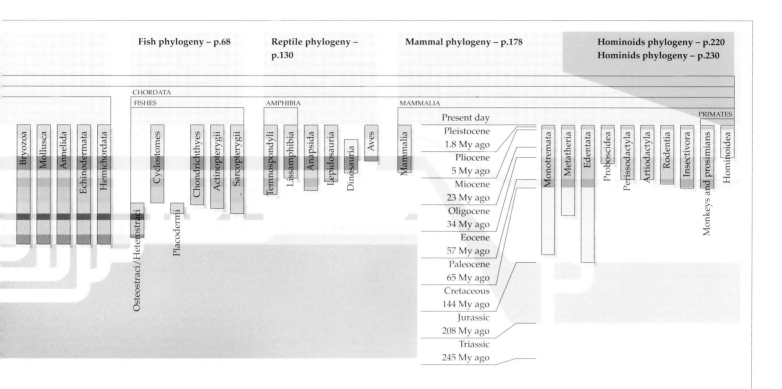

Fish phylogeny – p.68

Reptile phylogeny – p.130

Mammal phylogeny – p.178

Hominoids phylogeny – p.220
Hominids phylogeny – p.230

CHORDATA

FISHES

AMPHIBIA

MAMMALIA

PRIMATES

Bryozoa Mollusca Annelida Echinodermata Hemichordata Cyclostomes Chondrichthyes Actinopterygii Sarcopterygii Temnospondyli Lissamphibia Anapsida Lepidosauria Dinosauria Aves Mammalia Monotremata Metatheria Edentata Proboscidea Perissodactyla Artiodactyla Rodentia Insectivora Monkeys and prosimians Hominoidea

Osteostraci/Heterostraci Placodermi

Present day
Pleistocene
1.8 My ago
Pliocene
5 My ago
Miocene
23 My ago
Oligocene
34 My ago
Eocene
57 My ago
Paleocene
65 My ago
Cretaceous
144 My ago
Jurassic
208 My ago
Triassic
245 My ago

such as birds, mammals, vertebrates, mollusks, flowering plants, and ferns have also been easy to define in terms of their unique features. An obvious consequence of drawing up such groups is the hypothesis that they once shared a common ancestor.

Patterns of relationships are drawn up today using two independent sets of techniques. One, cladistics, is founded mainly on an examination of the anatomy of the organisms in question. Inclusive groups, such as "birds," "song birds," "thrushes," and so on, are defined by their unique shared character, or characters, and the patterns of relationships are shown by branching diagrams, or cladograms. They could equally well be shown by Venn diagrams, the basic way of showing the relationships of mathematical sets. It may ultimately be possible to draw a cladogram of all organisms, living and extinct, in which every species is linked into its correct place in the huge tree.

The second modern technique of analyzing relationships is by comparing molecular sequences of different species. All organisms share some basic proteins in their cells, and, in particular, all living things possess DNA (deoxyribonucleic acid), which codes the information that defines their development.

Proteins are made from amino acids arranged in particular sequences, and DNA is made from base pairs also arranged in particular sequences. Molecular biologists can now read off these sequences with some precision. The more closely related two organisms are (i.e. the later their latest common ancestor), the more similar are their DNA and protein sequences. Many human proteins are virtually indistinguishable from those of the chimpanzee with whom we shared a common ancestor some 5 million years ago. However, our proteins and DNA are very different from those of a lettuce, with which we shared a common ancestor more like 800 My ago. It is possible to draw up unique branching trees for any selected group of organisms based on measures of the shared similarities of their DNA or other protein.

The cladogram, or molecular tree, is converted into a phylogenetic tree by adding the time dimension. The information on time comes from the fossil record. The record of the history of life preserved in the rocks is surprisingly good, after 200 years of concerted study. It is possible to track modern groups back toward their times of origination, and to show with moderate precision the known distribution in time of the various groups .

The phylogenetic tree of mammals (above) is shown separately from the remainder of the tree because of the need to show some greater detail. This part of the phylogenetic tree of life has received an unusual degree of attention, both from cladists and recently from molecular biologists because of our interest in discerning the position of hominoids (humans and apes). Surprisingly, in view of all the research effort, however, the relationships of the main groups of placental mammals (Eutheria) are very poorly resolved. It could be that the concentration of research effort has confused the main problem. More likely the difficulty is that the main lines of mammalian evolution branched off very rapidly 65 My ago, after the extinction of the dinosaurs, and they are thus very hard to disentangle.

THE WORKINGS OF EVOLUTION

The theory of evolution by natural selection has not changed at all in its basic principles since it was formulated by Charles Darwin in 1859. Indeed, some of the best evidence is still based on the kinds of field observations which he had made over the previous thirty years before he published his famous book. Darwin noted obvious parallels between natural selection and artificial selection; it was well known that breeders could produce amazing effects in their plants, dogs, or pigeons simply by selecting carefully those parents that would be allowed to breed and produce the next generation. The same would surely happen to natural populations exposed to natural selecting forces in their environments.

Darwin noted that adaptations are superb evidence for evolution. He argued that plants and animals show features in their anatomy, physiology, and behavior that suit them for the lives they lead. The adaptations might not be perfect, but they are as good as they have to be. Darwin noted that many adaptations are convergent; fishes, reptiles, and mammals that swim fast all look the same, though they have evolved from different sources. Equally, the limb structures of a dolphin and a pig, now so different, can reveal a common ancestry.

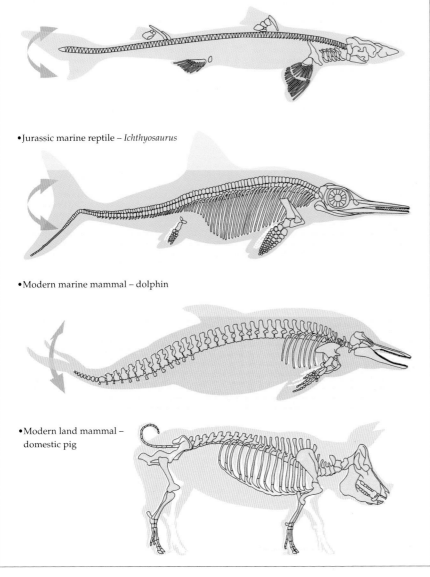

• Typical modern shark

• Jurassic marine reptile – *Ichthyosaurus*

• Modern marine mammal – dolphin

• Modern land mammal – domestic pig

sequences rich in well-documented fossils for paleobiologists to trace the rates and patterns of evolution in some species down long intervals of time. Some evidence now suggests that, contrary to the gradualist expectations of Darwin, evolution may, in fact, work in fits and starts. This may happen mainly during a brief phase of speciation (the formation of new species), and for much of its remaining career a species may then evolve very little. Niles Eldredge and Stephen Gould have put forward the idea of "punctuated equilibrium." It describes a pattern caused when part of the breeding population of a species becomes cut off from the rest, inside a different environment to which it adapts by evolving into a new species that need no longer change. "Phyletic gradualism" is the opposite proposition, based on evidence suggesting that most evolution occurs along the normal span of a species.

Identifying life's patterns

The huge diversity of life, both ancient and modern, is self-evident. Darwin's belief that it was a single system splitting and resplitting as each new species diverged from a single source is confirmed by all the findings of paleobiology and molecular biology. In order to make sense of what we know, we have to disentangle strands of development that we first have to re-create, possibly on the strength of bits of information separated by millions of years, thousands of miles. Which are the main highways, and which the minor routes into the past? Can we devise a language to describe their history, and all the traffic they have carried? That is the function of "systematics," which aims to identify life's patterns, and to understand and name the machinery of speciation and change through time. The branch of systematics that seeks to order these patterns is called taxonomy. Two of its main products are *phylogenetic trees,* which illustrate the pattern of branching relationships inside particular groups, and *classifications*, lists of species and subspecies within a given group, arranged in a descending order from basic categories to the lesser categories that make them up, and then to still smaller units.

The basic unit is the *biological species*. All of its members have to be able to interbreed, and to produce viable offspring. All domestic dogs can do that, no matter how varied their form, so they are grouped as the species *Canis familiaris*. The classification then expands into broader and broader features of animals' *morphology*, their physical form and structure. Dogs belong with the common wolf, coyote, and various forms of jackal in the genus *Canis*. This genus joins other genera of doglike animals in the family Canidae, which is placed alongside the Felidae (cats), Ursidae (bears), and other families of meat-eaters in the order of Carnivora. With other orders, this makes up the class Mammalia (possession of hair, production of milk), which is placed in the subphylum Vertebrata (backbone), of the phylum Chordata (notochord and tail), of the kingdom Animalia (ability to move), of the superkingdom Eukaryota (whose cells possess nuclei and organelles).

The German entomologist Willi Hennig realized in the 1950s that if this pattern of nested groups was logically organized it must reflect a hierarchy receding into time. He proposed a new method of analyzing the phylogenesis (evolutionary history) of species that has since grown into the science of cladistics. The most obvious features did not necessarily define a really close relationship (as when Cuvier lumped elephants together with hippos and rhinos as part of the now discarded order Pachydermata, because they had thick skins). Hennig argued that the key to success was to analyze all the characters of a proposed group and to leave out all features that did not mark a unique development in that group. In Hennig's system, a *clade* was the most logical type of group, because it contained the set of all the descendants of a given ancestor, but no others. Therefore the dinosaur group cannot be a complete clade unless it contains the birds, which descend from theropod dinosaurs.

Molecular biology has produced the most exciting recent advances in systematics by analyzing the proteins, and the DNA, of living organisms. The make-up of each protein in a species is assessed in terms of the sequences of amino acids along the protein molecule. Closely related species have similar protein and DNA sequences. The more distant the relationship, the more these sequences differ. That is because the common ancestor of two very distant relations lived much longer ago

than the ancestor of two close ones, and that gives much more time for their protein or DNA to have changed in the process of evolution. Molecular biologists are eager to sequence every protein and DNA molecule they can purify, so as to run comparisons through computer programs designed to reconstruct probable phylogenies, lines of descent. Sometimes their evolutionary trees will agree with the trees produced by cladistic programs based on morphological data. The arguments that follow when they disagree are hard to resolve because no one can assign a clinching priority to one or the other set of data.

The fossil record

Mostly it is living species that supply the kind of information that the morphologists and molecular biologists use in their search for a valid phylogeny, dependable pathways through the history of life. Only a century ago, it was the fossil record that seemed to offer the only available guidance. Fossils are the remains or traces of bygone life forms, preserved in sediments, coal, tar, oil, amber, or volcanic ash, frozen in ice, or mummified in arid sands or rarefied air. They are all we have left of 4 billion years of seething invention and exploration, performed by the trillions of individuals that make up the clade of Life. Are their voices so little to be trusted?

They are certainly biased. One of the key problems in perceiving, and interpreting, the history of life has been the incompleteness of the fossil record. Few fossils preserve all parts of the original organism. Soft parts usually decay and vanish. Since many organisms are all soft parts, it is utterly exceptional for jellyfishes, sea anemones, worms, most single-celled animals, and a legion of others to be preserved. The commonest fossils are those of plants or animals with hard parts – in tree trunks, the stiffening material lignin; in vertebrates, their bones (made from apatite, a form of calcium phosphate); in mollusks, their shells (made from apatite or calcite, calcium carbonate); in arthropods such as beetles, crabs, or trilobites, their exoskeletons (made from the protein chitin, or calcite).

Soft-bodied organisms are preserved only in rare conditions – for instance a quick burial, dead or alive, in conditions where oxygen is

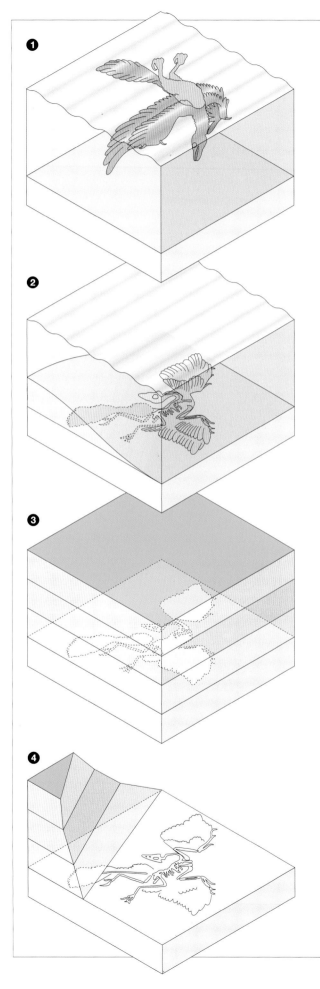

1

2

3

4

Fossils may be preserved in many ways, but the commonest is by burial underwater. The dead plant or animal may initially float for some time. Many larger animals are buoyed up at first by the gases of decomposition. At a certain point, the body wall ruptures spectacularly, and unpleasantly, and the carcass sinks to the bottom of the sea, lake, or river. If the water body is the sea or a lake, the remains may be moved about a little on the bottom before they are covered by fine-grained sediment. If the water body is a river, the carcass may be transported for some distance, then it may rot and break up. Scavenging animals may assist the breakdown by biting off chunks of flesh. Finally, bacteria will proliferate in the carcass and incorporate most of the protein into their own structures.

Usually all of the soft parts of the plant or animal are removed before, or soon after, burial. However, in deep seas and certain lakes, the bottom may have low-oxygen conditions, which inhibit scavengers and may allow near-perfect preservation of the whole organism. Pore spaces within its tissues may become filled with minerals carried in ground waters.

Millions of years later, changing circumstances may bring the fossil bed to the surface where it is exposed by erosion. Should a paleontologist chance upon it, the fossil may be collected, and it may contribute to our knowledge of the story of life on Earth.

absent and therefore scavengers and bacteria are excluded. Taphonomy is the science that studies the conditions in which organisms may be preserved or destroyed. Experiments carried out by taphonomists have shown that decay may set in very rapidly. A dead worm in normal seawater may completely disappear within a few days under bacterial attack. On land, even the carcasses of large animals like elephants may be so scattered by scavenging hyenas and vultures, and lesser flesh-eaters, and their bones so weathered and leached by heat and rain, that a few months of exposure leave hardly anything to be fossilized.

Suppose that an organism has upped its chances of preservation by its possession of hard parts and avoidance of fast annihilation by scavengers. Its best hope now is to have died in, or been quickly transported to, a place where sediments are being laid down, and will turn into rock. Nowadays this happens on scree slopes high on mountainsides, in alluvial fans spreading beneath mountains, on the inside bends of meandering rivers, in deep lakes, in lagoons behind coral reefs, or on the deep ocean floor. Lowlands and seabeds are best; hill- and mountain-dwellers are consistently under-represented in the fossil museum.

Safe in its grave, the fossil has only to survive its millions or billions of passing years. While the sediments are still soft, no flood must scour the lake or river, no storm churn up the tidal reaches, no deep turbidity current cut through the ocean floor. Once sandwiched in the crust, the fossil is vulnerable to earth movements, subduction and melting, uplift and destruction under pressure or high temperature. After all this, it may weather and crumble on the surface unless it is found by someone who knows what it is. It is no accident that so much of the history of life on Earth is focused on what is now North America and Europe. This is where most of the paleontologists live.

Several hundred thousand fossil species are known and have been named, possibly 0.001 percent of all life's species. So it is surprising that for some fossil groups there may be little new to find. New dinosaur species come to light in North America at the rate of fewer than five or six per decade now, in spite of intensive field-collecting programs. Major innovations – dinosaurs that would have to be placed in new families, for example – are found in regions previously little explored, in countries such as China and Australia, or in South America. Perhaps we really do have a reasonable sample of the diversity of dinosaurs, and of a few other groups of large fossil plants and animals. In general, however, paleontologists have only a very limited knowledge of Earth's former populations, though the very randomness of the sample probably ensures that few really basic groups remain to be discovered.

The oases in a mainly barren desert are the fossil *Lagerstätten*, a German term meaning roughly a fossil lode or storehouse. They are localities that yield exceptionally well-preserved fossils, possibly with soft parts showing. Over 100 key fossil *Lagerstätten* have now been identified, scattered across time. They offer us brief sequences that revive the fullness of an ancient habitat like frames intact on a reel of antique film. Here are the standards by which we can measure what has been lost in all the other, more typical fossil sites.

Sometimes the difference is sensational. The world-famous Burgess Shale *Lagerstätte* in British Columbia, Canada, has safeguarded a fauna of about 120 genera of previously unknown, often bizarre-seeming animals. If the soft-bodied animals are removed, as decay would usually ensure, only four or five remain, and these are much the same as the creatures that appear to represent the whole of the local fauna in the ordinary type of fossil deposit. Similar comparisons made by weeding out the fragile organisms from later *Lagerstätten* yield less startling results, but they remind paleontologists that 50 percent or more of all plants and animals found in any environmental setting would not normally make fossils.

One focus of *The Book of Life* is on fossil *Lagerstätten* as treasuries of unique information, keys to past life. The writers have aimed to describe both new discoveries and the latest thinking, working with the artists to produce original reconstructions that convey the profusion of life. If paleontology is "bones and stones," it is also fieldwork, research, and the people who do it. Naturally they argue. Inevitably they disagree. We hope to involve readers in the substance, flavor, and intellectual excitement of finding the evidence and tracing the rhythms that began when matter developed an unusual way to organize itself.

The inhabitants of Earth have some intensely practical reasons to investigate the mysteries of their origins and growth. We need to know whether life is the cosmic rule, or the exception. Does the universe teem with inhabited planets, as some scientists assume, or did something unique happen here? And if life itself is not unique, what about thought? Ever since the 1960s various groups of scientists have worked on SETI, the Search for Extra-Terrestrial Intelligence, a project that scans the most promising wavelengths looking for signs of galactic radio communication. If life is a natural extension of matter, where should we be looking for the rest of the family? If the absence of communication means that technological civilizations gain the power to destroy themselves, could such a pattern be predictable for life on Earth? Perhaps the biological past of our planet is the laboratory that can show us where we may be going, by exploring where we come from.

A packed scene in the Middle Cambrian of British Columbia, based on the fossils from the Burgess Shale. The giant arthropod Anomalocaris *swims above the shelled arthropod* Leanchoilia, *the sponge* Vauxia, *the shrimp-like* Waptia, *and* Pikaia, *a lancelet-like chordate. Below these, another* Anomalocaris *grubs in the mud, and the trilobites,* Olenoides, *scatter. At bottom right, the onychophoran* Aysheaia *creeps around a large sponge,* Vauxia, *and above it the enigmatic* Wiwaxia, *shows its double row of raised spines. At bottom left, a small shoal of the shrimp-like* Marrella *swim beneath three larger* Sidneya. *Some swimming sea cucumbers,* Eldonia, *swim above at top left.*

Chapter One

FOUNDATIONS

LIFE IN THE OCEANS

J. John Sepkoski Jr

Nobody knows how probable life is. Does our galaxy swarm with living creatures because matter has a tendency to form more and more complex structures whose properties include the potential to modify and reproduce themselves? Or does the simplest creature on Earth embody a chain of causes each so unlikely that it takes a cosmos, with its billions of planets, to permit it even once? We have no theory so comprehensive that it enables us to predict from Newton's mechanics, Maxwell's thermodynamics, or Einstein's relativity that one small solar system born less than 5 billion years ago in a suburb of the Milky Way would harbor life. Nor is there any formula that can account for the amount of change and almost inexhaustible variety that have written life's adventures all across our planet. That story is a mystery locked into deeps of time that are mostly beyond our reach. Even when discovery reveals some vital detail and solves a tiny fraction of the puzzle that we seem to confront, it is likely to give rise to a further mystery that points to dimensions of the puzzle not previously suspected. Our life dwells at the interplay between supreme cosmic forces and the ever-changing history of chemical and physical events sometimes too vast, sometimes too tiny, and mostly too chaotic to predict.

Every generation has to rewrite the book of life because it sees more of the story. There cannot be a final version, only the delight and illumination of the new perspectives that more knowledge reveals and that still more knowledge is bound to revise or replace. All of Earth's civilizations have needed to find explanations that make sense of their own beginnings and of the wealth of living things that teem through the waters, lands and skies of a solid Earth warmed by a burning Sun. Our fascination is no different, and it is no less intense for being founded on the same kinds of theories and the same gathering of facts that we rely on, say, to build new bridges or create new medicines.

It is no accident that some of the early history of life has become more accessible in a century which has begun to put together a picture of events involved in the early minutes and seconds of the universe, and of the formation of suns and solar systems. It needed the realization that our planet was once a very different place, to suggest what conditions would transform inanimate matter into the materials of life. The first act of the play took place on a stage that time and life itself have changed. The following account takes up a story that begins with elements forged in the nuclear furnaces of stars long dead, some of them produced and then scattered across the universe by the catastrophic events known as supernovas. They form the atoms that make up "interstellar dust."

About 4,600 million years (My) ago, gravitational forces acting in a cloud of interstellar dust that began to draw all the local matter inward towards the center in spiraling swarms.

The spinning disk of matter that resulted became so dense at the hub that it was able to fuel an atomic furnace. The Sun began to shine, and to blow other matter outward in the resulting "solar wind" of ionized hydrogen. Clusters of matter gathered further out to form spinning planetesimals, the cores of planetesmoids – rocky globes close to the Sun, gas giants deeper in space.

For millions of years the horde of bodies that formed within the inner solar system produced a riot of collisions and glancing blows before the forming planets swallowed most of them. Most of the survivors lay either in a stable zone inside the orbit of Jupiter – the asteroid belt – or outside the orbit of Pluto – the Oort comet cloud. The largest of the surviving asteroids is Ceres, which has a diameter of about 470 miles (750 km). Bodies much larger than that must have been involved in the violent bombardment that created the planets with impacts whose memorials are the craters strewn across the little-altered surfaces of Mars and the other terrestrial planets, as well as our Moon and the moons of the gas giants. The most severe episodes of planetesimal impact seem to have

petered out around 4 to 3.8 billion years ago, but occasional impacts continued, and we shall see that life may have been massively affected by far more recent collisions between the Earth and lesser bodies as little as 6 miles (10 km) in diameter.

The energy delivered by the asteroid bombardment raised the Earth's surface temperature high enough to vaporize water and melt rock. Gravity from the planet's growing mass pulled heavier elements into the core, where the radioative decay of those with unstable nuclei generated the heat that fuels volcanoes and drives the movement of whole continents today. Lighter elements, especially silica, magnesium, and aluminum, floated toward the surface of the new planet much like fat in a stew, to make the crust and outer mantle, gushing out of volcanos and other vents. Yet lighter molecules – hydrogens, oxygens, nitrogens, carbons – streamed into the primitive atmosphere and condensed into the earliest oceans.

The incoming bodies contributed extra masses of light material, including both carbon compounds and water (in the form of ice). Many of life's essential building stocks derive from the ability of carbon to form various (but energetically weak) chemical bonds with other elements; that is why the study of its compounds is called *organic* chemistry. At least some of the complex organic molecules were probably imported during the bombardment. They have been identified by Earth-launched probes investigating the most recent return of Halley's comet, and the atmosphere and surface of Titan, one of the moons of Saturn.

Any stirring of life more than 4 billion years ago would have been sterilized by the intense heat produced by the impacts of early planetesimals. And even when the surface cooled, we would not recognize our Earth. It was an alien planet, most of it ocean, with scattered chains of volcanic islands. The young Earth's day was less than ten hours long. It has been gradually slowed by the tidal drag of our Moon. A dim red Sun poured about 75 percent of its present radiance through a choking atmosphere dense with carbon dioxide and stinking of hydrogen sulfide and methane, with only traces of molecules of oxygen. Brown seas reflected a pinkish-orange sky whose greenhouse blanket of floating organic smog conserved the weaker solar heat. No modern multicellular organism could survive that poisonous world, yet such a world may have been the essential seed-bed for life.

Lift-off

Our human bodies record a sequence billions of years longer than human culture. Nested like painted Russian dolls are features that we share with more and more organisms the deeper we probe. As humans, we have uniquely large brains and upright posture; as primates, fingernails and stereoscopic vision; as mammals, warm blood and milk-fed young; as amniotes (the group that includes mammals, birds, and reptiles) internal fertilization and the ability to reproduce ourselves outside a watery environment. All animals have tissues, and all of us collect our carbon and energy from organic compounds derived from all those plants, algae, and bacteria that can use either chemical energy or light to manufacture their own supplies out of non-living raw materials. Like almost all organisms except bacteria, we have cells with nuclei and chromosomes, organelles, and an oxygen drive. In common with every living thing we have DNA, genetic blueprints, and metabolism – the equipment to absorb, dismantle, and exploit useful molecules.

This sequence is a journey proceeding out of time. Listed so simply, it is apt to sound like the flow of a single force with an inevitable outcome. Yet the long-range order that we see with hindsight is the outcome of a blizzard of causes blowing this way and that. It shows that evolution has not worked like a single steamroller headed in some purposeful direction, but as a set of improvisations and haphazard events, constantly reinventing itself. Rather than redesigning the fundamental machinery of life, evolution appears to have tinkered with the most variable and possibly least essential details. When such changes happen at the right time and place, they may prove useful as conditions change. Like our fellow species, we are a ramshackle collection of these useful adaptations, with layer tacked on to underlying layer, each one reflecting the successful inventions of bygone time. If the book of life could be written again from the very same first sentence, it might tell a very different story.

One key set of plot lines, and the process that writes them, will illustrate this claim. Each organism's genes contain the instructions for building its offspring, but biological reproduction is complex and prone

THE ANCESTRY OF THE HUMAN BODY

The body of any organism reads like an evolutionary chronicle. The more specific some feature is to the species and its close relatives, the more recently that feature evolved. Our own specific features, a large brain and upright posture, evolved less than 5 My ago. Fingernails, which we share with all primates, appeared around 55 to 65 My ago. Other features that we share with more and more animals appeared progressively earlier in time. Finally, when we examine the deep structure of our cells, we find universal aspects of life that date back to the origin of all living organisms.

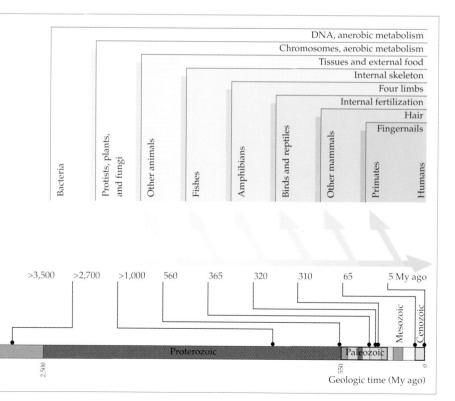

THE STRUCTURE OF THE CELL

One of the deepest divisions within life is between prokaryotic and eukaryotic cells. All life is composed of cells, which are biochemical factories surrounded by porous membrane and including genetic material (DNA) that codes cellular functions and reproduction. There are also structures (ribosomes) where the code is translated into proteins and other cellular components. The prokaryotic cell is rather simple, but the more complex eukaryotic cell has a genome comprised of strands of DNA that are enveloped within a membrane to form a nucleus. The cytoplasm outside the nucleus is crisscrossed by more membrane, termed the "endoplasmic reticulum." In this membrane lie organelles, mitochondria, which energize the cell, and plastids or "chloroplasts," which photosynthesize in autotrophic eukaryotes. Eukaryotic cells bear a whiplike "undulipodium" for movement.

Characteristics of typical prokaryote cell	Characteristics of typical eukaryote cell
Mostly small (0.2–10 micrometers)	Mostly larger (10–100 micrometers)
Genome resides in a single strand of DNA that is not bounded by membrane	Genome resides in 2–600 chromosomes, each in a combination of DNA, RNA, and protein within a membrane-bounded nucleus
No organelles	Most have mitochondria and photosynthetic cells have plastids
Enormous ranges of metabolic patterns	Almost all have a uniform style of oxidative metabolism
No complex membranes within the cell	Endoplasmic reticulum complexly folded within the cell
Simple flagella (whiplike structures for locomotion)	Complex undulipodia severing function of flagella

Schematic diagram of prokaryote cell

Small ribosomes
DNA
Cell wall
Cell membrane
Simple flagellum

Schematic diagram of eukaryote cell

Mitochondrion
Plastid
Large ribosomes
Endoplasmic reticulum
Nuclear membrane
Nucleus
Cell wall
Cell membrane
Undulipodium

to accidents. "Mistakes" must happen, as in the children's game of "Chinese whispers" – or "telephone" in the United States – where a message passed from one speaker to another accumulates random errors and may end as comic nonsense. If the sentence is still grammatical, which in the biological analogy means that the offspring can live and reproduce, it is passed on. Evolutionary change reflects a cascade of "mistakes." Perhaps they alter structure or behavior in ways that allow the organism to deal with its environment or its fluctuations better, or perhaps they provide a faculty that becomes useful in some new environment, or helps preserve its subject through a time of change.

When we think about the way such changes might accumulate through 4 billion years of life's history, we are dealing with processes that far outreach the scale of human experience, with its tendency to convert what we know to familiar terms. One built-in human bias is our "sizeism." Our eyes may overlook whole levels of life's diversity because we ourselves are large – larger than almost all living creatures except for the few creatures that we capture for our zoos or view on whale-watching expeditions. If we can sharpen our focus, we find that each of us is an ecosystem of small unseen organisms: pests such as skin mites or fungi, as well as essential guests. Our intestines hold bacteria whose cells outnumber the cells we call our own – the ones with our own genetic material. We share with all animals these residential caretakers that enable us to digest our food, housed deep inside our bodies, away from the oxygen that is for them a poison. The large "higher" plants are equally dependent on other bacteria that capture nitrogen and supply it to make proteins.

The division of life into plants and animals is a rough human classification that works well enough to describe various shapes and expected behaviors but has little to do with how life arose, or how early organisms made a living. Likewise, among living organisms, especially the small ones, the labels "plant" and "animal" represent two collections of features drawn from a much larger total range, and often found in combinations that ignore these simple categories. There are far more lifestyles than these two labels allow for, and some of them are practiced by organisms that live off what we would consider to be inorganic foods.

Working from definitions based on fundamental cellular machinery, we now recognize three basic types of living things. First are the archaebacteria, "ancient bacteria." Not many species now survive. They evolved on an alien Earth, and now live in extreme environments: hot springs where temperatures never fall below 131°F (55°C); salt flats where salinities are four times ocean levels; the intestines of cattle – or our own, for that matter. Second are the "eubacteria," many thousands of species with an enormously broad variety of lifestyles – fermenters, nitrogen-fixers, sulfur users, oxygen-producers, recycling specialists for all kinds of vital molecules.

Both archaebacteria and eubacteria are prokaryotes, "pre-seeds," so called because of the way they carry their DNA loose inside their outer cell membrane. The third fundamental division of life is the "true seeds" (eukaryotes, EKs for short), which package their genes in a "seed" or nucleus, a separate envelope inside the outer membrane (see phylogenies on page 60). They comprise all of what we consider to be "higher organisms," including ourselves.

There is one other problematic group, the viruses, which are particularly hard to define. They do not have cells, and probably stem from bits of DNA or RNA that somehow escaped from one or another of the three basic groups. Viruses certainly do not grow, and they feed only if we greatly expand our definition of feeding. In fact, they exist only because they can replicate, and the materials they use in order to replicate are what causes problems; these refugee bits of genetic material, when bound with proteins, can replicate themselves only by reinvading and subverting living cells.

Only over the past few decades, since Watson and Crick, have molecular biologists learned to scan the structure of the messages contained in various "informational molecules" – the nucleic acids, DNA and RNA, as well as complex proteins, such as hemoglobin and chlorophyll. With their techniques, they can measure how different these molecules are between species. Investigations quickly showed that species traditionally thought to be closely related

BACTERIAL METABOLISM

We tend to think of organisms as animals or plants: animals that eat other organisms, breathe oxygen, and fuel their cells; and plants that breathe carbon dioxide, capture energy from sunlight, and produce organic matter. These, however, are but a few ways in which life forms make a living. They are the ways of "eukaryotic" organisms. "Prokaryotic" bacteria have a much wider repertoire of lifestyles, including the use of chemical compounds that are noxious (or even lethal) to eukaryotes. This range of bacteria metabolisms is both interesting and important; it tells us about how simple early life exploited sources of chemical energy to maintain itself and reproduce, and it remains essential for the cycling of certain chemical elements, such as nitrogen, through modern ecosystems. A few of the ways bacteria make a living are diagrammed here. But there are many more, including the use of nitrogen (essential for synthesizing proteins) and the processing of metals (whose oxidation can release chemical energy).

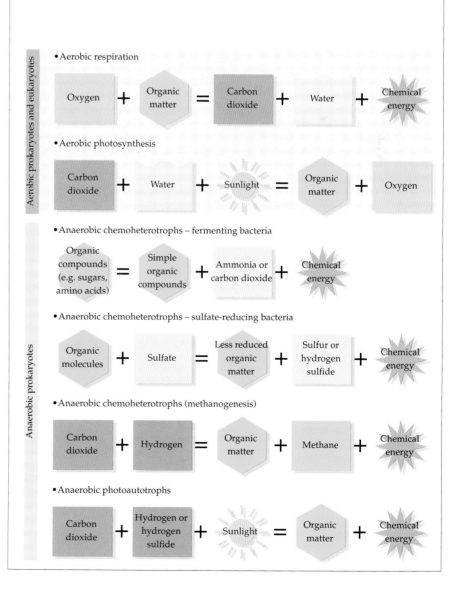

Irregular layers sitting one on top of another, stromatolites are traces of microbial communities, usually dominated by photosynthesizing cyanobacteria. Stromatolites occur mostly in carbonate and cherty sedimentary rocks, and can range in size from a fraction of an inch to tens of feet.

because they looked structurally similar in fact had DNA or RNA that differed only slightly. The more dissimilar the species, the less similar was the structure of their DNA, RNA, or complex proteins. Fossil evidence already suggested that these more dissimilar species had evolved away from each other earlier. Now we had another way to trace the evolutionary tree, even back to points where bones and external structure provided no clue. We could measure similarity in the structures of important organic molecules. This is how the three basic types of living organisms – archaebacteria, eubacteria, and eukaryotes (EKs) – have been distinguished.

One especially suggestive result of measuring molecular similarities among living organisms is the discovery that the bacteria with molecules most dissimilar from other life cannot live in the presence of free oxygen. These most primitive life forms on Earth – found in bubbling hot springs and stinking black muds – die immediately when exposed to oxygen. Their complex, carbon-based molecules quickly break apart. And this is precisely what we would expect for primitive organisms. Complex molecules of carbon, oxygen, and hydrogen cannot form in chemical reactions in the presence of molecular oxygen. It will burn them up – slowly if dilute, or rapidly in flame if the carbon-based molecules are concentrated (as when a pile of oil-soaked rags spontaneously combusts). But when there is little free oxygen and strong energy – lightning, ultraviolet radiation, or even kinetic energy released by impacting meteorites – complex organic molecules can form. These include carbohydrates (like sugars), amino acids (the building blocks of proteins), and nucleotides (the building blocks of DNA and RNA). Experiments simulating early Earth conditions have synthesized all these chemicals and more, ever since the work of Stanley Miller during the 1950s, when he was a student at the University of Chicago. But neither he nor anyone since has created life in a test tube, and there is much we do not understand about how organic molecules assembled into primitive life forms 4 billion years ago.

Without oxygen to breathe, how did life evolve? Modern bacteria again hold the key. Many versatile species of these tiny organisms find sources of chemical energy completely separate from the oxygen we require. Some can live on sulfur compounds and respire hydrogen sulfide, the gas that makes rotten eggs smell so bad, to extract chemical energy. Yet other bacteria live off the nitrogen in various organic compounds, and others ferment organic molecules to capture the chemical energy in carbon bonds. Early life seems quickly to have exploited ways of making a living in a world bathed by radiant energy from even a dim Sun, and free from the burning potential of molecular oxygen.

Another way to make a living that early bacteria seem quickly to have developed is using the energy of the Sun to drive chemical reactions. Even dim sunlight can be absorbed by large colored molecules called pigments, and the captured energy can then be used to break hydrogen loose from some simple compound, to be united with carbon dioxide and form carbohydrates. Primitive bacteria do not use water, H_2O, as their source of hydrogen. Instead they process safer compounds such as hydrogen sulfide, H_2S. (The oxygen released if they stripped hydrogen from water would burn their cells.) But water was an abundant resource, even on the primitive Earth, and through some series of genetic mistakes or mutations, several groups of bacteria began to synthesize complex molecules – enzymes – that could capture oxygen within their cells and then safely escort these lethal molecules outside their cell walls before any damage was done. Now they could exploit the most abundant source of hydrogen in the environment for photosynthesizing organic compounds. Cyanobacteria, one of the groups that evolved this capacity, are the most abundant fossils in Precambrian rocks and remain very abundant in the oceans today.

Carbon has two stable isotopes, ^{12}C and ^{13}C, the first very slightly lighter than the second. When living organisms use it, they tend to select the lighter isotope, which costs fractionally less energy to absorb. This means that when sedimentary rocks are found to contain more than the usual ratio of ^{12}C to ^{13}C, it was some form of life that must have altered the balance. Some of the oldest rocks on Earth form the Isua Group in Greenland, volcanic and sedimentary rocks laid down 3.8 billion years ago. Some chemical analyses suggest that there is a ^{12}C surplus in these rocks, and if this is true life must have put it there. As evidence for photosynthesis it would be startling because it comes a mere 200 My at most after the Earth grew cool enough to live on. It took fast travel to have come so far.

But what kind of photosynthesis were the organisms using? A vital marker is the estimated 600 trillion tons (tonnes) of iron ore composing the banded iron formations (BIFs) that are today Earth's key commercial source of iron. BIFs are marine rocks in which layers of iron-rich sediment alternate with iron-poor layers of other minerals. Only an alien Earth nearly free of molecular oxygen could lay them down. This is because reduced (oxygen-free) iron will dissolve in seawater, but oxidized iron – familiar rust – will not dissolve and will precipitate rapidly. BIFs indicate a cycle of activity, in which some possibly seasonal photosynthesizing agency turned up the oxygen supply, and rusted the iron dissolved in the primitive seas. Most of our BIFs were deposited between 2.5 and 1.8 billion years ago, but traces appear in the Isua Group – traces of photosynthesizing life.

Layered structures called stromatolites ("stone mattresses"), found in sedimentary limestones on every continent, are easily the most recognizable evidence of widespread life in the Precambrian. Sometimes they look like slices through the heart of a giant cabbage, sometimes like flat corrugations, or simple or branching pillars. In some places they appear as deposits over $1/2$ mile (1 km) thick and hundreds of miles across, built mainly between 2,000 My ago and the start of the Cambrian; but rocks from the Warrawoona Group in Western Australia, 3,500 My old, also contain stromatolites.

No one was quite sure about the biological nature of stromatolites until modern equivalents in Western Australia, Florida, and elsewhere were studied during the 1950s and 1960s. They turned out to be the work of bacterial communities dominated by photosynthesizing cyanobacteria, which form feltlike microbial mats covering the surface of shallow-water sediments. The microbes secrete a tacky gel that protects them from ultraviolet radiation and environmental contaminants. But the gel also causes sediment to stick to the microbes, and when the ensnared sediment grows so thick that it

dims the light, the community creeps sunward and a new microbial mat starts to build, and then another, and so on.

Stromatolites are trace fossils, byproducts of life. The first confirmed bacterial fossils from the Precambrian were discovered in 1953 when Stanley Tyler took samples from the Gunflint Chert, a rock unit 2 billion years old, associated with BIFs along Lake Superior in Canada. Tyler teamed with the Harvard botanist Elso Barghoorn, who studied the samples by cutting slices of rock, gluing them to glass plates, and grinding the slices so thin and translucent that their components could be studied under a microscope. He found tiny chainlike threads and spherical forms, the fossils of cyanobacteria and other prokaryotes. This pioneering work, along with that of B. V. Timofeev in Russia, who was discovering slightly younger fossil microbes, established the science of Precambrian paleontology.

Few sedimentary rocks survive from the oldest interval of Earth history, the Archean eon of 4.5 to 2.5 billion years ago, and those fossils they do contain are bacteria that were pickled in concentrated saltwater and fossilized in silica. Ancient cratons, the cores of early continents preserved in Canada, Australia, and South Africa, show that the Archean landmasses were small, and rich in volcanic rock. Earth's interior was hotter and more active than today, and faster heat-flows acted to break up any larger continents that began to form. When volcanic upsurges brought raw iron and other unoxidized minerals or gases to the surface, these would react chemically to pull any free oxygen out of the environment.

Around 2.5 to 2 billion years ago, early in the Proterozoic eon, a different Earth emerged as many of the radioactive elements present at its birth became expended, heat production tapered off, and crustal movements slowed. Larger landmasses began to form when smaller blocks collided and clung. Spreading expanses of shallow sea over these new continents provided extensive habitats where photosynthetic cyanobacteria could thrive. Life began to change the planet irrevocably.

The increasing populations of cyanobacteria on the sunlit bottoms of the new shallow seas relentlessly pumped out free oxygen, and the Earth rusted. From 2.2 to

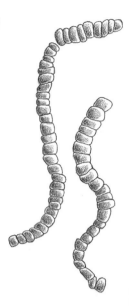

Warrawoona microfossils are the oldest known fossils, discovered in 3,500 My-old cherts in Western Australia. They are the remnants of filaments of cylindrical cells. This simple morphology is similar to some groups of living bacteria, especially cyanobacteria.

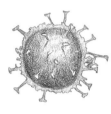

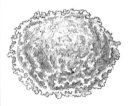

Precambrian acritarchs are the small, resistant cysts of unknown, single-celled eukaryotic organisms. Some are very simple spheres and ellipsoids, often found squashed with broken rims. Other forms are more elaborate with ornamented surfaces. The oldest acritarchs date from around 1,800 My ago.

1.8 billion years ago, huge volumes of banded iron formations were laid down. The massive amounts of reduced iron that had accumulated in the oceans rusted and precipitated to the bottom. Once the reduced iron ran out, there was no other chemical sink big enough to hold the continuing biological surge of oxygen, and this gas built up, first dissolved in water and then escaping into the atmosphere.

Tiny organisms that exploited a wealth of carbon dioxide, water, and sunlight had transformed the Earth. Their pollutant waste product, molecular oxygen, had turned the pink skies blue, the brown seas azure, sweeping out of the atmosphere the fumes of methane and hydrogen sulfide given off by other simple life forms. The stage was set for further change. More than half of geological time had flown.

Strategies for a changed planet

The oxygen boom must have brought a worldwide crisis with it. It drove older organisms that lacked oxygen-handling enzymes deep into airless habitats – stagnant waters, sediments, or dead organic material, which they could rot. Others it killed; some odd single cells in the Gunflint Chert do not appear again. But other bacteria stumbled upon cellular machinery able to use the power of oxygen to break down food into carbon dioxide and water. This new physiology yielded much more energy than older metabolisms, but it did not supersede them. Deep in the cells of oxygen-breathers the old machinery works to break down food into byproducts that it passes on to the newer oxygen-burning equipment for further conversion.

The origin of the "true-seed" EKs (eukaryotes) is a story so far from the traditional image of evolution through cutthroat competition that it has only come to notice since the 1970s. The American biologist Lynn Margulis observed that many living bacteria work together, exchanging biological services that help them to move about, or make food, or even reproduce. Some species live inside the cell walls of another, using its food but returning useful molecules.

Sometimes more than two species collaborate in this barter economy.

The EK cell, the basic unit of complex organisms, consists of a membrane wrapped around a drop of protoplasm, as in bacteria, but with a "true-seed" nucleus at the center. DNA in the nucleus combines with proteins to form chromosomes more complex than in bacteria, and carried in a membrane of their own. In almost all EKs, the cells also contain organelles – structures with special functions, such as the mitochondria that make and store chemical energy, or the chloroplasts that use light to make sugars in photosynthesizing EKs. These organelles are housed in their own individual membranes. Mitochondria resemble some oxygen-using purple bacteria, and most chloroplasts share many features with prochlorophytes, close relatives of cyanobacteria. Neither of these organelles is produced directly by the EK cell itself; instead they divide by simple fission controlled by their own DNA, much as bacteria do, when the nucleus divides to reproduce the cell. Most biologists now accept Margulis's evidence that the organelles are former partners of the ancestor cell – guests that had more to gain by trading off services than by going it alone.

When did they move in? It must have been after the oxygen build-up began, because the mitochondria are oxygen-users. Surprisingly, the host cell – the part with the nucleus – seems not to have been derived from the common eubacteria but rather from the more ancient archaebacteria. Containing both nucleus and organelles, the new EK cells were larger than previous bacteria. About 1.8 billion years ago, larger cells start to appear in the fossil record. They are termed acritarchs, tough organic coats or cysts, formed as a resting stage when single-celled EK algae or protozoans met with bad conditions, and shed when conditions improved.

EKs made a further and revolutionary jump. Differences in their genetic program enabled them to build bodies that used several cells instead of one, cells with a range of different structures and functions but which could still chemically communicate with each other. This was the turning-point that set life expanding out of the microscopic to exploit new resources by building complex structures that would range in size from mosses to sequoias, aphids to dinosaurs. Bacteria often have many-celled bodies, but each cell is almost identical. Only EKs can variegate their cell types into skin, bone, muscle, blood, leaf, bark, seed – a torrent of diverse shapes, tissues, and functions. From what we now know, EKs evolved multicellular pathways at least twenty separate times, far more than the three represented by animals, plants, and fungi.

Origin of complex life

We have little evidence for the early multicellular organisms. Their oldest fossil is *Grypania*, preserved as coiled films of carbon up to $1^1/4$ in (30 mm) in diameter on rocks in Michigan reported to be 2.1 billion years old – rather older than the earliest large fossil cells. Macroscopic carbon films become more common in younger Precambrian rocks but cannot be cited for sure as multicellular EKs. Much later comes a burst of evolution that gave rise to groups that include red and brown algae, chromophytes (such as diatoms and yellow-brown algae), green algae, fungi, and animals. Estimates based on the "molecular countback" system suggest that this surge took place 1 billion years ago, and the fossil record seems to confirm this.

The late Precambrian has a lot more sedimentary rocks intact, because the Earth's recycling machinery has had less time to bury, erode, or otherwise change them. In these rocks we start to see organisms that can be assigned to modern phyla. (Phyla are groups of species sharing some very basic features of their structure and biochemistry that are unique to the group and very hard to relate to other groups.) As well as multicellular red, brown, and green algae, there are other complex fossil structures not yet entirely understood.

What happened to release these cascades of change? They may coincide with the genesis of EK sex.

In advanced EKs, cells typically have paired chromosomes. Each parent contributes a randomly selected half of its own genetic material from each pair of chromosomes to create gametes, reproductive cells. No two gametes are the same. Each time that gametes combine to create a zygote – the fertilized cell

that will grow into a new organism – they share a different package of information that has been shuffled and redealt. Why reproduce in this way? In retrospect we can see that the constant reshuffling of genes has great evolutionary advantages. Mutations from different parental stocks can become variously combined in their offspring, leaving them with considerable genetic differences. Evolution is a constantly changing obstacle course that sorts through genetic combinations and selects those that allow individuals to best exploit resources and leave more descendants. With more genetic combinations, the sorting can be done much faster. Asexual organisms, which replicate nearly exact copies of their genome (their total genetic material) in their progeny, provide far less genetic variation for evolutionary sorting to work on.

But what favored two-parent sex in the first place? Obviously not advantages still hidden in the future. One appealing theory suggests, as the quick-fix gain, that sex confuses parasites. Retro-viruses that can penetrate cell membranes are stray bits of DNA with the ability to splice themselves into the sequence of the host cell's genetic program at a chemically recognized site. But if the host's program keeps changing with each generation, the sites are always being scrambled, and so may be harder for the viruses to find.

Perhaps it was an evolutionary invention to combat viruses that triggered the EK takeoff about 1 billion years ago, and drove the high rates of evolution that followed, or rather that must have followed. Yet no metazoans – multicellular animals – appear in the rocks of the fossil record until about 600 My ago, a silence of hundreds of millions of years – no fossils, no traces, no chemical indicators.

Were they too small? The bacterial mats that built stromatolites offered an ocean of easy pickings, but their fossils show no sign of the inroads made by grazing animals, small or not. If something in the environment blocked the rise of hungry metazoans, the simplest agent would be oxygen levels. While we know that oxygen began to accumulate in the Earth's atmosphere around 2 billion years ago, we do not know how much there was. There could have been plenty, as today, or only a little. Even with small amounts, single-

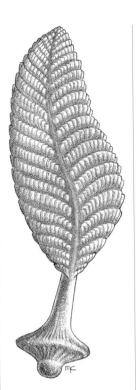

Charniodiscus, from the Ediacaran fauna of Shropshire. This frond-shaped animal, about 20 in (50 cm) long, lived tethered to the bottom by its bulbous holdfast. The frond on either side of the central axis has the quilted construction typical of many Ediacaran animals This animal could be a cnidarian, possibly related to living sea pens.

celled EKs can live, because their enzymes will carry oxygen the short distance from cell wall to mitochondria. But three-dimensional animals with numerous cells need a much richer supply, otherwise only their outer cells will thrive while the rest are starved. It is physically impossible for aerobic organisms with more than a few hundred cells to evolve in a low-oxygen environment.

The latest Precambrian was a time of many changes. A supercontinent, created earlier when plate movements welded smaller landmasses together, began to break up about 800 My ago. As it did, a series of glacial episodes ended with the harshest ice age in Earth's history. It left glacial deposits in Australia about 600 My ago when that continent was within 10 degrees of the late Precambrian equator! Photosynthesizers had drawn so heavily on the greenhouse gas, carbon dioxide, that they nearly caused the planet to freeze over. And in drawing down the carbon dioxide, they must have released much more oxygen into the atmosphere.

Yet it was this same period that saw a severe decline in the volume and variety of the stromatolites formed by those major photosynthetic agents, the cyanobacteria. Since about 1.5 billion years ago they had flourished in tropical seas almost everywhere. Their steepening decline after about 680 My ago, even at times when there were no glaciers, tells us that some new player had arrived onstage and was restricting or actively reducing those communities.

Earliest animals

Five-sixths of the story has gone by. Life has remolded its planet, but by slow persistence over the vastness of time, and through the work of organisms that paced their rates of evolution over hundreds of millions of years. A radical shift comes with the birth of the metazoans, animal giants by comparison, racing through evolution.

Right from the start, the rise of metazoans put evolutionary science in a quandary. Cambrian rocks no older than 550 My contained the earliest animal fossil record – arthropods, mollusks, brachiopods, and others. Below them there were no animal fossils. Darwin himself conceded that his

theory of evolution by natural selection required a history of previous populations for all these creatures to have descended from. Scientists hatched a brood of suggestions; all fossil-bearing rocks from the critical interval of animal evolution had been eroded or metamorphosed, or animals arose in freshwater lakes and only later entered the oceans. None proved satisfactory, and the wealth of animal fossils that defined the beginning of the Cambrian age remained an enigma.

Since the late 1940s, a series of discoveries made all over the world have enlarged the picture and lowered the time levels where more complex life forms first appear. They have also revealed far more life in the Cambrian itself, animals whose preservation would once have seemed impossible. The new story is more coherent than the old, but it raises new questions, and it still leaves hundreds of millions of years of metazoan ancestry during the late Precambrian without a fossil history.

A whole new period in Earth history has been recognized in recent years, known as the Vendian or the Ediacaran, after the Ediacara Hills in South Australia where the geologist R. C. Sprigg discovered its fossils in the late 1940s. Since then, similar evidence has been discovered in regions from Shropshire in England to Namibia, and from Russia to Newfoundland. The fossils date from 580 to 560 My ago, and are impressions or molds of a variety of animal forms, preserved in sedimentary rocks, without a trace of hard parts. They range in size from less than $1/2$ in (1 cm) to more than 40 in (1 m), and most are either disk- or leaf-shaped. The paleontologist Martin Glaessner (a refugee from Nazi Germany who settled in Australia) realized, when similar finds made in Shropshire turned out to lie underneath undoubted early Cambrian rocks, that Sprigg's fossils must predate the Cambrian.

Glaessner and his colleagues believed that they detected the roots of a number of modern phyla in the Ediacaran fauna – possible cnidarians (sea-pens, jellyfish), annelid worms, arthropods, and others – but the preservation is not sharp enough to identify clinching features. The paleontologist Dolf Seilacher offered a different view in the 1980s. He believed that true jellyfish, for instance, would have been preserved as dimples in the sand, as when they are stranded on a beach. The Ediacaran "jellyfish" appeared as bumps on the undersides of sandstone beds. To him, that indicated impressions of animals living on the bottom mud when the sand was deposited over them, not floating in the water. *Spriggina* had been viewed as a worm, tapering from a broadish head, but turn it vertical and this "head" becomes a base for a simpler animal planted on the sea bottom.

These and other features persuaded Seilacher that the Ediacaran fauna were a separate development, not related to modern animal phyla. He called attention to the curious structure that makes some of the animals look like quilted air mattresses, and he even speculated that they might have been large single-celled EKs, not metazoans at all.

We are still seeking a firm understanding of the Ediacaran animals, but it is possible that they represent an intermediate stage of simple animals that included various evolutionary "experiments" soon discarded. James Valentine of the University of California has argued that they could have built their bodies using eleven or fewer different types of cells, which is the maximum number in primitive animal phyla such as jellyfish. (Mammals may have more than 200 types of cells.)

These Ediacarans would have been static "suspension feeders" – animals that wait for the current to deliver sea-borne food particles, bacteria, and single-celled EKs.

The key theme here is the differentiation of cells, which is no simple task for evolution. It takes complex genetic machinery not only to grow a range of cells with specialized functions, but to have them work in harmony. Three billion years of evolution had selected for organisms best able to pass on their genes, yet most of the cells in an animal's body have jobs that require them to give up that goal.

Animals have an exclusive technique for managing their reproductive cells. Early in their development they set aside – in bio-jargon, "sequester" – a small team of cells destined for sexual reproduction and keep them in storage. Leo Buss of Yale University has suggested that this could be a precaution taken to protect the genome. Most of a body's cells replace themselves during its lifetime by

division, and do it so often that there are likely to be "mistakes" when it happens. The sequestered cells divide less often, and thereby minimize mistakes that could be fatal to the offspring they produce.

Other multicellular EKs use very different techniques. Plants, for example, keep cells capable of sexual reproduction at each growing tip, so that if one tip fails by mutation, others may still be available. Fungi have a "coenocytic" (cells-in-common) structure that contains few cell membranes and keeps a mass of nuclei in hollows in the body. Elaborate rules select which nuclei will be chosen to become spores for reproduction. One model for the origin of cell specialities in metazoans suggests that the ancestral animal might have been coenocytic, like fungi. Early forms could have cavities where many nuclei interacted with each other to settle which task they should perform. When these functions became more settled, cell membranes reappeared to protect the nuclei that performed them. Recent molecular studies indicate that animals are most closely related to the fungi on the broadest family tree. Perhaps some of the Ediacaran types will turn out to be holdovers from a coenocytic stage that experimented with sealed compartments dividing labor among a mass of nuclei.

But in any case, the Ediacaran animals were not alone at the end of the Precambrian. Other kinds of animals must have been present, because they left their mark, though not their bodies, at various Ediacaran locations – straight or wavy furrows on the sediment surface, or burrows extending just below it. The makers were small animals that were grazing microbes on the surface, or eating mud or silt so as to digest the organic particles and bacteria inside. (These are common behaviors to this day.) Whatever made the marks, it was not the fossil leaf and disk shapes, but some kinds of soft-bodied animals with a less preservable coat. Present-day

Spriggina, an Ediacaran fossil, 1½–4 in (4–10 cm) long. Some scientists have interpreted this to be an early annelid worm, or even a protoarthropod. Others think it is a colonial cnidarian-like animal with a bottom holdfast and a quilted, upright series of chambers.

Cloudina, *the oldest skeletal animal fossil. When broken apart, these fossils appear like stacked cones of calcium carbonate, about 1 in (2.5 cm) long. The animal is reconstructed as a simple tube-dwelling polyp that used its tentacles to capture food particles from seawater.*

OPPOSITE PAGE, TOP LEFT Opabinia, *a fantastic predator from the Burgess Shale. This streamlined segmented animal had swimming gills along its body, five eyes on its head, and a clawed, nozzle-like structure for capturing prey. Like so many Burgess animals, its evolutionary pedigree is unknown, but it is probably related to* Anomalocaris.

Cambroclaves, *a kind of sclerites known from Atdabanian and later rocks. The animal that wore these tiny spined plates is unknown and long extinct, but fused fossil clusters indicate that the cambroclaves must have covered the animal like a knight's coat of mail.*

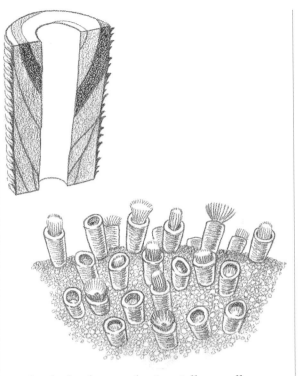

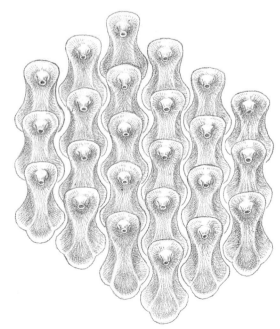

animals that burrow horizontally are all more complex, with more kinds of cell types, than jellyfish – the nearest comparable phylum to some of the Ediacaran fossils. These modern burrowers have more complex tissues, and also organ systems. They are *bilateria*, with the front end different from the rear and the left side a reflection of the right. They meet the environment "head-on" instead of sitting or floating in some roughly circular, directionless form. Nerves concentrate on the leading end, which also contains the feeding machinery. It was the early bilateria that gave rise to "advanced" animals.

One other fossil at Ediacaran localities stands out because it was unique for its time. *Cloudina* is the earliest known animal with a mineralized skeleton. It looks like a stack of miniature icecream cones, up to 1 or 1 1/2 in (3 or 4 cm) long – a structure that can be built by an animal perhaps no more complex than a cnidarian, that was secreting calcium carbonate out of its body.

Carnival of the animals

Where Ediacaran fossils lie directly beneath Cambrian rocks, they fade out some distance below the first appearance of Cambrian fossils. Dolf Seilacher has proposed that some sort of mass extinction, either gradual or catastrophic, wiped out the distinctive Ediacaran faunas. He argues that few Cambrian or later animals can be traced back to them. On the other hand, the Cambridge University paleontologist Simon Conway Morris has identified some later animals that may be descendants, including a middle Cambrian sea-pen rather like the leaf-shaped *Charnia*.

The wave of discoveries that rewrote the story of the earliest Cambrian began when the former Soviet Union mustered sizable teams of scientists to explore geological resources in Siberia after the end of World War II. There, above thick sequences of Precambrian sedimentary rocks, lie thinner formations of early Cambrian sediments undisturbed by later mountain-building events (unlike the folded Cambrian of Wales). These rocks are beautifully exposed along the Lena and Aldan rivers, as well as in other parts of that vast and sparsely populated region. A team headed by Alexi Rozanov of the Paleontological Institute in Moscow discovered that the oldest limestones of Cambrian age contained a whole assortment of small and unfamiliar skeletons and skeletal components, few bigger than 1/2 in (1 cm) long. These fossils have been wrapped in strings of Latin syllables but have been more plainly baptized in English as the "small shelly fossils" (SSFs for short).

In 1969, a 380-page monograph was published in Russian describing the unknown fossils, and paleontologists who now knew what to look for began to discover parallel

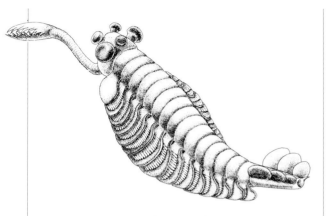

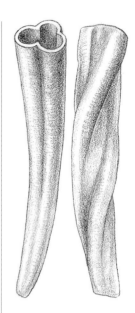

sequences scattered in sites from Meishucun in southern China, and India, to Newfoundland and Nova Scotia, and from Shropshire to southern Australia.

The study of SSFs has shown that they make up a temporal sequence of coherent sets of fossil animal species – in paleospeak, "faunal assemblages" – which can be recognized all over the world. The *Anabarites–Protohertzina* (*A–P*) assemblage lies lowest, and contains the oldest fossils with hard parts, other than *Cloudina*. *Anabarites* fossils are small calcium carbonate tubes with three ridges or keels, sometimes spirally twisted, the work of an unknown animal that probably lived inside the tube. Their three-part symmetry is unknown among living animals but appears earlier in some Ediacaran soft-bodied fossils. *Protohertzina* fossils are curved spines of calcium phosphate very much like the feeding graspers of modern arrow worms, which are predators. Thus, these fossils may be from the oldest carnivorous animals preserved in the fossil record. Vertical burrows dug several centimeters into the sediment appear for the first time in the *A–P* zone, either as single bores or else U-shaped, with both ends emerging at the sediment surface. Nowadays a number of different worms, arthropods, and other marine animals live this way, sucking in

particle-laden water at one end and expelling feces, indigestibles, and metabolites at the other. The makers of these traces were probably soft-bodied animals of several different phyla.

A more diverse set of SSFs appears near the top of the *A–P* zone and typifies the fauna of the overlying Tommotian stage, an interval of several million years dated around 550 to 540 My ago. Most are fossils less than $1/2$ in (1cm) across, a wide range of shells, tubes, and skeletal bits and pieces made of calcium carbonate, calcium phosphate, or silica. Here various primitive mollusks make their appearance. The single-shelled monoplacophorans moved like a snail on a muscular foot. The rostroconchs, now extinct, had folded shells that enclosed their body. Later in evolutionary time they probably gave rise to the bivalves (such as clams and cockles) that broke the fixture and hinged the shells. The inarticulate (hingeless) brachiopods had paired shells held together by muscle alone.

Many more tubular fossils appear in the Tommotian, their owners variously identified as extinct phyla or as simple foraminiferans, jellyfish, and wormlike suspension feeders, still plentiful today.

After billions of years of microscopic life, suddenly these early Cambrian seas were swarming with animals large enough to rise

Anabarites, a calcareous tube from the beginning of the Cambrian. This fossil has a very unusual three-point symmetry.

BOTTOM LEFT *Eopteria, a rostroconch mollusk. Although* Eopteria *comes from the latest Cambrian and Ordovician, it was descended from earliest Cambrian rostroconchs. These animals lived much like clams but did not have two valves.*

LEFT *Archaeocyathans. These solitary-to-colonial simple animals seem to have been related to sponges, filtering food particles from water brought in through their porous skeletons and expelled through the central opening.*

BELOW *Latouchella, an earliest Cambrian monoplaceophoran mollusk. The animal probably grazed on bacteria and algae, moving over the surface on a muscular foot.*

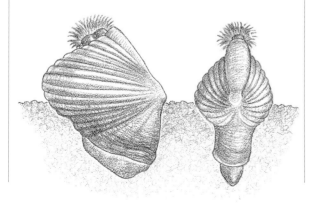

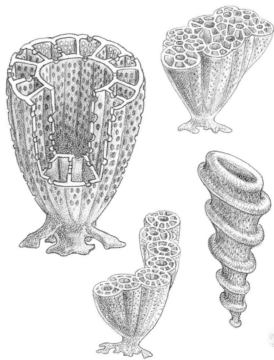

Chancelloria, composed of seven-rayed spicules. Although these resemble sponge spicules, they are hollow and constructed very differently. Still, complete fossils from the Burgess Shale show that the complete animal had a spongelike body.

Helicoplacus, a very early echinoderm. This unusual animal was shaped like a plumber's bob. Spiraling around its body were rows of calcite plates of typical echinoderm structure. Also in spirals were three feeding grooves rather than the typical five of the pentameral echinoderms.

Dailyatia, tommotid sclerites from the early Cambrian. These tiny plates are composed of calcium phosphate and are unlike the sclerites of any animal alive today. The reconstruction shows how the elongate tommotid animal may have appeared, with two rows of armor shielding its back.

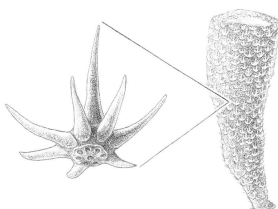

above the murky sea bottom, to channel water currents through their shells, or to burrow deep into mud and sand. There were foraging and tube-dwelling worms, all sorts of mollusks, brachiopods, and others. The conical-shelled hyoliths may be related to the living peanut worms (sipunculids). The spongelike archaeocyathids, with their porous calcium carbonate two-layered skeletons, one inside the other and connected by struts, were a lot bigger than most Tommotian animals. Some had conical skeletons and looked like vegetable strainers with a second strainer stacked inside. By the middle early Cambrian they formed dense thickets of individuals, and became the cores of small reefs. Evidence of the earliest true sponges is provided by various five- and six-rayed spicules, tiny silica needle clusters that function to hold their body shape together.

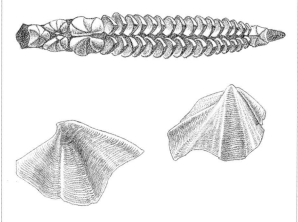

Various other spicules with different shapes and formed from different minerals were made by unknown animals, while equally unknown owners left the puzzling array of tiny fossils called sclerites, which are plates or spinelets that once combined to compose a suit of flexible armor called a

scleritome. Molluskan chitons still wear this kind of chain mail as they graze over intertidal rocks in modern oceans. When their owner dies, the very diverse sclerites lose the connective tissue that held them together, and fall apart like an unstrung puppet with no recognizable components. Rarely, the whole animal turns up and we can see how it wore its sclerites. Others remain as perplexing assortments of puzzle pieces that cannot be put together to shape their owners.

Above the Tommotian assemblages lies the Atdabanian stage, dated around 540 My ago, where many surviving Tommotian groups are joined by a squad of new recruits. These include scleritome-bearers that offer just as few clues as their predecessors, but some of the groups that now enlist are launching much more familiar careers. The Atdabanian introduces the earliest bivalve mollusks, articulate brachiopods with a tooth-and-socket hinge to stabilize the shells, and the oldest echinoderms, relations of modern sea urchins and starfish.

Arthropods make their fossil debut here, though Tommotian rocks bear trace fossils, made by legs scraping the sea bottom, possibly engraved by older species with skeletons too fragile to endure. The Atdabanian fossils include trilobites, an arthropod class that left the most common fossils of later Cambrian rocks, though we shall see that they were not necessarily the most common members of the complete Cambrian fauna. Trilobites were shaped like wood lice or pill bugs – oval, and divided into three segmented sections, head, thorax, and tail, the exoskeleton thick with calcium carbonate. Later species grew up to 2 ft (60 cm) long. Some fossil arthropods were ostracodes, tiny crustaceans living within pairs of phosphatic shells. Other Atdabanian arthropods were bizarre forms, difficult to relate to any later members of their phylum.

Inventing the skeleton

What we see in this sudden profusion of animal fossils is a vast exploration of lifestyles available at a wide range of sizes. But large size often requires support, even in the buoyant medium of water. The fossils found in early Cambrian rocks represent skeletons for strengthening bodies and attaching muscles; shells for encasing feeding chambers where particles could be extracted from captured water; and hardened devices for rasping and cutting food items.

Finally, skeletons also shielded their wearers from environmental contaminants, microbial parasites, and animal predators. Soft bodies would have no protection against the sharp spines of *Protohertzina*. Holes in some Chinese specimens of *Cloudina* look like the work of a predator that bored into its skeleton to eat its tissue. But weapons provoke defenses. Soon after Ediacaran times we find both EK algae and cyanobacteria secreting regular skeletons of calcium carbonate over their organic surface – the first photosynthesizers to do this. Were they adapting to the feeding of mobile grazing animals that roved the seabed?

A stream of new designs – tubes, shells, spines, supports – created the mineralized animal parts whose graveyard marks the end of the Precambrian, the dawn of the Phanerozoic age, the age of visible life, and the most conspicuous change in the visual language of the fossil record in the whole history of Earth. From then onward, a riot of mineralized shells and skeletal structures used calcium carbonate, calcium phosphate, and silica. Today, almost every animal phylum secretes not only one or more of these three staple materials, but also exotic compounds such as iron oxides, sulfates, and halides (such as in the form of magnetite, gypsum, fluorite and other minerals).

No obvious single factor explains that rapid evolution. However, the birth of this ability to buttress the body and put on armor may have been an evolutionary accident. Components of all the minerals used to make hard parts can affect cell activity – for instance, too much calcium is poisonous, and phosphate makes the valuable compound ATP, but is scarce in ocean water. In both cases, an organism can benefit if it pumps these substances to the cell wall, either to control their concentration or to store them for future use.

We really do not know why animals rapidly evolved hard skeletons, but Heinz Lowenstam of the California Institute of Technology has suggested that once animals began to isolate masses of minerals outside cell layers, those masses could quickly be co-opted for other needs: support, muscle attachment, feeding, defense. Furthermore, simply shifting an available mineral would be a cheaper way to make hard parts than trying to synthesize tough organic molecules such as the chitin that arthropods make. If the first crude efforts worked, animals with cells that specialized in mineral applications might be differentially selected. But this could happen frequently only after animals had evolved more than the dozen or so cell types we think Ediacaran animals had. A cell that specialized in secreting minerals must be backed up by other cell types performing the more essential roles of metabolism, movement, sensing, reproduction, and the like.

This exploitation of an existing biological process or piece of kit to do work it was never "designed" to do is a typical feature of evolution. It does not waste its assets. However, all sorts of other factors may have contributed to the sudden uptake of minerals by so many different species. For example, a permanent rise in the levels of oxygen dissolved in the water might enable specific organs in the body to take over the job of delivering this richer supply to the inner cells. Then animals could afford to use the rest of their outer surfaces in different ways, instead of having to use the whole area to capture an adequate supply. Organisms must play many roles, and changing environments are complex. Evolutionary understanding is never simple, so that all explanations citing only one factor must be as suspect in the book of life as they are, say, in human history or biochemistry.

The Cambrian explosion of animal life

Animal life today is phenomenally diverse, more so than any other of life's six kingdoms. Over the past three centuries, scientists have described an estimated 1.5 million species of living animals, but so many more species have not yet been studied – particularly small ones in the tropics – that true totals of 5 or even 50 million have been guessed at. Most of these species (mostly arthropods and parasites, 75 percent of all species) live on the land. Far fewer species live in the oceans

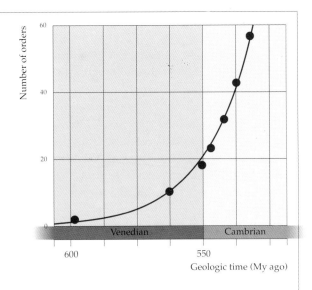

The Cambrian explosion of animals was rapid but not irregular. Counts of numbers of fossil taxa (orders, families, genera) at various times around the Vendian–Cambrian boundary exhibit a geometric rise, as illustrated in the graph to the right. This is the kind of rise expected when organisms radiate into a world of unlimited ecological opportunity.

THE SUDDEN MULTIPLYING OF SPECIES

The phenomenal speciation and taxonomic differentiation of animals during the early Cambrian may be unique only in scale. Animals appear capable of very rapid speciation when new habitats appear. Five million years ago, a brackish sea formed over southeastern Europe, and within a few tens of thousand years more than 30 genera of bivalve cockles evolved in this new habitat. At around the same time, the freshwater Lake Tanganyika formed in East Africa, and since then 34 genera and more than 100 species of cichlid fishes have differentiated there. Some of these cichlids are illustrated below.

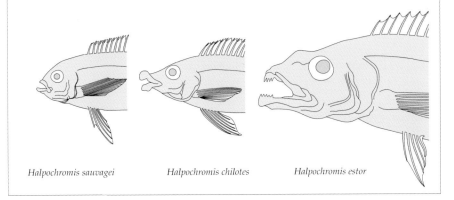

Halpochromis sauvagei *Halpochromis chilotes* *Halpochromis estor*

(about 295,000 have been recognized). Yet it is the ocean that contains more of the main divisions of the animal kingdom, the phyla – almost every one of them – because it was the ocean that gave birth to them all. They lie at the root of life's diversity.

Scientists divide the world's surviving animals into anywhere from 25 to 35 phyla, classed according to basic features such as the number and kinds of tissues and organs, body symmetry, and the presence and nature of body cavities. It is customary to list them in their order of complexity, putting sponges and jellyfish before annelids and mollusks,

and arthropods before chordates. Some phyla comprise very obvious and distinct groups of species, while others spread their definition wider. The phylum Chordata interests us because it is our own. As well as the core group of vertebrates, it contains two other less obvious groups. The tunicates – sea squirts – produce larvae that have dorsal nerve cords and a stiffening rod, the notochord, in common with ourselves, although the adults lose them. In the cephalochordates – lancelets – the adults retain notochords but develop no bones.

What is extraordinary about animal phyla is that nearly all of them seem to have evolved around the Cambrian–Precambrian boundary. Furthermore, that evolutionary explosion also produced a horde of unique and sometimes outlandish-looking animals whose fossils cannot be assigned to any living phylum.

How are we to explain this phenomenal burst of natural creativity, precedented only in the staggering array of bacterial metabolisms seen at the beginning of life and the flush of EK phyla at the invention of sex? And does it show animal life itself erupting into a superb and radical phase of new designs, roles, and strategies, or are we simply seeing species previously invisible adopting larger sizes and new skeletal fashions? Perhaps these were ancient lineages of soft-bodied metazoans, with sources deep in the Precambrian. Two lines of research have tackled this truly vital question in recent decades.

First, detailed investigations of trace fossils have found evidence of many more activities, and many more new actors to perform them, in lowest Cambrian rocks. Feeding burrows now meander, or spiral, or spread into branching chambers. Some are much larger than in Ediacaran versions, and delve much deeper into the sea bottom. A greater variety of trails and scratch marks engraves the sunlit seabed. Animals with skeletons made some of these traces, but far more of them, and especially of the burrowing kind, were the work of soft-bodied animals – "worms" of various kinds, and various they are, seeing that this catch-all term applies to practically any animal that is long, thin, boneless, and preferably legless. We have no label of equivalent vagueness – except perhaps "bug"

– for the harder, more compact animals that we tend to find easier to tell apart. But in any case, these multiplying traces prove that the Cambrian outburst affected all the animal kingdom, with or without a skeleton.

A second line of evidence comes from kinds of animals only rarely preserved in the fossil record. These are animals either with delicate skeletons or with no skeleton at all. To preserve them, it takes a special kind of accident, a statistical marvel, except that with world enough, and time, marvels must happen. They come as *Lagerstätten*, rare windows on the fullness of ancient life, sites of exceptional preservation that entomb the fragile remains of soft-bodied animals and expand our knowledge even of hard-boiled contemporaries more easily fossilized.

Of all the wonders of the Paleozoic world, it is the existence of the Burgess Shale that most transfigures our perspectives on early animals. Stephen Gould has described it as the most important fossil deposit ever found. The discovery of its 120 or more animal species allows us to study the alphabet of metazoan life while it was still being designed.

The Burgess Shale is a rock unit about 200 ft (60 m) long and at most 8 ft (2.5 m) thick that crops out on the flank of Mt Stephen, in British Columbia, Canada. It dates from the mid-Cambrian of about 520 My ago, and was discovered in 1909 by the then dean of American Cambrian paleontology, Charles Doolittle Walcott, secretary of the Smithsonian Institution. He noticed that shiny black films preserved on surfaces of stray slabs were the flattened fossils of an ancient fauna that included many soft-bodied and lightly clad animals. The Burgess fauna had lived on banks of mud in shallow seas below a plunging reef wall. Occasional mudslides had dislodged them into a deeper basin where they died but were saved from decay by its lack of oxygen. Between 1910 and 1920 Walcott collected and partially described about 65,000 fossils, which he identified as arthropods, worms, jellyfish – familiar living phyla. His labels stuck.

Reopening of Walcott's Burgess Shale Quarry in the late 1960s provoked Professor Harry Whittington of Cambridge University to re-examine Walcott's collections. During the 1970s, he and two doctoral students, Derek Briggs and Simon Conway Morris, were forced by what they found to discard a century of Cambrian assumptions, including many of Walcott's. Their insights sprang in part from Whittington's discovery that although pressure had flattened the Burgess

Wiwaxia and Halkiera. Wiwaxia is a 1–2 in (2–5 cm) long animal, known from the Burgess Shale. It was completely covered by plated armor, with two rows of projecting scales. It may be related to Halkiera, a sclerite-bearing animal of the early Cambrian. A recently discovered complete fossil from Greenland shows that halkieriids lacked projecting scales but had curious shell-like plates at the front and back. Their relationship to other animals is unknown.

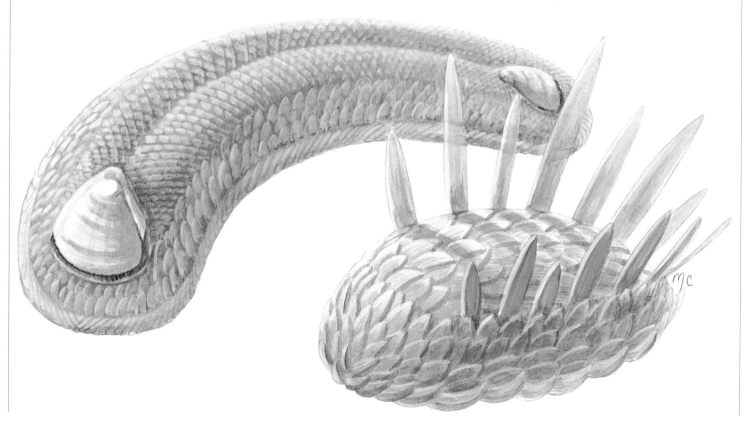

animals, their fossilized compressions were just thick enough to contain a wealth of details that could be bared by delicate probing and the imaginative power to charm these faunal smears back into three-dimensional forms that no one had seen before.

Their work – since reinforced by finds in China, Greenland, and elsewhere – has resurrected a Cambrian world so much fuller, more lively, and more varied than it once appeared that it has revolutionized our vision of the scale and sheer potential of animal life in one of its most crucial phases.

Most of the Burgess fossils are arthropods, but only a fraction are the usual durable trilobites. The rest are lightly skeletonized species vanishingly rare among fossils, and most of them cannot be fitted into modern groups. Some lie along recognized pathways – sponges, mollusks, echinoderms – and some are ancient members of modern phyla rarely preserved – priapulid and polychaete worms, for example. That leaves at least twenty species bizarre to any eye and seeming to belong to phyla deleted before they could produce a clan.

The treasure is too rich to catalog here. We can select a few broad trends, and some of the species that illustrate the exuberance and sheer inventiveness of Cambrian life.

In modern times, arthropods are the seas' most diverse phylum, as they are the land's. In equatorial Canada, only 20 My after their Cambrian debut, they achieve an equal range of forms. *Marrella* is the standard Burgess species, the most primitive ever discovered, less than 1 in (2.5 cm) long, with a simple segmented body, at least two dozen pairs of legs and gills (the latest way to collect oxygen and filter food), long antennae, and two pairs of long spines curving backward from its headshield. Remove the ornate horns, and all other biramous ("double-branched") arthropod forms can be derived from the basic design. Other arthropods play variations on multi-segmented bodies, all sorts of appendages and graspers around the mouth, chitinous carapaces, variously joined legs, and plain or fancy gills.

The Burgess Shale has trilobites familiar from other Cambrian deposits, but preserving soft parts previously unknown. *Naraoia* is a

Xenusion *and* Microdictyon, *lobopod animals known from extraordinarily well-preserved Atdabanian fossils. They may be related to living onychophorans. Each had many paired legs with stiffened shoulder pads above.* Aysheaia *from the Burgess Shale may also be a member of this group.*

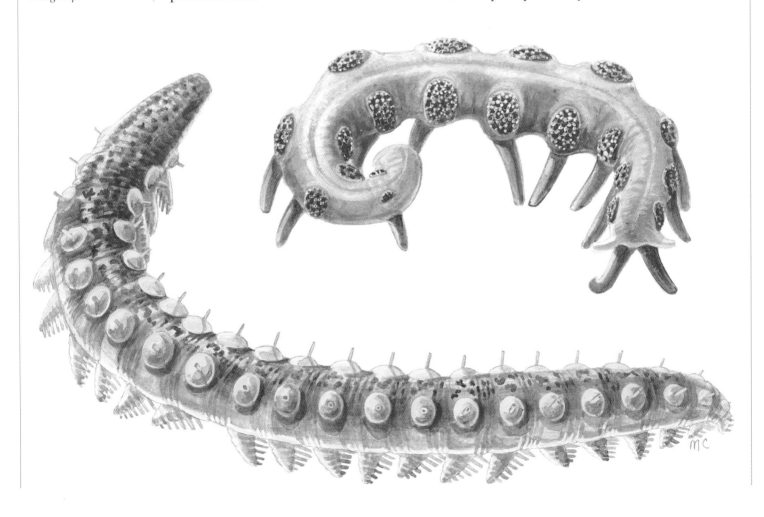

primitive trilobite with a carapace divided into two sections, instead of the three in the usual model. Many more of the Burgess arthropods are extinct forms with designs now lost, thrown up by genetic "mistakes" that must have worked for a while in Cambrian seas. In different conditions, would their designs have made sense, and the ones we know now have been eliminated, to be recalled as curious "failures"?

Wiwaxia is $3/4$–2 in (2–5 cm) long, and is shaped like a single walnut shell covered by flattened plates with some projecting spines. Stefan Bengtson in Uppsala and Simon Conway Morris at Cambridge have deduced from the arrangement of its scales that sclerites called halkieriids in the early Cambrian were worn by a similar animal. When some were found intact in Cambrian rocks of northern Greenland, the animals looked as forecast – only with a shell at each end of their bodies, setting a further puzzle.

A particularly bizarre animal from the Burgess is *Hallucigenia*. As Conway Morris first reconstructed it, it was a dream of a creature, 1 in (2.5 cm) long, standing on seven pairs of spiny stilts, with a row of seven soft snorkels rising from its back. But neither he nor other workers were entirely comfortable with this view. Its mystery was finally solved with the discovery of an Atdabanian fossil, *Microdictyon*, half a world away in south China. This animal had been caterpillar-shaped, with plump short limbs, each surmounted by a minute mesh of phosphate like shoulder pads. Another related animal, *Luolishania*, had short spines in the same position. What if the spikes on *Hallucigenia*, and another early Cambrian fossil, *Xenusion*, took the design still further? Lars Ramsköld of the Swedish Museum of Natural History carefully dissected some specimens, and found that each of the seven "snorkels" had a hidden twin. These must be the legs, and the stilts were really shoulder spines. *Microdictyon*, *Luolishania*, and *Hallucigenia* probably belong to the phylum Onychophora, along with another Burgess animal, *Aysheaia*. The onychophorans, close relatives of the arthropods, survive as about eighty species of terrestrial velvet worms in the tropics.

Although *Hallucigenia*'s wildest dream has ended, other Burgess species have arrangements of tissues, organs, symmetry, or segmentation definitely unique in their time and to this day. *Dinomischus* is a stalked suspension-feeder that looks rather like an unopened daisy whose head has a palisade of tentacles (the petals) that captured floating particles. *Amiskwia* is a mobile wormlike swimmer. Its stout body has lateral fins and a paddle at the end, and a pair of tentacles sprout from its bulbous head.

Opabinia is a predator. Its segmented body, up to $2^{3}/4$ in (7 cm) long and flanked by a fringe of lobes and gills, tapers to a tailpiece mounted with three pairs of staggered blades, set as broad Vs (see illustration on page 49). The head sports a group of five eyes, two to either side of the central one, and a long and flexible clasping organ that protrudes from the front like the nozzle of a vacuum-cleaner.

Then there is *Anomalocaris*, much the largest Burgess animal, nearly 2 ft (60 cm) long, a giant predator whose streamlined body was propelled by the beating of paired lateral flaps. Its strong pair of jointed spiny graspers fed prey to a circular mouth that looks like a slice of pineapple, able to dilate and contract toward its open center to hold and crush its food. *Anomalocaris* represents another Burgess mystery that had to be unraveled. Before Briggs and Whittington discovered them together, different parts found as separate fossils had been named as parts or wholes of three animals variously identified as a shrimp, a sea cucumber, and a jellyfish (graspers, mouth, body, respectively).

Before the restudy of the Burgess Shale, predators were thought to have been rare at the dawn of metazoan history, and the scene quite placid. Few of the known shelly fossils seemed capable of grabbing and consuming other animals. The new school of mobile predators revised that version, and they were not alone. Fossils of the arthropod *Sidneyia* and of soft-bodied priapulid worms contain their final meals of trilobites, brachiopods, or hyoliths.

A full cast of predators adds the links to the food chain that create a whole working ecology active near the reef now known as the Cathedral Escarpment, in seas as busy as the land was silent. Stalked suspension-feeders sway in the current as they filter out their food, and shelly or spiny grazers eat photosynthetic organisms on the sea floor. Deposit feeders crawl and scuttle to swallow

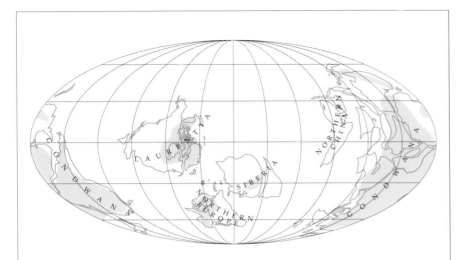

THE CAMBRIAN WORLD

This paleogeographic reconstruction illustrates the positions of continents and their land areas where they are estimated to have been during the early Cambrian. Laurentia (Paleozoic North America) straddled the equator at a rotation of 90 degrees to its modern orientation. To the south, Siberia lay in the subtropics, and North Europe in the southern temperate zone. Most of the other modern continental masses were amalgamated in the supercontinent Gondwana, shown at both margins of the map extending from high northern to high southern latitudes. South China was positioned in the tropics off Gondwana's western coast.

food on the sediment surface, or force soft bodies through the mud below, delving for grains rich in bacteria. There may also have been macroparasites that fed on animal tissue but did not kill their hosts, if the *Aysheaia* species turns out to have lived on the sponges it is found with. Predators and/or scavengers hunt or lie in wait for the newest food source, the carcasses of other animals.

One more face in the Burgess crowd draws special attention: *Pikaia*, a wormlike swimmer 2 in (5 cm) long. The V-shaped bundles of muscles along its flanks are structures known only among chordates, our own phylum. The dense mass reaching back from its head along the top of its body could be a notochord supporting the dorsal nerve column – again unique to chordates. *Pikaia* seems to be a cephalochordate, related to living lancelets, but that identification is not certain.

Conodonts were other early swimmers, named after minute toothlike fossils that first appear in mid-Cambrian deposits. A soft-bodied animal preserved from the early Carboniferous and certainly a chordate has the same sharp conodonts lining its mouth, so it looks as if chordates included predators soon after the Cambrian explosion began.

Decoding the Cambrian

The Cambrian had always been a mystery, a highway starting in a desert. Now we have pathways converging to feed it, and it turns out to be a broader road than we knew. Yet it is no gradual story that our discoveries reveal, but a revolutionary episode, crammed with new animals, some big and some small, some with skeletons and some without. Can we account for the sudden appearance of so many fundamentally different *kinds* of animals – the basic phyla? And why did the variety of animals *within* each phylum rocket so steeply upward? Scientists would like to know what gave "higher" animal life the functions and structures it started with, and whether the evolutionary rules have changed since then.

There is no doubt that animals are capable of bursts of evolution, or that they have undergone them many times. Five million years ago, it took no more than some tens of thousands of years for more than thirty new genera of cockles to fill a brackish sea that had formed over the present Balkan region. Around the same time, the freshwater Lake Tanganyika emerged in East Africa, and it has since given rise to thirty-four genera and more than a hundred unique species of cichlid fish. The formula is plain: an empty environment, untapped resources, and easy access to potential benefits. Fast evolutionary sorting pushes and pulls the best new talents till all resources are exploited, all niches filled, and the upsurge stabilized.

Apply this formula to the Cambrian explosion. Suddenly there were creatures whose size and mobility enabled them to exploit whole oceans full of resources never previously vulnerable to capture and easy to digest. Here was virtually limitless abundance at a number of levels: for the smallest metazoan, the ability to wolf down microbes living on the sea bottom or floating in the water; just a bit larger, and the power to push through the sea bottom and dig into the feast of microbes living in the sediment; next step, crawling over the surface rasping photosynthesizers; and finally the option to prey on any of the grazers and browsers swarming on the first and second levels.

What happens when animals flood into empty ecological space is a sequence well known in life's history. Not only individuals, but also species, begin to multiply, and at a geometric rate: 1, 2, 4, 8, 16 . . . As Geerat Vermeij puts it, new species are basically accidents that happen as populations of individuals adapt to local environments. The easier the access to new and diverse environments, the more likely the accidents that lead to speciation, the birth of new species.

We can't date rocks of the early Cambrian with a margin of error less than 5 to 10 My, but in the longer term the number of fossil taxa does seem to increase geometrically. This happens after the apparent extinction event – perhaps the work of more complex metazoans – that finished the simple Ediacaran animals. The increase flagged in less than 20 My, and after that diversity leveled off until the end of the Cambrian. These early animals seem quickly to have exhausted their options for exploiting the conditions.

That pattern accounts for the growth of populations and the diversity of *species* over time, but it does not explain why so many of the broad *groups* recognized as phyla and classes – animals with fundamentally different designs – should have evolved quite so fast.

Here we have data for a close comparison with an analogous event. Three hundred million years after the Cambrian explosion, a colossal extinction occurred at the end of the Permian, when no more than a few thousand species survived. The situation at the dawn of the ensuing Triassic was similar to the start of the Cambrian. Yet as Doug Erwin, Jim Valentine, and John Sepkoski have noted, throughout the burst of diversity that restored numbers of species during the Triassic, not a single new animal phylum evolved. There must have been something primally different about the animals on the brink of the Cambrian.

That difference may center on the way the genetic commands were read. The DNA in humans and all other animals controls how fast and in what order we make the proteins and other organic molecules as we grow from zygote to adult. Our cells issue the building instructions in a cascading network; some genes switch on or off the work of others, which do the same for others, and so on. In some cases, a lot more than one gene will issue the same command, and more than one stands by to receive and obey it. There are plenty of fail-safe procedures in this cascade. One faulty switch – a "mistake" or mutation – is not certain to halt or disorder the sequence; another will do the job. Half a billion years of systems testing have evolved this multiple redundancy that stabilizes the genetic transmission machinery by enabling it to ignore its own mistakes and still produce a fully working animal.

Early in animal history, the first command modules may have been simpler. Instead of being straitjacketed by back-up messages, the error in a mutated gene may have been more likely to be accepted and acted upon. Mostly it would kill the bizarrely developing animal, and waste the reproductive efforts of the parents. But occasionally, the freak might survive and flourish, able to exploit resources unavailable to its parents. In a brave new world that still had food and space to spare, its less than maximum efficiency might not condemn it. Then adaptation would set to work on its offspring to improve what they had.

This scenario provides a mechanism for promoting the uniquely fast build-up of very different animals with all sorts of odd equipment during the early Cambrian. (Every one of them was an "oddball" to begin with, and what look like orthodox prototypes today may have beaten long odds to survive successful Cambrian cousins now perceived as freaks. Nothing is pre-programmed.) But the pressure of an expanding population and monopolization of available resources would give a "bad gene" less and less breathing space, and would add a premium to any parent that could breed true.

In the long run, and working in the fairly stable conditions that our planet has often provided, selective forces would reward those genetic assembly lines that held true, and remove the more erratic. It would preserve the systems with more fail-safe switches (those whose products worked in the prevailing conditions), and discard the ones with fewer (untested products, more at risk). These precautions would accumulate in the

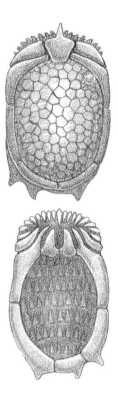

A ctenocystoid, a middle Cambrian echinoderm. Echinoderms experienced an immense evolutionary radiation, producing nine classes during the Cambrian and another thirteen during the Ordovician. Ctenocystoids are related to a bizarre early Paleozoic group informally known as "carpoids," which some workers argue may have been ancestral to the vertebrates.

Gogia, an eocrinoid echinoderm. Eocrinoids appeared during the Atdabanian and left abundant fossil plates in later Cambrian rocks. These animals were sedentary suspension feeders, capturing food particles with their arms and passing the particles to the mouth at the top of the body. They are thought to be ancestral to the many kinds of stalked echinoderms common in the post-Cambrian Paleozoic.

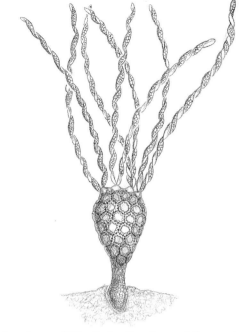

structure of life. So by the time of the next great wipeout at the end of the Permian, the survivors carried 300 My of genetic conditioning to breed true; no major aberrations would be commonplace; and therefore nothing so deviant that we might class it as a separate phylum was likely to crop up.

Much of this account has to be speculation, until we learn more about the functions and history of the systems that develop and pass on the genome. We also have a lot more to learn about the family connections among all the early Cambrian animals, both weird and familiar. It is as if we were scanning Victorian family photos, trying to distinguish real resemblances from a general fashion for stays, stiff backs, whiskers, or stern expressions.

Orthocone nautiloid cephalopods from the Ordovician. These nautiloid mollusks were among the groups that inherited the marine world as the Cambrian fauna dwindled. The early nautiloids had straight to slightly curved shells in front of their tentacled bodies; but quickly closed spiral shells, like that of the modern Nautilus, *evolved.*

New waves of life

In many respects, we have passed the most important events in life's history. Bacteria had oxidized the Earth, one-celled eukaryotes had stumbled on two-way sex, and animals had entered the wider scale of size and complexity. Life had passed through its major turning-points. Yet a slow walk along the modern shoreline measures our distance from the Cambrian scene. Those clams and mussels, snails and crabs stranded above the tideline were minor players in ancient seas; that glint of silver is a beached fish; and above the splash zone grow weeds, shrubs, and trees, still uninvented half a billion years ago.

Subtle changes set in toward the end of the Cambrian. Some lesser previous groups such as articulate brachiopods expanded in some habitats, and new groups such as gastropods and cephalopods evolved. The gastropods now include snails, winkles, and other single-shelled animals that creep on a muscular "foot." Cephalopods were different kinds of single-shelled animals that extended and tapered their shells and then partitioned them, living in the outer chamber but able to pump gas in and out of the inner ones, so as to float. By squirting jets of water they could propel themselves as roving predators, scavengers, and browsers. Now they appear as the pearly *Nautilus*, and as the squid and octopus, which have reduced or lost the shell.

Starting about 500 My ago, these groups and others exploded in a scramble that marks the end of the Cambrian and the start of the Ordovician. That burst of growth went on to triple the diversity of marine animal life in the space of 50 My – in numbers of species, an increase greater than the Cambrian's. The living fabric changed as the first wave of evolutionary fauna gave way to the second, the Paleozoic fauna, with its new cast of dominant groups.

Articulate brachiopods channeled currents through their two hinged shells and used their tentacled lophophore to snaffle food particles. Crinoid echinoderms ("sea lilies") stood on permanent stalks to capture these particles up above the level worked by the lower-lying brachiopods. The technique worked so well that stalked relatives, blastoids, cystoids, and others, unfurled their feathery arms above the seabed. Tabulate and rugosan cnidarians, corals now extinct, captured tiny zooplankton. Bryozoans ("moss animals"), tiny colonial organisms that might spread out as encrusting mats or stick up as plantlike growths, were the only verified new phylum (Ectoprocta) to appear, as yet unknown from Cambrian rocks.

Not all of the dominant animals were tethered to the sea bottom. Cephalopod mollusks branched into an assortment of big beautiful shell forms from curved, horn-shaped cones to elegant spirals. Starfish evolved from some armless echinoderm ancestor, probably to prey on sedentary shellfish. Calcareous ostracods, tiny crustaceans, swarmed into diverse forms and lifestyles as scavengers and particle feeders.

With these new groups of Paleozoic fauna, the ocean ecology reorganized. The key animals were mostly filter feeders living on the seabed and with more specialized habits, including ways of feeding at higher levels than before, tiered systems to trawl a greater volume of water. Cambrian species had eked out their living through a number of environments; their successors tended to specialize for a single environment, probably adapting to use resources more efficiently. Somehow in these changed lifestyles, Paleozoic species lasted longer, and extinction occurred less frequently than among their Cambrian antecedents.

A habitat that they themselves constructed was the tall massive blocks of tropical reefs, cemented clusters of algal and animal skeletons that grew above the bottom toward surface waves and currents. Lured to these stable, more solid quarters above the sift of mud and scratch of sand, many other animals thronged round for food and lodging – crinoids, snails, shellfish, bryozoans, worms. Corals joined in, but the busiest builders for 300 My, till the end of the Triassic, were calcareous sponges that laid down sheets of solid skeleton.

The older groups dwindled, though trilobites and hyoliths clung on till the Permian. Mere remnants of the Cambrian fauna survive in modern times – hingeless brachiopods in the muds of shallow seas or on the undersides of rocks, and a few deep-sea species of single-shelled mollusks, monoplacophorans, only rediscovered a few decades ago.

Five bouts of mass extinction culled early animals during the 50 My of the Cambrian, and each time they bounced back to their previous levels. Yet the Paleozoic faunas that eventually squeezed them to the margins were no less vulnerable to extinction. John Sepkoski has calculated that at least 70 percent of ocean species were erased at the end of the Ordovician period in what may have been the second largest of all mass extinctions, at least for bilaterian ocean-dwellers. It temporarily wiped out all reef-forming communities, and many families of brachiopods, echinoderms, and ostracods, as well as the beautiful agnostids, little floating trilobites evolved during the early Cambrian.

Vast continental movements seem to have been the cause of these convulsions, which came in a three-pronged attack. North America straddled the equator as a reef-fringed tropical continent. But Gondwana, a supercontinent that welded together Africa, Antarctica, Australia, India, and South America, had drifted to lie over the South Pole. Glaciers grew over this massive supercontinent and the oceans cooled. Polar faunas now expanded toward the equator, and much of the tropical fauna became extinct. Meanwhile the glaciers on Gondwana locked water away from the seas, which lowered their level and drained North America. The reefs died, along with all the species living in interior seas behind them. Finally, as glaciers melted and seas warmed at the very end of the Ordovician, lower ocean levels stagnated and much of the fauna there appears to have choked.

Revival came quickly. Reef communities reappeared with newly evolved residents that embarked upon a long expansion toward a Devonian peak. On muddy bottoms, the new species belonged to the same main groups – brachiopods, corals and sponges, bryozoans, and echinoderms. Mobile predators evolved to unprecedented sizes. Eurypterids were arthropods that looked like giant scorpions (a close relative) and would reach 6 1/2 ft (2 m) in length by the late Devonian. Some nautiloids also became giants, growing up to 10 ft (3 m) long.

The luxury of the Devonian did not last. Late in that geologic period another massive extinction, whose causes are poorly understood, racked the Paleozoic fauna.

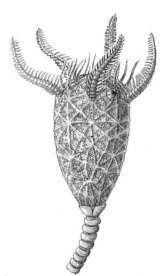

Eumorphocystis, an Ordovician stalked echinoderm. This animal is a diploporite cystoid. With the crinoids, blastoids, coronoids, paracrinoids, and parablastoids, the cystoids formed gardens of stalked suspension feeders in the post-Cambrian Paleozoic seas.

PROKARYOTE AND EUKARYOTE PHYLOGENY

These phylogenies are based upon similarities in informational molecules, namely subunits of ribosomal RNA (rRNA; RNA = ribonucleic acid). In each tree, groups that are connected have the most similar rRNA, and the length of the lines between connections, as well as of the terminal lines, reflects relative genetic differentiation. The phylogeny for the prokaryotes was assembled in the laboratories of Karl Woese and James Lake; the tree for the eukaryotes comes largely from the work of Vincent Sogin's laboratory.

Only a sampling of the numerous phyla of bacteria and protoctists have so far had their rRNA analyzed. Thus, these phylogenies are incomplete, and some of the illustrated relationships may change slightly as new phyla are analyzed and added. Some of their characteristics are listed below:

1 Thermotogales: primitive anaerobic heterotrophs, hot-spring dwellers.
2 Green non-sulfur bacteria: anaerobic chemo-autotrophs and photoautotrophs, also in hot springs.
3 Flavobacteria: anaerobic heterotrophs, including our intestinal symbiont *Escherichia coli*.
4 Cyanobacteria: important aerobic photosynthesizers.
5 Prochlorophytes: like cyanobacteria but with both chlorophyll *a* and *b*.
6 Purple bacteria: anaerobic photoautotrophs that utilize hydrogen sulfide.
7 Gram-positive bacteria: heterotrophs that break down organics into simpler compounds, including ethanol.
8 Halobacteria: pink archaebacterium living on salt flats.
9 Methanogens: anaerobes that are uniquely capable of producing methane.
10 Eocytes: anaerobic chemoautotrophs living in hot springs.
11 *Giardia*: facultatively aerobic heterotroph with nucleus but no mitochondria.
12 Microsporidians: also unicellular parasites lacking mitochondria.
13 Kinetoplastids: facultatively aerobic unicells with mitochondria.
14 Euglenids: aerobic unicells with both plant- and animal-like functions.
15 Entamoebas: amoeboid unicellular parasites.
16 Cellular slime molds: freshwater heterotrophs that aggregate into multicellular colonies.
17 Red algae: mostly multicellular photosynthesizers, some reef-forming during the Phanerozoic.
18 Ciliates: mostly unicellular heterotrophs covered by cilia for locomotion and feeding.
19 Dinoflagellates: planktonic or symbiotic unicellular photosynthesizers, cause red tides.
20 Plasmodial slime molds: amoeboid cells that aggregate into multicellular stalked reproductive bodies.
21 Sporozoans: unicellular heterotrophs.
22 Oomycetes: "water molds" with fungus-like lifestyles.
23 Labyrinthulids: heterotrophs that form colonial networks.
24 Brown algae: brown-pigmented photosynthesizers, including the giant kelp.
25 Xanthophytes: mostly freshwater, yellow, multicellular photosynthesizers.
26 Diatoms: unicellular photosynthesizers that secrete silica skeletons.
27 Chrysophytes: photosynthesizers with golden-yellow pigments.
28 Acanthamoebas: familiar heterotrophic amoebas that have lost undulipodia.
29 Green algae: diverse unicellular to multicellular photosynthesizers.
30 Choaenoflagellates: unicellular to colonial heterotrophs with strong undulipodia.

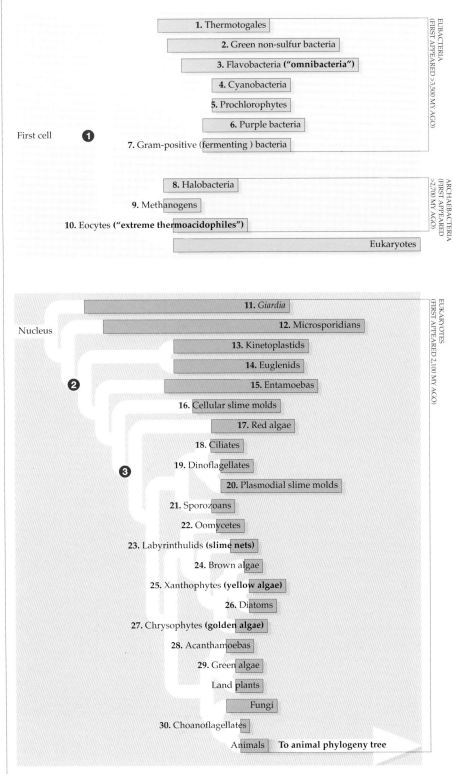

First cell

EUBACTERIA (FIRST APPEARED >3,500 MY AGO)

1. Thermotogales
2. Green non-sulfur bacteria
3. Flavobacteria (**"omnibacteria"**)
4. Cyanobacteria
5. Prochlorophytes
6. Purple bacteria
7. Gram-positive (fermenting) bacteria

ARCHAEBACTERIA (FIRST APPEARED >2,700 MY AGO)

8. Halobacteria
9. Methanogens
10. Eocytes (**"extreme thermoacidophiles"**)

Eukaryotes

Nucleus

EUKARYOTES (FIRST APPEARED 2,100 MY AGO)

11. *Giardia*
12. Microsporidians
13. Kinetoplastids
14. Euglenids
15. Entamoebas
16. Cellular slime molds
17. Red algae
18. Ciliates
19. Dinoflagellates
20. Plasmodial slime molds
21. Sporozoans
22. Oomycetes
23. Labyrinthulids (**slime nets**)
24. Brown algae
25. Xanthophytes (**yellow algae**)
26. Diatoms
27. Chrysophytes (**golden algae**)
28. Acanthamoebas
29. Green algae
Land plants
Fungi
30. Choanoflagellates
Animals | **To animal phylogeny tree**

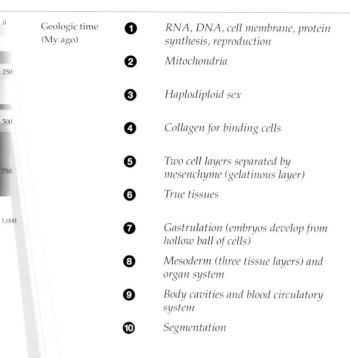

Geologic time (My ago)

1 *RNA, DNA, cell membrane, protein synthesis, reproduction*

2 *Mitochondria*

3 *Haplodiploid sex*

4 *Collagen for binding cells*

5 *Two cell layers separated by mesenchyme (gelatinous layer)*

6 *True tissues*

7 *Gastrulation (embryos develop from hollow ball of cells)*

8 *Mesoderm (three tissue layers) and organ system*

9 *Body cavities and blood circulatory system*

10 *Segmentation*

ANIMAL PHYLOGENY

This phylogeny is also based largely on molecular evidence (ribosomal RNA), assembled in the laboratory of Rudolf Raff. The rRNA of only a limited number of animal phyla has been studied, and many long-extinct phyla are not shown here. Among those exluded are the Ediacaran animals. Other phyla not represented are the oddball animals of the Cambrian explosion. If these many extinct groups could be accurately added to the animal phylogeny, the tree would look very different: broad near the base and rapidly winnowing upward toward the top.

Animal phylogeny

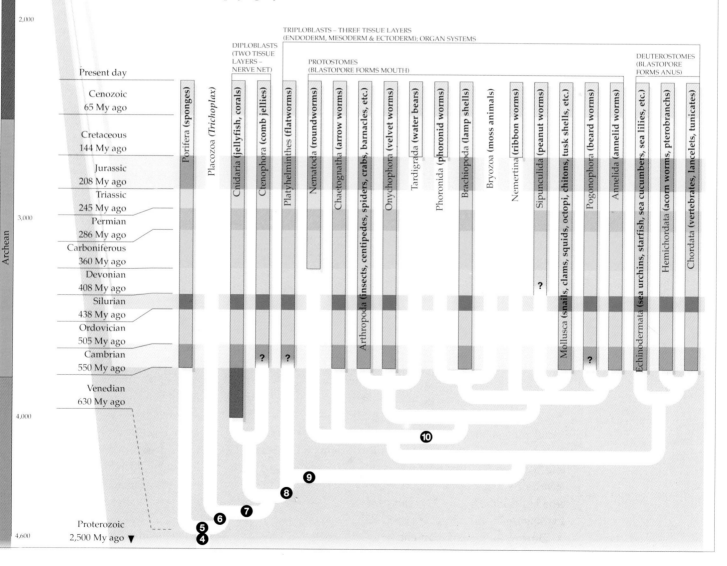

Again, in the ensuing Carboniferous, the fauna recovered and flourished, but now with subtle differences. Animal groups like bivalved clams and snails, sea urchins and marine vertebrates, including sharks and bony fishes, had increased their numbers proportionately. They continued to do this slowly for another 100 My of evolutionary time, until animals in the oceans suffered the great Permian mass extinction, another poorly understood event that devastated the fauna, extinguishing at least 95 percent of ocean species.

In the ashes of the Paleozoic era, those animals that had fared disproportionally after the Devonian now took over the oceans, building the third and modern evolutionary fauna. Its citizens include the clams and snails, sea urchins and fishes, as well as myriad crustaceans and new kinds of bryozoans – the animals we pick up along the strandline today. They seem even more specialized than their Paleozoic predecessors, with many more predators and many more sediment dwellers that burrow deeper into muds and sands for food and shelter. Their sharp division of resources in the environment allowed them to evolve to unprecedented diversities by the Recent (the Holocene epoch), at least doubling the peak enjoyed by the Paleozoic evolutionary fauna during the luxurious millennia of the Devonian.

Throughout this tumble of faunas, an elementary and recurrent sequence appears to govern the origin of new animal groups in the oceans. They tend to originate in very shallow habitats along the shoreline, and then spread out into the deep. The shoreline habitat is a restless one, washed by tides, storms, and freshwater floods, with ups and downs of temperature, algal blooms and die-offs, scouring currents, and rains of sediment in anything from a trickle to an avalanche. These harsh conditions will favor any species that for some reason is hardy, and resistant to extinction. During the march of evolutionary time, any such species will cling on and increase while neighbors fail. And sometimes they or their descendants will move into nearby deeper reaches vacated by shorter-lived species. The result is slow offshore expansion by the more resistant faunas.

During the Ordovician changes, the early members of the Paleozoic fauna first made inroads in very shallow water. As they expanded toward the deep, new animal groups occupied the shallows – the ancestors of the modern fauna. By the Silurian, many of these animal groups were beginning to explore life in middle depths. But more interestingly, some were exploring in the opposite direction, into freshwater and even on to land.

Moist habitats just above sea level must have offered access to very early explorers. Ultraviolet radiation from the Sun does not harm cyanobacteria that exude their insulating mucus, a very efficient sunscreen. Recent studies of some Precambrian soils have indicated higher than random concentrations of ^{12}C, the lighter carbon isotope that photosynthesizing organisms use for their work in preference to ^{13}C. Photosynthesizing bacteria and perhaps algal microbes are the likely source. They were the advance guard of an invasion that would green and change the land, create new habitats to shelter and feed new animals, and eventually produce half of the world's supply of oxygen, constantly renewed.

Photosynthesizers had lived in the sea for billions of years. Multicellular EK algae evolved as seaweeds kept stable by holdfasts on the rocks and firm sediment. They were descendants of cells that had invented sexual reproduction, except that their's retained a rhythm of alternating generations that all land plants still employ. One generation became a non-sexual form – the sporophyte – able to produce spores. Each spore then grew into a new and different stage, the gametophyte. It produced gametes, sexual cells, that had to pair off in order to give rise to the next generation of sporophyte plants, which continued the cycle. The gametes were mobile cells that united by following each other's chemical signals.

The ancestors of all land plants are green algae. These photosynthesizers had certainly adapted to freshwater rivers, lakes, and ponds before the end of the Ordovician, and perhaps much earlier. Freshwaters contained the dissolved carbon dioxide they needed, as well as phosphates and nitrates essential for constructing necessary biomolecules. The major problem faced had been controlling cellular water; cells are salty, and water will tend to flow outward into the fresh unless effective molecular barriers prevent it.

Once this problem was solved, it was a simple step to exploit other moist environments like seasonal ponds and spring seeps that soaked the soil. The standard trick is to grow fast and reproduce quickly so that some of the many progeny can land in the ephemeral habitat when it appears again. But there is another solution that evolutionary sorting can favor. If cell walls toughened by the rigors of freshwater life become even more impenetrable, so that they can shut out the drying of fresh air and the harm of ultraviolet radiation, the adult photosynthesizers can wait out the dry season and grow again during the rains. Now they can produce multiple generations of offspring.

To meet these challenges, plants acquired new equipment: first, a coat that kept the water in, but would also admit air through small breathing pores (stomata); then hardier spores to protect their propagation; and eventually special cells (xylem) to conduct what little water might be available. Roots were a subsequent invention. The earliest land plants were confined to a rhizome, an underground stem that could put out shoots but did not sink a solid grip into the ground.

Land plants have followed two main strategies. The plants that developed xylem, roots, and efficient systems for circulating water are the vascular plants, the abundant tracheophytes. In them, the sporophyte generation (weed, rose, oak) is large and dominant, and may grow as a long-lived structure that has miniaturized the gametophyte generation, retained it on the larger plant, and found ways to unite the gametes without shedding them into water.

A different route was taken by the bryophytes, which survive as mosses, hornworts, and liverworts. In them, the dominant generation is a gametophyte which has to stay close to the ground in shady, damp conditions because it still sheds gametes that must travel to find each other. Their union produces a short-lived sporophyte generation visible as a capsule of spores on top of a short bare stem. These asexual spores grow into the next gametophyte, which is green and possibly leafy, but cannot be tall. These are primitive habits, closer to the behavior of algae in the sea, and it is generally believed that plants may first have come ashore as bryophytes, living along shady water margins.

Spores of a kind that only land plants produce have been found in late Ordovician terrestrial sediments, and could belong either to bryophytes or to very simple vascular plants. Soils of a similar date also contain structures that look like the burrows of arthropods – U-shaped trace fossils. We have a lot more to learn about the animal landings. Various kinds of arthropods derived from coastal relatives appear to have evolved air-breathing systems independently of each other, but it is not possible to list the order of arrival of the mites, springtails, scorpions, and millipedes, all of which have extraordinarily diverse descendants living among modern vascular plants. Erosion and decay on land allowed far fewer animals to fossilize than could be kept intact beneath sediments in seas.

The earliest known vascular plants are named *Cooksonia*, and come from the upper Silurian. These vinelike plants, only about 4 in (10 cm) tall and lacking both roots and leaves, sent their short shoots upward to capture sunlight and to release their spores into the wind. Varieties of *Rhynia* – plants of a similar design, but larger and more complex – appear in a key *Lagerstätte* of the lower Devonian. The Rhynie Chert of Scotland preserves them as part of a plant–arthropod community that has a lot to tell us about the early ecology of the land. The arthropods include scorpions, pseudo-scorpions, mites, and even spiderlike trigonotarbids. A few million years later, there are terrestrial *Lagerstätten* that include tracheates, arthropods that breathe through branching tubes supplied from external holes – the group that includes centipedes, millipedes, and insects. In the millions of years before the vertebrates came ashore, the early terrestrial world was an arthropod proving ground.

Here was a new laboratory of forms and lifestyles, but the ocean had a start of 3 billion years, and it still had wonders to perform.

The large predatory lungfish Dipterus *lurks among the water weeds, waiting to lunge at one of the slender spiny acanthodian fishes swimming in a shoal in the middle. Some of the water plants in this scene from the middle Devonian of the Orcadian Basin, northern Scotland, 380 My ago, live on land, but with a toehold in the water. Sunlight streams in from above, revealing early spiders, insects, and worms, and a world not yet explored by the vertebrates.*

Chapter Two

THE RISE
OF THE FISHES

Michael Benton

One new kind of animal drove a small armored tank of a body through Ordovician seas. *Astraspis* is an agnathan (jawless) fish, 4 in (10 cm) long, found in the Ordovician of North America. It has a large, oval-shaped plate of bone over its back and belly, and the plates are linked along the side by small scales that surround a series of gill slits. This chunky outer skeleton forms a rigid phosphatic box that protects the head and body region. There are no fins. Only the tail sticks out behind the box, covered with bony scales and able to flex from side to side.

From *Pikaia* to *Astraspis* is a leap of upward of 30 million years, from Cambrian to Ordovician, from agile darter to defensive plodder, and in particular from chordate to vertebrate. The fossil record offers no obvious chordate stage before *Pikaia* and nothing that links *Pikaia* or some lost related species to the very different form of *Astraspis*, whose heavy armor looks likely to have fitted it to a life on the seafloor trawling for food with a jawless mouth that might suck in water but was unable to bite. The only earlier evidence comes in the form of small black fragments of bone found in marine sediments laid down during the late Cambrian in Crook County, Wyoming. Their discovery was announced by Dr J. E. Repetski in 1978, but their owner, if it was a fish, has still not surfaced.

The phylum Chordata includes all animals with a notochord, the flexible stiffening rod running down the middle of the back. The subphylum Vertebrata, containing more than 99 percent of all chordate species, has vertebrae as well as a notochord. These bony elements link to form a backbone that is the main stiffening element in adult vertebrates, and more or less replaces the notochord, which is usually recognizable only in the earliest stages of development. The backbone provides an attachment site for the skull, for numerous muscles, and for various appendages including limbs.

History is written by the victors, so it is

One of the oldest fishes, Astraspis, *lived during the Ordovician in North America. This jawless fish has a simple mouth opening with which it sucked small-scale detritus from the sea-floor sediment. Its head and body region was encased in thick bone shields, pierced by an eye opening near the front, and a row of gill slits behind. The fish could swim by beating its less heavily armored tail portion from side to side.*

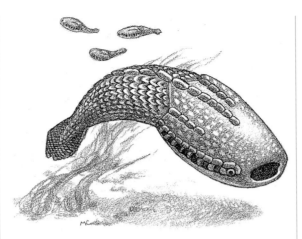

said. In the story of life there are no victors, only provisional survivors. Of these survivors, only one species is able to write the story, and this is partly because the vertebrate design lends itself to the development and protection of a brain. This organ is present in other animals, but there are limits on its growth – one of them imposed very early in the history of life, when animals were first developing basic equipment like a front and a back, sense organs, and the ability to use information from the sense organs to issue simple instructions like stop, start, forward and backward.

The cluster of nervous tissue where the data arrive and the orders depart may be only one of several such clusters in an animal's body, but it is the most important, and it occupies the most useful site, which is somewhere near the front. In the vertebrates, this cluster (ganglion) and the main nervous system are housed in the skull and backbone, separated from the rest of the body. In the arthropods and mollusks, the central ganglion is built as a ring of tissue that surrounds the gut. Any tendency for this tissue to grow is likely to squeeze the tube of the gut and constrict the supply of food. This is a contradiction that the arthropod design has never resolved, though with 80 percent of the world's animal species arthropods might not think themselves any the poorer – supposing that they could think. Still, the contrast in design has made enormous differences to its vertebrate owners, and paleontologists have a vested interest in tracing chordate origins.

With so little fossil evidence, we are forced to look for clues among the living non-vertebrate chordates that differ less obviously from our common ancestors of about 570 My ago. These living chordates are a mixed bag,

and bags are what some of them resemble. The adult sea squirt (subphylum Urochordata) is a floppy, slightly lumpy animal shaped like a crudely made flask or coffeepot with a short fat spout placed fairly close to the offset tubular hole that forms the lid. It is fixed to the seabed, grows up to 6 in (15 cm) tall, and feeds by pumping water in through one siphon and out through the other (the lid and spout). Small particles of food are filtered out as the water is drawn through numerous gill slits in the wall of the pharynx, the large inner sac. All chordates have gill slits in the region of the throat or pharynx at some stage of their life cycle, if only in the embryo; but it is the juvenile sea squirt that explains why this sedentary bag with a cellulose hide should be viewed as a chordate. The larval stage is a tadpole with a large head bearing a "mouth" at the front, small gill slits on the side, a primitive "eye" that is no more than a spot of light-sensitive pigment, and a long thin tail containing a notochord.

This mobile larva is only a dispersal agent. When it hatches, it swims off in search of a site for attachment – this is a common way for fixed seabed animals to populate new areas. Within a few days, without ever feeding, the sea squirt tadpole chooses its spot and settles head-first on the seabed. Tail and notochord wither away; gill slits and pharynx expand; and the animal rotates its internal organs to suit its changed form.

It might seem simple to see how a middle Cambrian animal like *Pikaia* could become a fish by adding a mineralized skeleton, but what sort of animal was it that evolved into *Pikaia*? Where did the chordates come from? Evidence from anatomy, embryonic development, and molecular biology suggests that the closest major phylum to the chordates is the echinoderms – starfishes, sea cucumbers, crinoids, and sea urchins. A group of fossils called the calcichordates, now extinct, may provide the link between echinoderms and chordates, but the evidence is not straightforward. Calcichordates have an external covering made from calcite plates, as sea urchins do, so they have generally been seen as true echinoderms. But they also have a strange flexible "arm," and here the views conflict. Dick Jefferies, of the Natural History Museum in London, maintains that this appendage is not an arm but a tail – a chordate feature – and that all other features place the calcichordates along the line from echinoderms to chordates.

Early fishes

Fishes are rare in the fossil record all through the Ordovician and in much of the Silurian period (505 to 438 and 438 to 408 My ago respectively). Only in the Ordovician of Australia, North America, and South America do isolated fish forms first appear, and they remain apparently quite unimportant animals such as *Astraspis* and the more streamlined *Arandaspis*, from Australia. A larger array of jawless armored fishes is known from the late Silurian, ranging in size from 1 to 8 in (2 to 20 cm). Some had heavy armor sheaths, while others had more flexible scale coverings like chain mail. All of them had a pair of small eyes near the front and, where it can be seen, a small, rather featureless and jawless mouth. Feeding may have been the key to the long-term fate of these earliest agnathan fishes, even though they did become extremely diverse during the Devonian period (408 to 360 My ago). It seems that the evolution of jaws in other groups at this time put a limit on agnathan prospects, and the only living agnathans are several species of lampreys and hagfishes – all of them with long, unarmored, eel-like bodies and round, sucker-like mouths. They feed parasitically by attaching them-selves to other fishes and rasping at the flesh, or by wrenching bits off dead or dying fishes.

The Orcadian Basin

The wealth of fossil fishes in the Devonian became clear to collectors as long ago as the 1820s and 1830s. One of the richest areas lies to the north of Scotland, on the site of the former Orcadian Basin, a large lake that covered the area from Inverness and Elgin, northward over Caithness, Orkney, and the Shetland Islands, and east of the present Scottish coast, over the Moray Firth and into the North Sea. A thickness of more than 3 miles (5 km) of sediments was deposited in Caithness, and possibly as much as 6 miles (10 km) in Shetland.

Collectors in the 1830s found that certain layers of fine sandstone and siltstone, generally red, and known as Old Red Sandstone, could be split open to reveal dozens of perfectly preserved skeletons of fishes appearing as shiny black apatite. These shoal-bearing seams could be found by looking for the telltale cross-sections of fishes in broken faces of rock – thin layers of black material squeezed between the layers of sand grains. In other localities the fishes occurred in limestone nodules. Here the fish had acted as a nucleus for deposits of calcium carbonate to accumulate. Collectors became so expert that they could judge which nodules were likely to contain a fish, and some could even predict the type of fish, and its attitude, from the shape of the nodule.

These dramatic discoveries in northern Scotland drew international interest, and particularly from the famous Swiss

Specimen of the lobe-finned fish Gyroptychius agassizi *from Orkney, northern Scotland. This typical fish of the great Orcadian Basin swam in warm waters in search of small prey.*

FISH PHYLOGENY

The evolution of fishes is represented here by a phylogenetic tree based upon recent cladistic analyses of the relationships of the various living and fossil groups. The time scale indicates present best estimates of the times at which each new evolutionary line branched off.

The oldest uncertain remains of fishes are late Cambrian in age, and the record is sparse throughout the Ordovician. During the Silurian and especially the Devonian, fishes diversified greatly in the seas and in fresh waters. The jawless fishes (agnathans) and placoderms were dominant. These gave way to cartilaginous and bony fishes during the late Devonian and Carboniferous.

Bony fishes radiated in several bursts. They gave rise to the amphibians in the Devonian. Then, new groups, the "holosteans," radiated in the Triassic and Jurassic, and the teleosts in the Jurassic and Cretaceous. The teleost radiation was huge, giving rise to tens of thousands of species, from goldfish to tuna, and salmon to sea horses.

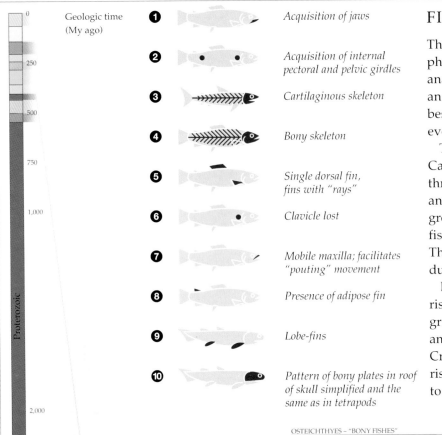

1 Acquisition of jaws

2 Acquisition of internal pectoral and pelvic girdles

3 Cartilaginous skeleton

4 Bony skeleton

5 Single dorsal fin, fins with "rays"

6 Clavicle lost

7 Mobile maxilla; facilitates "pouting" movement

8 Presence of adipose fin

9 Lobe-fins

10 Pattern of bony plates in roof of skull simplified and the same as in tetrapods

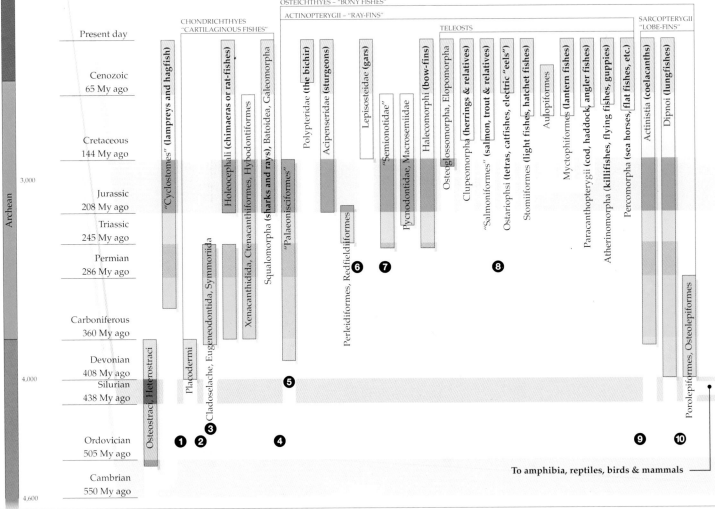

To amphibia, reptiles, birds & mammals

THE FISH IN THE SCOTTISH SANDSTONE

An army surgeon, a stonemason turned accountant, a professor of medical jurisprudence, and an ambitious naturalist: these were the founders of modern paleoichthyology, the study of fossil fish.

In 1830 the Swiss naturalist Louis Agassiz (1807–73) decided that he would describe and illustrate all the world's known fossil fish. His

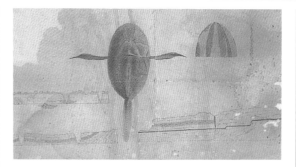

A fossil of Osteolepis macrolepidotus *from Orkney.*

John Malcolmson, a surgeon home from army service in India, corresponded with Agassiz and sent him drawings of specimens collected by his fellow Scotsman, Hugh Miller, the former stonemason.

Miller soon turned to popularizing his own discoveries, particularly in the lavishly illustrated book he published in 1841 under the title *The Old Red Sandstone*.

twelve-year task depended on a community of amateur enthusiasts who were partly competitors and partly collaborators in the quest for knowledge. Between 1833 and 1844 he published five volumes of his massive work on fossil fish, announced discovery after discovery, and made his name as one of his century's greatest paleontologists.

That reputation rests on the work of enthusiasts now mainly forgotten. The immensely generous Professor Thomas Traill of Edinburgh University made his collection of Devonian fossils available for copying.

FAR LEFT *Louis Agassiz (1807–73).* LEFT *Hugh Miller (1802–56).*

LEFT Pterichthyodes milleri, *one of the fossils named by Hugh Miller.*
FAR LEFT Coccosteus cuspidatus.

paleontologist and anatomist Louis Agassiz (1807–73), whose vast five-volume work on fossil fishes, published between 1833 and 1844, documented everything that was known about its subject at that time. Agassiz later moved to the United States, where he helped to establish the Museum of Comparative Zoology at Harvard. He visited Scotland twice to see the Old Red fishes, which he described and illustrated in a large volume in 1844. This showed the world an astonishing range of armored agnathans and placoderms, tiny spiny acanthodians, and streamlined, heavily scaled, lobe-finned fishes.

The other famous Old Red figure of those times was Hugh Miller (1802–56), a very different kind of enthusiast, who made his living as a stonemason and in his spare time wrote poetry and collected fossil fishes. In 1841 he published *The Old Red Sandstone*, an account of the geology and the fossil riches of his native area. This and later books had a great success in popularizing their subject among a broad Victorian readership that enjoyed their weighty yet vivid style. Here is Miller in *Footprints of the Creator* (1847) describing the bony head shield of the placoderm *Asterolepis*:

> The head of the largest crocodile of the existing period is defended by an armature greatly less strong than that worn by the *Asterolepis* of the Lower Old Red Sandstone. Why this ancient ganoid should have been so ponderously helmed we can but doubtfully guess: we only know, that when nature arms her soldiery, there are assailants to be resisted, and a state of war to be maintained. The posterior central plate, the homologue, apparently, of the occipital bone, was curiously carved into an ornate massive leaf, like one of the larger leaves of a Corinthian capital, and terminated beneath, where the stem should have been, in a strong osseous knob, fashioned like a pike-head.

This is the style of a man who enjoyed his subject, and was likely to hold strong views. In particular, he waded into one of the great scientific controversies of his day, about the possibility of evolution.

It is often assumed that Charles Darwin invented the idea of evolution in 1859 with the publication of *On the Origin of Species*. In fact evolution in the sense of change through time had been a serious topic among scientists in Europe for a century. The first broad proposals came from French naturalists of the eighteenth-century Enlightenment, and these ideas were championed in various English-language publications of the 1840s. Miller insisted that the fossil record actually denied evolution – a view that may seem odd today, when it is usual to see in the fossil record a progression through time from simple to complex forms. Miller argued that the Old Red fishes were *more* complex, and so more advanced, than their degenerate living relatives. They had effective armor, which modern fishes lack; the skeleton as a whole was more complex then than now; and the diversity of major fish groups found in the Old Red lakes was greater than today's.

This argument was forgotten later in the nineteenth century. It dates from a time when ornamental decorative styles were much admired, and perceived as an advance on the "primitive" crudeness of plain functional forms. Later on, the interest in Old Red fishes shifted to expanding knowledge of their anatomy and overall diversity. More intensive collecting brought many new species to light. Careful preparation, and examination under the microscope, revealed anatomical details that had never been seen before.

The twentieth century has expanded our knowledge of the Devonian world as a whole. The new science of taphonomy aims to define the entire set of circumstances – physical, chemical, and biological – in which fossils are deposited and preserved. We can now describe the general scene around the Orcadian lake, and begin to understand why it is that the Scottish Old Red fishes are so well preserved.

The north of Scotland has not always had a winter landscape of snow and howling gales. In Devonian times, Britain lay on the equatorial belt; and in fact when all the scattered Devonian fish sites are plotted on to a map of the world they occupied, they turn out to lie around the equator, where now they are in Greenland, Canada, the Baltic, China, Australia, Russia, and Britain. The Orcadian lake appears to have been freshwater, although other Devonian sites may have been marine. These warm waters were surrounded by low land plants that still had their roots in the water in the late Silurian and early Devonian, growing above the surface like short reeds. By middle Devonian times, some plants had moved away from the water's edge, and by the late Devonian large plants – trees – had evolved.

New patterns of growth included more complex branching systems, spines, tiny leaves developed by flattening the spines, and larger leaves produced by growing shoots in the same plane instead of three-dimensionally, then joining groups of these together with new tissue growth. Most of these strategies were means to collect more light. One way to outstrip rivals was to achieve greater height, and this required stronger support; secondary xylem strengthened some plants with woody growth, and more extensive root systems both anchored and fed the first trees, which could grow up to 33 ft (10 m) tall. The first land plants, the rhyniophytes, and their successors, the trimerophytes, rose, declined, and became extinct in the Devonian, though not before giving rise to the horsetails, club mosses, and ferns, some of them giant varieties, that would dominate the Carboniferous world.

Other key advances among the plants brought improvements in reproduction. In a useful division of labor, in some plants the spore-bearing (sporophyte) generation began to produce two kinds of spore, one of them large and female, the other small and male. The larger developed into a female gametophyte plant that carried a small food store; the smaller produced a tiny and short-lived male gametophyte plant that had nothing to do but release the sperms that would fertilize the female and produce the next sporophyte generation. It was now a shorter step for the sporophyte to retain the female spores instead of dropping them, and to find a way to introduce the male spores to them in a moist environment, and then plants could carry out reproduction without depending on moist soil conditions. This was the beginning of the move toward making true seeds, which could protect and feed the fertilized female spore or ovule, and which could wait for the right conditions before germinating. The various plants that developed the first seed forms toward the end

of the Devonian were the progymnosperms. They could begin to colonize both inland and upland areas not previously available. The differential spore production that seems to have begun this process is called heterospory.

It seems likely that by the end of the Devonian most of the land-dwelling invertebrates had come on shore, even those flatworms, leeches, and earthworms that do not appear in the fossil record. Snails may have arrived; insects had evolved from among the myriapods (centipedes and millipedes); spiders, mites, and scorpions were proliferating. Fungi had been around at least since the Silurian, and some paleobotanists believe that they derive from red seaweeds unable to compete with the first plants that lived and died half in and half out of the water, leaving masses of dead material available for an organism that could insert filaments inside these nourishing corpses. The fungi did not need to photosynthesize energy from light if they could live as saprophytes on their richer

Small plants grew up around the sides of lakes and rivers in Devonian times. These include the fernlike Archiopteris hibernica, *here preserved beautifully in the yellow Devonian sandstone of Kilkenny, Ireland. The specimen is 10 in (25 cm) long.*

relations. It has been suggested that they might actually have helped these relations on to the land by supplying nutrients to roots and helping to prevent them drying out. Their presence helped to break down dead matter, as well as providing food for small organisms.

The development of soils, together with the litter immediately above them, provided new habitats for small animals, as well as more food for larger plants. These plants in turn were ladders to the upper air, where a new ecosystem in the treetops is reflected in the tropical forests of the present day. Life was expanding around the Orcadian lake.

Fishes of the Old Red Sandstone

Jawless (agnathan) fishes are not common in the typical Orcadian middle Old Red Sandstone beds, but *Cephalaspis* and *Pteraspis*

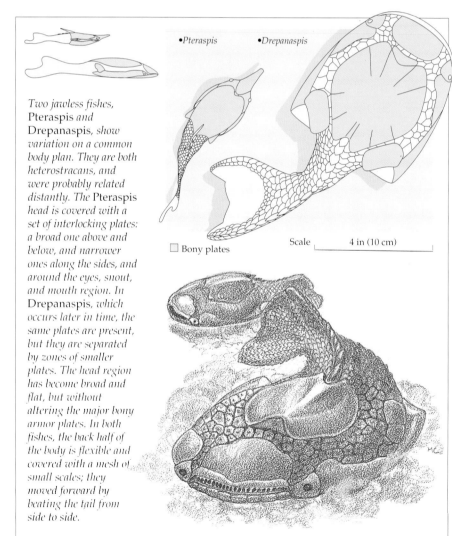

Two jawless fishes,
Pteraspis *and*
Drepanaspis, *show variation on a common body plan. They are both heterostracans, and were probably related distantly. The* Pteraspis *head is covered with a set of interlocking plates: a broad one above and below, and narrower ones along the sides, and around the eyes, snout, and mouth region. In* Drepanaspis, *which occurs later in time, the same plates are present, but they are separated by zones of smaller plates. The head region has become broad and flat, but without altering the major bony armor plates. In both fishes, the back half of the body is flexible and covered with a mesh of small scales; they moved forward by beating the tail from side to side.*

•*Pteraspis*   •*Drepanaspis*

☐ Bony plates   Scale ├─── 4 in (10 cm) ───┤

appear in the lower Old Red. *Cephalaspis* had a broad armored head cover shaped like a low horse's hoof, rounded at the front, flat underneath, and rising gently to a point in the midline of the back. The slit of a mouth could only have been used to grub for food trapped in sediments on the bottom. There are two eye sockets on top of the head shield near the midline, together with three unusual depressions, filled with small scales, one in the midline between and behind the eyes, the others running along each side of the head. These may be sensory organs, possibly able to detect slight movements in the water – predators? – by electrical impulses, a technique used by many modern fishes. The slender body was covered in fine scales. There was a short fin on the back, and a long tail fin underneath the tail. *Cephalaspis* swam by beating its tail from side to side.

There were many other groups of jawless fishes at this time, with a worldwide distribution. One group consists of partially "exploded" flattened forms. *Drepanaspis* is like a *Pteraspis* that has been firmly squashed in a press. The body is flat and all the head-shield plates are there, but spread apart so that upper and lower, side, eye, mouth, and snout plates no longer meet but are separated by spaces filled with "tesserae," small scales. Their flattened shape suggests that they lived on the bottom of the ocean.

The second great fish group represented in the Old Red Sandstone is the placoderms ("plated skin"). At first sight they do not look very different from the armored agnathans, but there are several new features, including a neck joint that enabled the head shield to be lifted to tilt the head, and a key new piece of equipment: jaws. The placoderms appear all over the world, but in time they were almost wholly restricted to the Devonian period – the only major fish group to have so short a span.

One of the most striking Old Red forms was *Pterichthyodes*. It had a peaked head shield pieced together from several bony plates, a heavily scaled body, and pointed, scale-covered arms or fins sticking out at either side of the head. These arms were totally enclosed in bone, and were movable at the "shoulder" joint as well as at a second joint halfway along. They may have been used to lever the animal along the bottom, or possibly to throw sand over its back so as to hide it, but they

cannot have helped much with swimming.

Later placoderms tended to lose body armor. The Old Red *Coccosteus* had a much reduced head shield that was divided into two parts, one over the head itself and the other just behind, over the shoulder region. These two shields hinged together on a pair of ball-and-socket joints, one at each side. The jaws opened in a way that seems unusual to us, by lifting the head up. The lower jaw also dropped a little, and the double hinging action gave a wider gape, which helped to make the order of placoderms known as arthrodires into the most ferocious of Devonian marine predators, and easily the largest vertebrates yet to evolve. *Dunkleosteus*, from the late Devonian, grew up to 33 ft (10 m) long. Their movable jaws enabled the placoderms to grasp and manipulate prey, and to cut and grind – functions that were impossible without jaws.

No series of fossils shows the stages of jaw development, but the theory most widely accepted is that they evolved by modifying slender bones in the throat region that supported the gill arches of jawless fish. The front three sets of bony supports may have migrated and fused to become parts of the brain case, more firmly attached to the skull, and of the upper and lower jaw. The gill slit supports consist of a major upper and lower element which are hinged, so that the step to becoming hinged jawbones is not too great, especially because the hinge would also be useful if it helped the fish to pump more water through its gills, and so obtain more oxygen, by flexing strongly.

Another class of jawed fishes was the acanthodians, spined fishes, much more modern-looking than either the agnathans or the placoderms. They appear to be related to the modern bony fishes – the majority of fishes – although some paleontologists place them in the line of the cartilaginous fishes (sharks and rays). Acanthodians swam in huge shoals in the Old Red lakes and seas. Fairly small fishes – no more than 8 in (20 cm) long – they were active mid-water swimmers, unlike most of the agnathans and placoderms, which lived on the bottom. The fins bear long spines, and there are several more pairs of spines along the belly region.

The sharklike fish Cladoselache *pursues a shoal of tiny-spined acanthodians during the late Devonian.* Cladoselache *had a slender body, clearly designed for speed; its deep, broad front paddles presumably helped it to steer and to maintain position as it swam, just as in modern sharks. Its deep tail fin was supported on an upturned fleshy tail, and its dorsal fin was broad. The acanthodians lived in great abundance at this time, and they relied on their spines, and their shoaling behavior, to avoid capture.*

This living fossil of the coelacanth fish Latimeria chalumnae *(bottom) was fished up off the east coast of Africa and preserved in a tank; the specimen is about 6¹/₂ ft (2 m) long. Its ancestor of 150 My ago,* Coccoderma suevicum *(top), is very similar but smaller. This superb specimen, 1 ft (32 cm) long, from the Lithographic Limestone of Bavaria, Germany, shows all the fine bones and the overall outline of the body.*

Acanthodians have eyes placed well forward, which indicates that they used their sight in murky waters. They also had lateral line canals, detector systems that can identify small disturbances in the water. Some acanthodians were toothless, and are likely to have fed on food particles filtered from the water, but most had a fine array of teeth. These were voracious predators, and one fossil acanthodian has been found with a bony fish lodged in its body cavity, swallowed whole. They survived into the Permian period (286 to 245 My ago).

Cartilaginous fishes, class Chondrichthyes, represented today by the sharks and rays, arose in the late Devonian, and *Cladoselache* from Ohio is the best-known early form. Already this is an obvious sharklike animal: small head with long jaws, pointed dorsal fins, deep pectoral (shoulder) and tail fin, and

streamlined body, without armor. *Cladoselache* reached a length of 6¹/₂ft (2 m) and was plainly a powerful predator, driven by the deep tail that beat from side to side, and able to use its vast pectoral fins to control its stability in the water and make rapid twists and turns in pursuit of its prey.

Class Osteichthyes, the bony fishes, are today's most common fishes, represented by thousands of species. They arose in the Devonian, and a typical early form is *Cheirolepis*, from the middle Old Red Sandstone of Scotland, 10 in (25 cm) long, with a slender body equipped with the usual set of fins seen in modern species – midline dorsal, tail and anal fin, and two sets of paired fins, pectoral and pelvic, attached to the shoulder and "hip" regions.

Compared with its modern descendants, *Cheirolepis* has a heavy bony skull. Teeth were carried on three bones, the maxilla and premaxilla in the skull, and the dentary in the lower jaw, just as in later vertebrates. Detailed studies carried out in the 1970s by David Pearson and Stanley Westoll of the University of Newcastle upon Tyne showed that the bony head was extraordinarily kinetic, or mobile, and that various sets of bones were able to move freely in relation to each other. In fact the entire skull could come apart when the jaws opened. The lower jaw dropped in the usual way; the cheek unit and skull roof went up and back; the gill region expanded and moved down and back; and the shoulder girdle moved down. All this convertible machinery enabled *Cheirolepis* to expand its gape to swallow larger prey fishes – up to two-thirds of its own length.

Cheirolepis and its associated species, past and present, are actinopterygians – "ray-fins." A second group of bony fishes represented in the Old Red Sandstone are the sarcopterygians or "lobe-fins," so called because their paired fins, the pectorals and pelvics, had fleshy central portions instead of the thin blades with bony struts and paper-thin skin found among the ray-fins. The fleshy lobe was composed of mobile bones and muscles that allowed the lobed fin to be used in swimming and a kind of "walking" on the bottom.

Osteolepis lived alongside *Cheirolepis*, *Pterichthyodes*, and other fishes in the Scottish middle Old Red Sandstone and other Devonian sites. It was about 8 in (20 cm) long, slender, and rather like *Cheirolepis* in its bony head, regular suit of scales, and fins designed for rapid swimming. It happens that the lobed fins of *Osteolepis* and its relatives, the "rhipidistians," show striking similarities with another group of vertebrates, and contain clues to one of the major evolutionary steps, the origin of the amphibians. These heavier fins needed a second anchorage on the skeleton, which is also a requirement for making an efficient limb. This possession of a feature that will be useful in some quite different situation is called "pre-adaptation." But the fins made their owners better at living in the water; they did not evolve because the lobe-fins were somehow programmed for promotion on to the land.

Another major Devonian lobe-fin group, the lungfishes, appear in the Scottish middle Old Red Sandstone in the shape of *Dipterus*,

also about 8 in (20 cm) long, its pointed fins mostly crowded well back on its slim cigar-shaped body. *Dipterus* did not have rows of teeth on the margins of its jaws, though it had some smaller teeth in front. Like other lungfishes, it was equipped with a pair of large dentine-covered grinding plates in the middle of its palate, which indicate a diet of tough hard food that had to be crushed. The lungfishes reached their peak in the Devonian and have dwindled in diversity since then, but they survive in the Southern Hemisphere.

The third lobe-fin group, the coelacanths, also appeared in the Devonian, although they are not typical Old Red forms. They survived, mostly at low diversity, for many millions of years, but were thought to have become extinct around 70 My ago, until in 1938 a modern coelacanth was fished up off the eastern coast of Africa, and was named *Latimeria* the following year, after its discoverer, Marjorie Courtenay-Latimer.

The Orcadian lakes

Why are the Scottish Old Red fishes so consistently well preserved? Taphonomists have investigated the formation of the sediment and the physical, chemical, and biological conditions in which the carcasses died and came to rest. Their studies show that the red, brown, and gray sandstones, siltstones, and limestones of the Orcadian Basin were laid down in a large subtropical lake, fed from surrounding highlands. Rates of erosion were rapid because these uplands were not yet clothed in vegetation, and the particles eroded were often red or purple because of the quantity of iron oxide they contained – the signs of a hot atmosphere that assisted oxidation. Lake levels rose and fell in time with the annual wet and dry seasons, but they also followed longer cycles. The annual cycles are shown by varves, layers of siltstone about 1/25 in (1 mm) thick, whose regular laminations in 40 in (1 m) of rock record the passage of a thousand years. The longer cycles, typically 33 ft (10 m) thick, consist of repeated sequences as follows:

1 dark-colored siltstones and mudstones, containing black organic material, and deposited in anoxic phases;

ORCADIAN LAKES

The Devonian fishes from the north of Scotland are world-famous, not only for the fact that they were discovered in the 1820s, and provided material for some of the early classics in paleontology, but also because of their diversity and superb preservation. Recent studies of the paleogeography and sedimentary environments of the Old Red Sandstone of Caithness and Orkney have shown what life was like in the subtropical lake system of that time. In particular, it has turned out that the gray- and red-colored sequences of sandstones, siltstones and mudstones have recorded, almost year by year, the changing conditions in an ancient lake. One in particular, at Achanarras, has been studied in detail by Nigel Trewin of Aberdeen University. He found evidence for a gradual shallowing of the water, beginning with dark, fish-preserving muds, progressing through sediments that had been laid down in more turbulent waters, to rocks that had experienced very shallow conditions, where the mud had cracked in the drying sunshine.

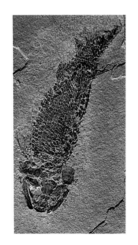

Fossil fishes from the Old Red Sandstone of the Orcadian lakes are often exquisitely preserved. Fishes then all had heavy bony scales, and many of the fish beds show very little sign of damage to the specimens. It seems that whole shoals of fishes were killed instantly, and fell to the bottom, where they were preserved.

Devonian paleomap

▨ Mountainous regions
☐ Inland seas

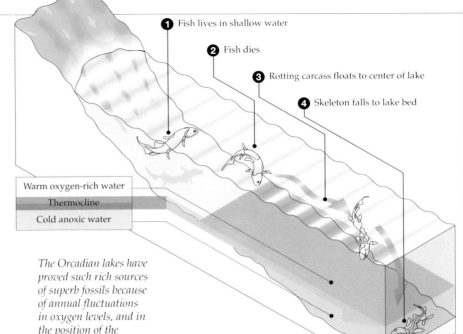

1 Fish lives in shallow water
2 Fish dies
3 Rotting carcass floats to center of lake
4 Skeleton falls to lake bed

Warm oxygen-rich water
Thermocline
Cold anoxic water

The Orcadian lakes have proved such rich sources of superb fossils because of annual fluctuations in oxygen levels, and in the position of the thermocline.

Deep lakes today have a thermocline, above which waters are warmed a little by the Sun, and below which the water is icy cold and deoxygenated. Fishes live in the top layer. At certain times of the year, there may be vast algal blooms, when surface-living plants flourish, and use up all the nutrients and oxygen in the water. This is usually followed by mass death of the plankton, *and of most of the other life in the lake. Everything falls to the bottom, into deep cold oxygen-poor waters, where the organic material and the carcasses are shrouded in mud.*

Northern Scotland lay in equatorial latitudes during the Devonian period. The red rocks speak for hot climates, and there is evidence in some detail that the lakes dried out. The sedimentology of the lakes and surrounding river systems shows that sediment was draining down off the proto-Highland mountains. A great deal of the rock exposure may be seen onshore, but much of the mapping has now been done with information from offshore boreholes drilled in the search for North Sea oil.

NORTH SEA
SCOTLAND
BRITISH ISLES
EUROPE

Location of fossil beds in Scotland

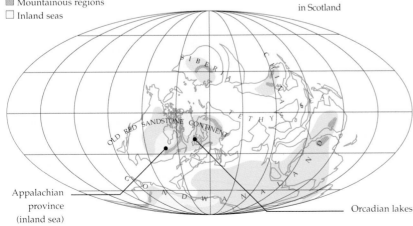

SIBERIA
CHINA
OLD RED SANDSTONE CONTINENT
TETHYS
GONDWANALAND

Appalachian province (inland sea)

Orcadian lakes

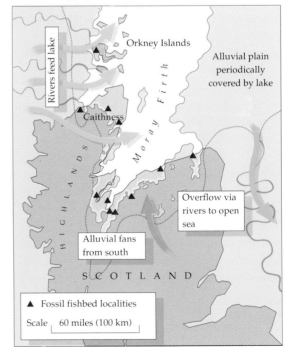

Rivers feed lake
Orkney Islands
Alluvial plain periodically covered by lake
Caithness
Moray Firth
HIGHLANDS
Overflow via rivers to open sea
Alluvial fans from south
SCOTLAND

▲ Fossil fishbed localities
Scale | 60 miles (100 km)

2 sandstones deposited by gravity currents, slides triggered by storms or earthquakes (in some lakes only);

3 sandstones and siltstones with wave ripples, which represent shallowing phases and fairly rapid deposition;

4 green mudstones and sandstones with mudcracks, formed in drying lakes.

The fishes are found as scattered fragments all the way through the sequence, but mainly in "fish beds" crammed with whole specimens. These beds correspond to anoxic phases when the lake was at its deepest, and they seem to have been deposited every ten years or so, as indicated by painstaking work done by Nigel Trewin of Aberdeen University. The anoxic conditions may have been caused by an algal bloom when thriving algae removed oxygen from the water, by decaying plant matter, or by severe storms fetching deep anoxic waters to the surface. Rapid changes in salinity or cold shock may also have contributed to catastrophic fish kills.

The heavier-armored jawless fishes and placoderms probably died at the bottom and were moved slowly into deeper water by gentle bottom currents. The free-swimming acanthodians and bony fishes would have floated at the surface after death. Gases of decay blew up their stomachs like balloons, until they burst and the carcass could sink to the bottom, and many of these fishes have their rib cages disrupted. Once on the oxygen-free bottom, there were no scavengers, and the carcasses were covered over by the fine rain of sediment from above.

Obviously the fishes did not live in the suffocating muds that preserved their bodies. Food chains were based in the warm oxygenated surface waters around the lake shores. *Dipterus* and *Osteolepis* probably fed on reedy waterside plants and possibly on worms and arthropods. The acanthodians may have fed on plankton and small animals, while they themselves used their impressive arrays of spines to discourage larger predators. The jawless fishes and heavy placoderms like *Pterichthyodes* scavenged on the bottom. The medium-sized unarmored fishes were fed on by *Cheirolepis* and *Coccosteus*, which have been found with acanthodians and *Dipterus* in their stomachs. These predators were preyed on in turn by a large lobe-fin called *Glyptolepis*, 40 in (1 m) long, which may have lurked like a pike among the waterside plants to charge at passing fishes.

Life had been exploring the water for something like 3.5 billion years. Already it had dispatched the bacteria, algae, fungi, plants, and an army of invertebrates on to the land. Eurypterid trace fossils show these "sea scorpions" making rushes on to shoreline areas from the Silurian onward, and had they managed to gain a foothold they would have been giants in a world that teemed with Lilliputian prey. Animals of the lowest waters had developed elaborate tiering systems with different sets of feeders partitioning their resources from deposits in and on top of the soft sediment up to levels 40 in (1 m) above the seafloor. Each layer would have its characteristic occupants, from deep-delving bivalves all the way up to builders such as bryozoans and corals. The Devonian was an age of immense coral reefs, built up over millions of years, sometimes hundreds of miles long, rich in faunas and floras, the largest structures life had ever built. Devonian seas and lakes swarmed like a modern metropolis, awash with competition. For the larger animals, land could be a well-stocked larder and a safe haven, even if they could only commute there.

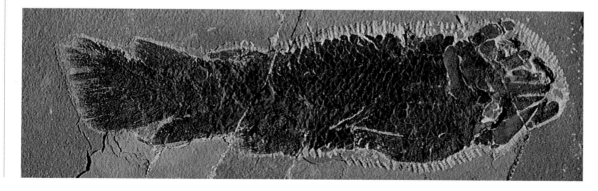

This specimen of the lobe-finned fish Glyptolepis *from Achanarras quarry, northern Scotland, shows typical Orcadian lake preservation. The skeleton is nearly complete and has not been scavenged; it is preserved in a dark-colored, fine-grained sediment deposited in cold, oxygen-poor water.*

The Carboniferous forests of Nyrany, Czech Republic, some 320 My ago, were rich in terrestrial life: great trees, seed ferns, horsetails, and the like. Giant insects, including dragonflies as large as seagulls, patrolled the damp forests in search of prey – in this case a cockroach. Amphibians were common. Some spent most of their lives on land, while others remained largely aquatic. All had to return to the water to deposit their spawn (eggs) from which the tadpole-like larvae hatched.

Chapter Three

FOUR FEET ON THE GROUND

Michael Benton

There is some debate about when the vertebrates ventured out of the water, and why they did it. There is no doubt, however, that amphibians – able to breathe air, but dependent for their reproduction on water – had arisen by late Devonian times.

Some of the disputed evidence for a much earlier arrival consists of footprints. There is a slab bearing twenty-three footprints reported from late Silurian or early Devonian rocks in Australia in 1977. The prints are not clearly marked, and unfortunately the slab was not found in place, but in the floor of a courtyard. More definite evidence comes from skeletons in the late Devonian: *Ichthyostega* and *Acanthostega* from Greenland, *Tulerpeton* from Russia, and *Metaxygnathus* from Australia.

Why move on to the land? The classical theory put forward by Alfred Sherwood Romer in the 1950s was that fishes moved on to land in order to escape from drying pools. The Devonian of Scotland and elsewhere was a time of seasonal droughts, he thought, and to be able to survive in those conditions must have conferred a great advantage. The fishes could either estivate – bury themselves in the mud and "sleep" through the dry season with their systems shut down, as modern lungfishes do – or move across land in search of another pool. So, according to Romer, walking on the land evolved as a means of staying in the water!

This theory has been criticized because there is only limited evidence for droughts, and because it would explain only a minimum amount of land adaptation, not the complex amphibian limb. The simplest hypothesis is that the lobe-fins moved on to land in order to tap new resources: not only food, in the form of plants and small animals that were not being exploited by anything else, but also oxygen, vastly more plentiful in air than it is in water, if only they could absorb it safely.

Problems of life on land

Animals, particularly larger animals, have a special problem with gravity on land, just as the plants do. Like the plants, they also had to devise ways to breathe, to minimize water loss, and to keep reproducing, when their equipment was designed for water.

Breathing was the least of the barriers that the Devonian lobe-fins had to overcome in order to evolve into amphibians. The main difficulties were support and movement. In water the body is practically weightless, which means that there is little need for special adaptations to maintain its shape, keep its organs in place, support its weight, or enable it to walk.

First, support. The backbone of a fish is adapted for swimming, and especially for sideways bending. In a four-legged amphibian the main stress is downward. The backbone and associated muscles are modified to prevent the back from sagging when the animal stands up. New muscles are needed in the head, to hold it up, and in the belly region to prevent rupture by the weight of the internal organs.

Locomotion on land works very differently from swimming. When you watch a salamander or a lizard walking, it sways its body into sweeping sideways curves – a legacy of its swimming ancestors – but the main propulsive force comes from an entirely new source. Instead of a smooth gliding motion, now it was a series of jerky steps that took the amphibians where they wanted to go. The paired fins of the lobe-fins can produce simple "walking" movements, but these needed radical improvement in the first amphibians.

There is a broad resemblance between the bones in the paired fins of rhipidistian fishes, such as *Osteolepis* and its relative *Eusthenopteron*, and in the equivalent limbs of amphibians. In the pectoral fin skeleton of *Eusthenopteron* it is possible to recognize the humerus (upper arm bone), radius, and ulna (forearm bones), and there is a less reliable match with some of the wrist and finger bones.

How well could the rhipidistians walk? Mahala Andrews of the Royal Museum of Scotland, in Edinburgh, attempted a detailed reconstruction of the forelimb of *Eusthenopteron* in the 1960s, and produced a model of its walking abilities. She concluded that *Eusthenopteron* could swing its forelimb back and forward through only 20–25 degrees, with most of its movement at the shoulder joint. The elbow could bend very little, though the elements at the end of the paddle were more flexible. So *Eusthenopteron* must have flopped its front fins to and fro and dragged itself along through soft mud. It could not hoist its body off the surface for proper walking.

Once again, paleontologists can only guess at the stages in between. The earliest amphibian skeletons already have fully developed shoulder, elbow, and wrist joints, bulky limb bones, well-developed muscles for support and locomotion, and clearly formed hands and fingers. The limbs were held in a sprawling posture, with the elbow and knee sticking out sideways and the humerus and femur, upper arm and thigh bones, held roughly horizontal during walking. During each stride, the body dips up and down as the limb moves up and along from its almost horizontal start position. The same "stride-bouncing" effect can be seen in all land vertebrates, including humans. Imagine a person walking along behind a wall that just reaches standing height; the head bobs in and out of sight.

Land life led to many more skeletal changes. For example, the shoulder and hip girdles grew and strengthened, and connected more firmly to the skeleton, in order to stabilize the limbs and provide attachments for the new and enlarged walking muscles. In fishes the hip girdle is tiny, and can "float" in the muscles of the belly; in land vertebrates it is fused to the backbone. The shoulder

The rhipidistian fish Eusthenopteron, *from the Devonian of Canada, had a flexible body covered with broad, circular, bony scales. Its head was invested with heavier bony plates. The strange forked tail fin, and the other midline fins on the back and belly, assisted in swimming and steering. The pectoral fins may have functioned in steering, but since they were fleshy and muscular, they could have performed a simple walking maneuver.*

EVOLUTION OF THE TETRAPOD LEG

Tetrapod legs must have evolved from the fins of precursor fishes. But how could a swimming fin be translated into a walking leg? The clue seems to lie in the bony skeleton within the paired fins of certain Devonian fishes.

The osteolepiform fish Eusthenopteron *may be close to the ancestor of tetrapods. This large fish, 40 in (1 m) long, had muscular lobed fins. It is a sarcopterygian, the group of lobe-finned fishes represented today by rare lungfishes and coelacanths. The skeleton of the front and rear paired fins, lying behind the head and beneath the back part of the body respectively, is complex. Individual bones of the front fin, for example, match closely those of tetrapod arms; there are an upper arm bone (humerus), two forearm bones (radius and ulna), and, arguably, wrist bones (radial, intermedium, ulna). These elements may be seen clearly in the arm of one of the first tetrapods,* Ichthyostega, *which had seven fingers. It is likely that* Eusthenopteron *could "walk" with its forefins at least, swinging them back and forward to drag it over the mud of drying pools toward nearby water. It was still a major shift of function to make this into a walking move-ment, mainly because* Ichthyostega *had to be able to hoist its body clear of the ground and support its own weight.*

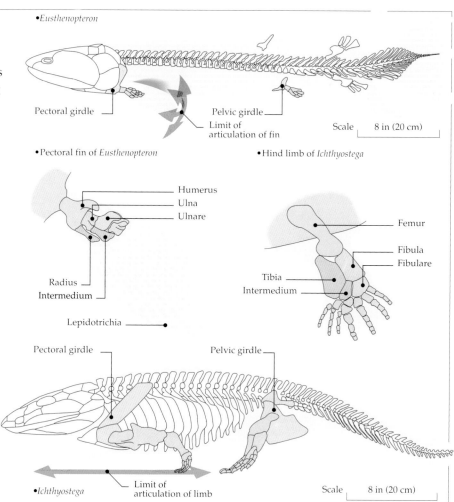

• *Eusthenopteron*

Pectoral girdle

Pelvic girdle

Limit of articulation of fin

Scale 8 in (20 cm)

• Pectoral fin of *Eusthenopteron*

Humerus
Ulna
Ulnare

Radius
Intermedium

Lepidotrichia

• Hind limb of *Ichthyostega*

Femur
Fibula
Fibulare

Tibia
Intermedium

Pectoral girdle

Pelvic girdle

• *Ichthyostega*

Limit of articulation of limb

Scale 8 in (20 cm)

girdle in fishes is part of the back of the skull, but in a tetrapod this arrangement would transmit the shock of each step directly to the head, and scramble what few brains it had. The shoulder girdle became divorced from the skull and associated with the side of the rib cage.

The head became modified in other ways. The first amphibians had lost the remarkable kinesis of the bony fishes' skulls, its usefulness canceled by the need for heavier skull bones to prevent the collapse of the head under gravity or by collisions when stumbling about on land. The jaw joint in the first tetrapods was a simple hinge between the lower jaw and the skull. The rhipidistians had already developed paired nostrils.

The first amphibians probably dealt with the threat of drying out by staying near the water. Ways of giving birth had to change, but again the amphibians seem only to have gone halfway. Just like modern amphibians, the earliest tetrapods must have laid their eggs in the water, where they hatched into

the tadpoles that later metamorphosed into the adult land-living form. This is confirmed by a few rare finds of fossil tadpoles complete with the feathery gills they must have used in order to breathe underwater.

Ichthyostega

The most completely known early tetrapod is *Ichthyostega*, from the late Devonian of Greenland. It was about 40 in (1 m) long, and retained several obviously fishlike features: a streamlined body, a rounded skull as in *Osteolepis*, the shoulder girdle close to the back of the skull, a large tail fin, and lateral line canals on the skull.

The ribs were unusually broad, and overlapped down both sides of the animal. This may have been partly to support the internal organs, but it is unique to *Ichthyostega*, and would have put a severe limit on the vertebrates had it been taken much further and produced a thoracic cage

One of the first amphibians, Ichthyostega lived during the late Devonian of Greenland. Its skull was nearly identical to that of the rhipidistian fish Eusthenopteron and, like its close fish ancestors, it retained a deep tail with fins. This amphibian had four land-going limbs and its head was clearly separated from its body by a neck (short as it may have been). Its backbone and rib cage were designed for weight support on land.

more like a breastplate. The head and limbs are heavy, which shows that *Ichthyostega* did not move fast on land, and in fact it probably had to stop from time to time on any expedition, and rest its head and belly on the ground. The long sharp teeth show that it fed on flesh, perhaps with a few land-living arthropods and worms to supplement its likely diet of fish.

For many years it was assumed that *Ichthyostega*, like all tetrapods, had five fingers and five toes on each of its hands and feet. This is the basic five-digit ("pentadactyl") model on which every tetrapod limb was supposed to be based: whale's paddle, bird's wing, horse's hoof, human hand. New specimens investigated at Cambridge University by Jenny Clack and Michael Coates showed in 1990 that this standard number (the basis of the human counting system) was not fixed at all. *Ichthyostega* had seven toes, and *Acanthostega* eight. The "extra" digits were perfectly well formed and not at all irregular. This new discovery confirms earlier Russian accounts of the late Devonian amphibian *Tulerpeton*, which also had more than five digits.

These discoveries have caused a major rethink of models for the evolution of limbs, and of other skeletal structures. It turns out that all Devonian tetrapods had more than five digits before the number came down to the standard five, or fewer. If fingers can be lost so easily, clearly the vertebrate hand

and foot are not stuck to the number five. Here was an exploded paradigm. Paleontologists had assumed that each finger and toe was programmed into the cell from the start of the embryo's development, and that it would take a big evolutionary shift to change the numbers.

Newer theories suggest that forming a structure like the vertebrate limb is a procedure that leads along a path which has a series of decision points. The entire path, the developmental program, is coded genetically, but how far it goes can be decided by aspects of its environment. Only the upper limb bone, the humerus or femur, is present in the early limb bud. It develops as far as a decision point where the two lower limb bones are generated. The next decision point comes at the wrist, when a variable number of bones are laid down in cartilage, and finally the fingers are generated in an arc that involves a series of decisions running from the inside out. To grow more than five fingers or toes just requires the generating arc to stay active a little longer.

Experiments in developmental biology by Père Alberch of the Madrid Museum, among others, have identified miniaturization as a cause of digit loss. Dozens of salamander species have four fingers instead of the standard five – a transition quite common in the evolution of these modern amphibians – and what happens at this decision point is always the same: the fusion of the fourth and fifth fingers, furthest from the "thumb." Clearly there is a latent fixed potential to lose digits, and the theory put forward by many experts is that if the limb bud is smaller than a given size at a given stage of development then only four fingers will be produced, but if it is larger the number will be five.

Diversity of Carboniferous amphibians

The amphibians are often said to have "ruled the Earth," whatever that means, during the Carboniferous period (360 to 286 My ago). Certainly they spread and diversified, and the key lines of later tetrapod evolution were all laid out during that time.

In Carboniferous times, the continents were coming together to form a smaller group of more or less continuous landmasses, with connections from Europe to North America, and from Africa to South America, Antarctica, and Australia. Fossil amphibians have been found in plenty, especially in the coal-bearing districts of Europe and North America which lie in an arc around the Carboniferous equator.

The late Devonian trees were now the most striking aspect of the landscape: giant club mosses such as *Sigillaria* and *Lepidodendron*, 100 and 130 ft (30 and 40 m) tall; horsetails like *Calamites*, growing up to 50 ft (15 m); equally tall tree ferns like *Psaronius*. The pteridosperms, seed ferns, were flourishing, together with other seed plants like the cordaites, now extinct, and the conifers. These seed-bearing plants go by the collective name of gymnosperms. Paleobotanists often find their seeds in the fossil record, but may be unable to pair the seed with the plant. Trees like the conifers, upland dwellers, still harked back to the massive sporing strategies of marine plants by releasing millions of pollen grains to fertilize the female cone. These wind-blown vectors required heavy expenditure in pollen to get results, and would work much better where the trees were massed together without much competition.

Much of the lush Carboniferous forest grew in lowland swamp and estuarine areas that were periodically flooded by the sea. The forest phase produced thick layers of leaves and rich soils, while the flooding knocked over the trees to produce organic layers that broke down and compressed in the course of millions of years into coal seams often several yards thick. The cycles of forest growth and inundation continued to lay down more and more layers for more and more millions of years, and vast coal deposits were accumulated in sediments over 3,000 ft (900 m) deep.

These forests were stalked by giant millipedes, some up to $6^1/2$ ft (2 m) long; and dragonflies with seagull wingspans flew through the upper branches. The insects seem to have developed flight by the end of the Devonian, perhaps when tree-dwellers with small lateral appendages that may have evolved to gather heat found it useful to make gliding descents from trees, and natural selection went to work on the appendages. Other insects in the Carboniferous forests included springtails, stoneflies, and cockroaches, while the spiders, scorpions, and centipedes continued to thrive in the detritus of the forest floor.

One of the best early Carboniferous faunas was found in the 1980s at East Kirkton, near

The Carboniferous plant Neuropteris, *a fernlike frond found commonly in Carboniferous coal deposits.*

The amphibian Crassigyrinus *from the early Carboniferous of Scotland had reverted to an aquatic life, nearly losing its limbs. The tiny arms and legs might have assisted in swimming, but they would have been virtually useless in moving such a blimplike body on land.* Crassigyrinus *was probably a lurking predator, like a pike, hiding in the dense plant life around the edges of ponds and seizing fishes in its massive jaws.*

Edinburgh, Scotland, by the professional collector Stan Wood. A prominent amphibian here was the grotesque *Crassigyrinus*, an animal with a huge head and mouth, a long fishlike body, and the limbs of an animal about a quarter of its size. *Crassigyrinus* fed on fishes and was clearly a swimmer rather than a walker. It may have lurked among the tangle of aquatic vegetation and fallen trees on the bottom, and varied its diet by taking large invertebrates. *Crassigyrinus* was a reptiliomorph amphibian (tending toward a reptile form), an early representative of one of the two great lines of amphibian evolution that arose in the Carboniferous.

The Nyrany amphibians

Carboniferous amphibians appear most commonly in a number of European and North American localities, some of them famous for the exquisite preservation of their fossils. One of the best known is Nyrany, a former mining town in the northwest of the Czech Republic. Hundreds of specimens were collected between the 1870s and 1900, when commercial mining ceased.

The fossil amphibians nearly all come from a layer of coalified shales and mudstones, around 1 ft (30 cm) thick, called the Plattelkohle. The sediments were laid down about 300 My ago in a swampy area, an intermittent lake about 5 miles (8 km) long and 1 1/4 miles (2 km) wide.

The amphibians clearly lived in and around the lake, and their carcasses either sank straight to the bottom or were only carried a short distance before being buried in a fairly undecomposed condition. Comparisons with similar lakes of today, where reeds grow in less than 40 in (1 m) of water, suggest that the 1 ft (30 cm) of the Plattelkohle took 300–700 years to build up. The main amphibian layer probably covers less than a century, and the animals were not all killed by a single event but occur on different laminae (thin layers) in the sequence.

The amphibians are so well preserved at Nyrany because the bottom waters seem to have been anoxic, as in the Orcadian Basin fish beds. Even tiny animals are often preserved complete, and traces of skin and guts may be seen. There is often a carbonaceous film around the white or brownish bones, which paints in the former body outline, as well as soft structures like gills. With no oxygen in the bottom waters or the mud below, there were no bottom-living or burrowing scavengers to recycle the corpses. In addition, the conditions of sedimentation were very low-energy – no strong currents, tides, floods, storms, or massive animals passing through – so that the bodies in their lakebed graveyard were not swept about and broken up. Swamps are often acidic, like the peat bogs they sometimes become, and in such conditions bone dissolves, but the Nyrany swamp appears to have been neutralized by salts contained in the incoming sediment. The only real damage seems to be explosion of the gut by the build-up of gases of decomposition – again as in the Orcadian fishes. Many of the larger Nyrany amphibians have disrupted rib cages with loss of the associated skin and scales.

Andrew Milner of Birkbeck College, London, has carried out a census of the 700 specimens of Nyrany amphibians that he was able to see in museums. They belong to twenty amphibian and four reptile species, though other finds include rare fishes and small shrimplike creatures. The tetrapods fall into three main ecological categories: open water, swamp and lake, and land.

Three species of larger swimming amphibians occur as very rare examples. They are an eogyrinid anthracosaur (a reptiliomorph group) called *Diplovertebron*, and two loxommatid ("long-eyed") batrachomorphs, *Megalocephalus* and *Baphetes*. These amphibians were all about 2 ft (60 cm) long, with broad powerful skulls, short limbs, and long, deep-sided tails. They would have spent most of their time in the water, where they fed on fish. They appear in the deposit because they have strayed from the middle reaches of the lake toward the swampy shore.

The commonest finds are naturally the animals that are preserved where they lived, on the shallow swampy fringe of the lake. Mostly they were small and partly aquatic, and scuttled about among the plants and debris on the shallow bottom.

Branchiosaurus, for example, is a strange little animal about 3 in (7.5 cm) long, a fascinating temnospondyl amphibian that seals the high quality of the Nyrany assemblage. Because it has fleshy feathery gills extending from the sides of its body it has often been interpreted simply as a tadpole form. But some paleontologists have suggested that at least some of the *Branchiosaurus* material may belong to adult animals that have matured in a juvenile body. The process is still common among modern amphibians, and is known as pedomorphosis.

There may be evolutionary advantages for an amphibian adult to take over the ecological role normally played only by the tadpoles and remain forever larval in form, feeding on aquatic arthropods.

Microsaurs teemed in late Carboniferous and early Permian times, most of them on land, but with a range of aquatic cousins. *Microbrachis*, was a thin, newtlike animal, 1 ft (30 cm) long, with a small head and short limbs. It fed on small floating organisms in the lake.

The most diverse part of the Nyrany fauna is the lakeside animals: thirteen species, and none of them particularly common, which tends to support the theory that they were washed in by chance from time to time. They include the temnospondyls *Gaudrya* and *Amphibamus*, the anthracosaurs *Gephyrostegus* and *Solenodonsaurus*, the aïstopod *Phlegethontia*, the microsaurs *Hyloplesion*, *Sparodus*, *Ricnodon*, and *Crinodon*, and three primitive reptiles: *Brouffia*, *Coelostegus*, and *Archaeothyris*.

Anthracosaurs were a small group of land- and water-dwelling animals that survived from the early Carboniferous to the late Permian. Both forms are represented at Nyrany. They had deep narrow skulls, unlike the temnospondyls, and they lay along the reptiliomorph line of amphibian evolution – on the way to the true reptiles.

The lakeside microsaurs were clearly terrestrial animals, small (4–6 in, 10–15 cm, long) and superficially lizard-like. The same goes for the three Nyrany reptiles, and all of

- ◼ Tillites (glacial deposits)
- ◿ Coal deposits
- ◗ Vertebrate fossil sites
- ▦ Humid conditions
- ▢ Glaciated cold
- ▢ Arid desert

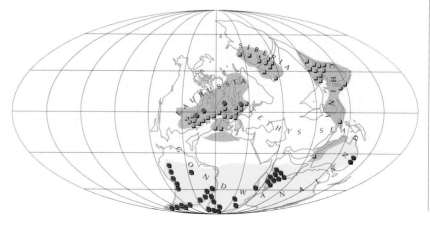

CARBONIFEROUS WORLD
Continental reconstructions for the Carboniferous are disputed. Some geologists find evidence for an early Pangaea, or supercontinent, quite distinct from the one that occurred later during the Mesozoic. The reconstruction here shows a different view, with parts of eastern Asia separate as major island continents.

There is dramatic evidence in Carboniferous rocks for major climatic belts. Glacial deposits, including tillites (soils produced by glacier movement), are found abundantly in southern continents; these prove a major ice sheet over the South Pole, and extending northward. Coal, for which the Carboniferous is famous, indicates a major humid warm climatic belt running from North America over much of Europe. The great forests were home to the early amphibians and reptiles.

Through much of the Carboniferous there appear to have been three major continents: Gondwanaland in the Southern Hemisphere, Laurussia (North America and Europe), and Siberia. The two northern continents combined to create Laurasia.

these animals probably fed on insects and other terrestrial prey. Despite the resemblances, none of them is at all related to true lizards.

The Nyrany amphibians are mirrored in other late Carboniferous deposits of similar age in Britain and North America, and there seem to be common patterns among the ecological groupings. It seems reasonable to conclude that the associations between these animals are not merely chance sets of fossils but are a rough sample of true ecological communities – that is, groups of plants and animals that have fairly consistent compositions and interactions.

Carboniferous seas

Much of Europe and North America lay around the equator in Carboniferous times, and their marine fossils reflect these tropical conditions. Vast thicknesses of limestone were deposited by the usual agents, the marine invertebrates building up huge reefs. These sites were occupied by tabulate and rugose corals, giant brachiopods, crinoids, and algae. Starfishes, gastropods, and sea urchins roamed among the reefs, while great coiled nautiloids, ammonoids, and fishes swam

above. The fishes had lost several of their most prominent early groups, and looked very different from those of the preceding Devonian period. The placoderms, armored agnathans, and rhipidistians had all gone, and the acanthodians and lungfishes were in decline. Sharks and bony fishes dominated, and the sharks had developed all sorts of strange equipment.

The eugeneodontid sharks seem to have had bodies like *Cladoselache*, according to the scanty fossil record. Most of the skeleton is poorly known, with one striking exception: the great tooth whorl that was mounted in the lower jaw. This was a coil of teeth that unrolled during growth, and kept producing new teeth as the old ones wore away, like a dental conveyor belt. In order to accommodate this spiky spiral, the snout and lower jaw grew longer.

Stethacanthid sharks are known from Ireland, Scotland, and the United States, and have a specially freakish look to a modern eye. *Stethacanthus*, about 40 in (1 m) long, had a thick, prominent spine mounted just behind its head and topped by dozens of small teeth. It looks as if the handle of a shaving brush has been jammed into its back with the bristles uppermost. The patch of teeth at the top is matched by a similar patch growing straight

Strange sharks inhabited the early Carboniferous seas of Scotland, Ireland, and North America. Stethacanthus, on the left, had a radar-like turret over its shoulder region, surmounted by a platform bearing rows of small sharp teeth. Another patch of teeth graced its forehead. Falcatus, on the right, had a projecting bony spine in the same position, pointing forward like the gun barrel on a tank. This supernumerary bone was also equipped with rows of teeth, as was the top of the head. The function of these spines in the stethacanthid sharks is a mystery.

out of the forehead, and this has produced an ingenious suggestion from Rainer Zangerl, who found the new material in the Bear Gulch Formation in Montana in 1984. When *Stethacanthus* was partially buried in the sand, the patches of teeth on the crest and forehead would have mimicked a vast gaping mouth on the seafloor. Zangerl proposed that their function was to scare off predators.

Another stethacanthid, *Falcatus*, has an L-shaped spine with the L tilted forward in the same shoulder region so that the lower bar sticks up and the upright points forward over the head. This toothed spine is found only in sexually mature males (you can tell the sex of a fossil shark because the males have "claspers" underneath, extensions of the pelvic fins). The spine above the head may have been used in pre-mating courtship displays, rather as stags use antlers. It is not clear whether or not it also had a more practical clasping function. One specimen has been found with the female locked on to the spine, but with the female above and the male below, with his back turned to her; clearly they needed a lot more practice.

Carboniferous sharks like *Xenacanthus* and the hybodonts, fast movers with no unusual adornments, were closer to the main line of shark evolution. They heralded a major radiation of the sharks from the middle of the age of dinosaurs to the present day.

The bony fishes of the Carboniferous also showed advances over their Devonian forerunners. The Carboniferous and Permian varieties, like *Cheirolepis* before them, had simple hingelike jaws, an asymmetrical tail, and heavy bony scales. They lived in hordes in some localities, and developed a wide variety of body shapes. Most were typical cigar-like designs, but others, such as *Cheirodus* from the Carboniferous, were deep-bodied and very flat and narrow. This deep body form has been repeated several times during the later evolution of the fishes, arising independently, obviously useful, but for reasons not explained.

By the end of the Permian the ocean's phytoplankton and the land's thriving plant life had pumped enough oxygen into the atmosphere to bring it close to its present level. Vast forests supported a teeming animal population that could draw on this plentiful energy supply to find places in expanding food chains. The later Paleozoic is a scene of consolidation and development, rather than radical new creations. Vertebrates might range from 1 or 2 in (2.5–5 cm) to lengths of 15 ft (4.5 m) and more on land, 30 ft (9 m) and more at sea, but these were variations on an established theme.

From now onward, the basic cast had been assembled. Inventions and innovations would have to come from them, or not at all. On the other hand, this was a cast of extraordinary versatility, with the potential to change physique and shape far beyond any limits so far reached, and with a wealth of new scripts and new roles as yet unwritten.

It has been suggested that one reason for the huge quantities of plant debris that have survived as coal is that the breakdown specialists, the bacteria, fungi, and invertebrate vegetarians, were not yet fully equipped to tackle the chemistry of cellulose and lignin in the plants' materials. Any improvements in this direction would be hard to follow in the fossil record. But certainly the tetrapods found a new role in the consumption of plants. Where they had begun by eating other animals, and each other, changes in the teeth of some of the larger newcomers identify them as herbivores; and their bodies broadened to make room for larger digestive tracts that needed to contain a lot more bulk and a longer gut, presumably with new varieties of internal flora.

Divorce from the water: the reptile solution

The amphibians had developed all sorts of mechanical and load-bearing technologies to hold them up and move them about on land. Changes in the middle ear enabled them to pick up airborne sound – an essential requirement of life on land. Their next invention was the most dramatic of all; it freed them from their dependence for reproduction on fair-sized bodies of water, and enabled them to start exploring the furthest reaches of the land, far from ponds and rivers, and from the seas that had created their ancestors.

What turned the amphibians into a new class of vertebrates was the cleidoic ("closed") egg, also called the amniote egg, after the membrane that protects the embryo inside it. The cleidoic egg has a semi-permeable shell that holds a complex of membranes, fluid, and enough food to allow the embryo to develop fully in a safe environment before it hatches. The shell of the egg is usually hard and calcareous, although snakes, some lizards, and some turtles have eggs with leathery shells. (Lizards lay a long way in the future at this stage, when some reptiles would develop the ability to raise their snouts as well as lowering their jaws, and would specialize in the flexible skeletons also possessed by their descendants, the snakes. A lizard is a reptile; a reptile does not have to be a lizard.)

In the cleidoic egg, the developing embryo takes its food from a yolk sac and passes waste materials into another sac, the allantois. In order not to pollute the egg's environment, wastes are stored as insoluble uric acid, instead of the soluble urea produced by animals whose eggs can be irrigated by water (mainly fishes and amphibians). The membrane around the embryo is the amnion; both embryo and yolk sac are additionally surrounded and protected by a second membrane, the chorion. Gases for breathing pass readily in and out of the shell by way

AMNIOTES AND THE CLEIDOIC EGG

The first tetrapods – four-legged land animals – were amphibians, animals that lived on land but had to return to water to breed. The break with the water was made by the amniotes (reptiles, and their descendants, the birds and mammals), during the early Carboniferous period. The key to amniote success was the cleidoic egg (right).

The cleidoic egg allowed the early reptiles to move away from waterside habitats, and to colonize dry regions. Other adaptations assisted in this widening of the hold of vertebrate life on land. Reptiles have scaly waterproof skins, to prevent evaporation of water through the skin, as can happen in amphibians. Many reptiles also have remarkable abilities to conserve water. For example, most produce near-solid urine, an ability inherited by the birds.

The cleidoic ("closed") egg has two features. First is the semi-permeable outer shell. The shell is usually a hard mineralized coating, but can be of leathery construction in some lizard and snake groups. The shell protects the fluids inside from evaporation, and provides some protection for the developing embryo from physical damage. The second feature is the possession of amniotic membranes within the egg, the various fluid-filled sacs that protect the embryo (amnion), enclose the food supply (yolk sac), gather the waste (allantois), and surround all of this (chorion).

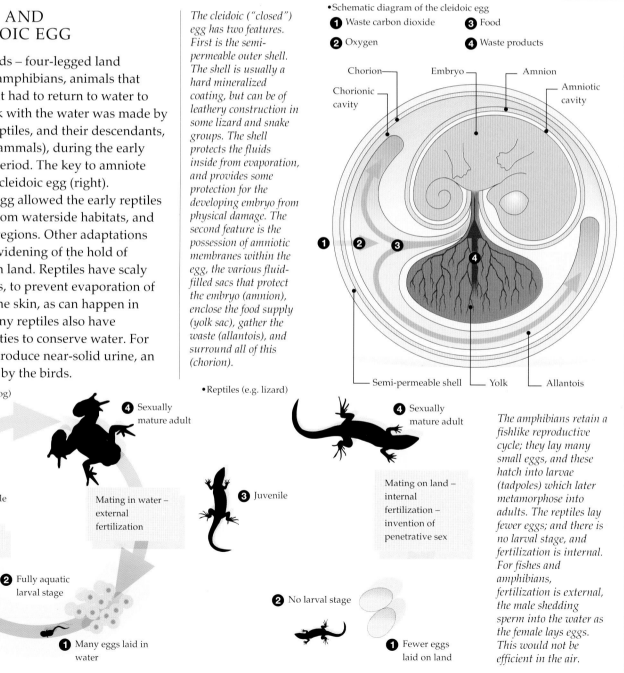

•Schematic diagram of the cleidoic egg

❶ Waste carbon dioxide ❸ Food
❷ Oxygen ❹ Waste products

Chorion — Embryo — Amnion
Chorionic cavity — Amniotic cavity
Semi-permeable shell — Yolk — Allantois

•Amphibians (e.g. frog)

❹ Sexually mature adult
❸ Juvenile
Metamorphosis
Mating in water – external fertilization
❷ Fully aquatic larval stage
❶ Many eggs laid in water

•Reptiles (e.g. lizard)

❹ Sexually mature adult
❸ Juvenile
Mating on land – internal fertilization – invention of penetrative sex
❷ No larval stage
❶ Fewer eggs laid on land

The amphibians retain a fishlike reproductive cycle; they lay many small eggs, and these hatch into larvae (tadpoles) which later metamorphose into adults. The reptiles lay fewer eggs; and there is no larval stage, and fertilization is internal. For fishes and amphibians, fertilization is external, the male shedding sperm into the water as the female lays eggs. This would not be efficient in the air.

of the allantois, but fluids are retained, and so the embryo does not dry out.

The eggs are not laid in water, as happens with amphibians, so this reproductive tie to a fishy ancestry is decisively broken by the reptiles. Further, the eggs are larger, and well stocked with food, so that the aquatic larval stage, the tadpole, is also omitted. Reptiles lay fewer eggs than the fishes and amphibians do, because so much more reproductive energy has to be invested in each egg, both as food store and as living accommodation. By contrast, a frog must produce huge amounts of spawn in order to give just one or two of her tadpoles the chance to survive.

Reptile reproduction takes place on dry land, so that internal fertilization is essential. Male fishes and amphibians usually shed their sperm in the general direction of a mass of freshly laid eggs, which is a wasteful technique, but works well enough in the water. When mating takes place on land, the sperm are delivered economically, straight into the female.

Until recently, the oldest known reptiles dated from the start of the late Carboniferous, and we have also met reptiles in the Nyrany deposits, sometime after that. A recent find in Scotland has totally altered our view of reptile chronology. The fossil was discovered in 1988 by the professional collector Stan Wood, during his excavations at East Kirkton. It seems likely that it is a reptile, although we cannot say which amphibians were its near ancestors, or which reptiles its direct descendants. The fossil received its scientific name of *Westlothiana lizziae* in 1991, commemorating both the region where it was found and its nickname, Lizzie.

The East Kirkton deposits date from the early Carboniferous, about 350 My ago, and this possible reptile predates by 40 to 50 My the rather more fully known reptiles, *Hylonomus* and *Paleothyris*, which had previously seemed to open the reptile record. Even after so many years of collection and study, a single survivor can still rewrite whole chapters in the story of life.

Hylonomus and *Paleothyris* are slender animals, 8 in (20 cm) long, with the comparatively small heads that are typical of reptiles – about one-fifth of the trunk length, instead of the one-third or one-quarter found in most amphibians. The skull is high – a

feature inherited from the reptiliomorph amphibians – and the complex of bones seen in amphibians at the back of the skull is much reduced. On the way to the cleidoic egg, it is likely that some amphibians started to lay fewer but larger non-amniote eggs on land, and the advantage to the hatchling would have been proportionally greater in a small animal. Both of these reptiles have lightly built skeletons, but retain the sprawling limbs and small shoulder and hip girdles of amphibians. The hands and feet have very long digits, as modern lizards do.

It is likely that these early reptiles used their sharp teeth to pierce the cuticles of insects. The high skull, and some new muscle groups that pulled the lower jaw inward and upward toward the palate, would enable them to grapple with struggling insects, some of them probably quite large. Most of the amphibians lacked a strong grip, owing to their low skulls and weak jaws.

These two early reptiles are beautifully preserved in remarkable conditions. They were found *inside* fossilized tree stumps in Nova Scotia, Canada. This area was covered by lush forests of *Sigillaria*, up to 100 ft (30 m) tall. Water levels periodically rose, and the forests were flooded. Trees were knocked down, and this left just the roots and the base of the trunk, which was often enclosed in layers of storm sediments. Soon the hearts of the club-moss stumps rotted away, and they were colonized by millipedes, insects, and snails. Small reptiles and amphibians may have fallen in by accident, caught in these natural pitfall traps, or climbed in on purpose, in search of food. Many of the reptiles seem to have lived in the stumps for some time, feeding on the insects and snails, since their

One of the oldest-known reptiles, Hylonomus lies trapped in a hollow tree stump in mid-Carboniferous times. Exquisitely preserved skeletons of these small reptiles have been found inside fossilized tree stumps in Nova Scotia, Canada. When the tall seed ferns and club mosses were brought down by floods, their interiors rotted, providing a haven for beetles and other detritivores. As the reptiles moved in to feed on the insects, they became trapped.

fecal pellets are preserved. Sooner or later more floods topped up the stumps with sediment, entombing and preserving the delicate reptile skeletons.

Now, fossil eggs are well known from certain dinosaurs, but remains are sparse to non-existent in the Carboniferous and Permian. The oldest supposed fossil reptile egg comes from the early Permian of Texas, about 270 My ago, but it is not an entirely convincing specimen. If the key characteristic of a reptile is the cleidoic egg, and not even one such egg survives the Carboniferous period, how is it possible to identify as reptiles animals like *Westlothiana*, *Hylonomus*, and *Paleothyris*?

The simplest answer derives from what we know about the phylogeny of the reptiles – that is, the evolutionary sequence in which they developed. The fact that all living reptiles, ranging from lizards and snakes through to crocodiles and turtles, have the same kind of egg suggests that the cleidoic egg arose only once. It is utterly unlikely that so complex and integrated a structure could have evolved independently in different amphibian lines.

As well as the living reptiles, we also know that various extinct groups, such as dinosaurs, also had the same cleidoic egg, as far as we can tell from details of the shell microstructure. Hence it is possible to backtrack on the evolutionary pathways of all these groups, with a range of known physical features, and to say with confidence that their single common ancestor must already have had the necessary egg. *Hylonomus* and *Paleothyris* belong to a line of reptile development with features that cannot possibly come directly before various other known lines of development. Therefore all these lines must have diverged from an ancestor that came before them all, and which originated the cleidoic egg.

The amniote radiation

We have seen the phylum Chordata, the chordates, develop a line that led toward the fishes, and then a branch line that emerged from the sea as the amphibians. Now it had branched again, and we have been referring to this new branch as the reptiles. Certainly we still have a class Reptilia, but it does not include all of the descendants of the common reptilian ancestor, because it omits the birds and the mammals (classes Aves and Mammalia).

In technical terms, a group that is descended from a common ancestor but does not include all of the ancestor's descendants is called "paraphyletic." Scientists find it useful to seek groups that do include all of the ancestor's descendants, which are said to be "monophyletic." This still leaves us with a gap, however. Scientists require a name for the larger monophyletic group when they need to take an overview of the evolutionary history that comprises the entire lineage of reptiles, birds, and mammals. These make up the superclass of the Amniota.

The outlines of amniote evolution took shape in the late Carboniferous, when two major lineages developed among the reptiles. One of these led eventually to the mammals, leaving behind it various "mammal-like" reptiles, all of them extinct. Another group led ultimately to the present-day reptiles – lizards, snakes, and crocodiles – and also branched out to produce the birds. A third group may also have developed during the same period, but the fossil record is empty where we most need it; this group survives as the tortoises and turtles (order Chelonia, also called Testudines).

The members of these groups can be readily identified by the presence or absence in their skulls of features called temporal fenestrae: openings located behind the eye socket in the cheek area at either side. The function of these openings is uncertain, but they may simply be a means of economizing on bone construction in places where there is no everyday stress on the skull. This would save energy both in building the skull and in moving it about.

There are three main patterns of temporal fenestrae, corresponding with the lineages mentioned above. They are:

1 *Anapsids*: reptiles with no temporal openings. This is the primitive condition seen in fishes and amphibians as well as in the earliest reptiles such as *Westlothiana*, *Hylonomus*, and *Paleothyris*. The anapsids also include the turtles and their extinct relatives, first known in the late Triassic, as well as a whole variety of early reptiles whose sequence and evolutionary relationships are uncertain. An early anapsid group contains *Hylonomus* and *Paleothyris*, with various others, under the forbidding name of protorothyridids. These are thought to show developments toward the diapsid lineage that appears below. All of the anapsids are now extinct, except for the turtles.

2 *Synapsids*: reptiles with only a single pair of temporal openings, each placed low behind the eye socket. This monophyletic group includes all of the mammal-like reptiles, now extinct, that gave rise to the mammals. The earliest synapsid group in the upper Carboniferous was the ophiacodontids, described below.

3 *Diapsids*: reptiles with two pairs of temporal openings, one above the other behind the eye socket. This monophyletic group with a glorious evolutionary history includes the lizards, snakes, dinosaurs, crocodiles, pterosaurs, birds, and the upper Carboniferous petrolacosaurids.

Another much later reptile group introduced a fourth pattern of fenestrae. The *euryapsids* have a single pair of facial openings, but smaller than in the synapsids, and located higher behind the eye sockets. This is a mixed set of extinct marine reptiles – nothosaurs, plesiosaurs, placodonts, ichthyosaurs – that probably evolved from a number of diapsid ancestors, possibly by losing the lower pair of openings.

Of the three main groups of late Carboniferous reptiles, we have already met the anapsid contingent, the protorothyridids. Most of the synapsids fall into the family Ophiacodontidae. The oldest is *Archaeothyris*, which was found not only in the Nyrany lakeside fauna but also in the Nova Scotian tree stumps, along with *Paleothyris*. The skeletons are not complete, but the narrow high-sided skull has the typical single lower opening behind each eye.

The earliest known diapsid group is the petrolacosaurids. *Petrolacosaurus* comes from the late Carboniferous of Kansas. It is slender, about 16 in (40 cm) long, and superficially similar to *Hylonomus* except for the smaller head, longer neck, and longer, agile-looking legs. The sharp teeth around the edges of the jaws, and supplementary rows of teeth on the palate, show that it fed on insects and other small animals.

The early diapsid reptile Petrolacosaurus *moves rapidly through the forest-floor debris in mid-Carboniferous times in search of insect prey. One of the first reptiles,* Petrolacosaurus *was the ancestor of crocodiles, dinosaurs, birds, lizards, and snakes.* Petrolacosaurus *fed on large insects in the leaf litter, puncturing their tough chitinous exoskeletons with its sharp teeth and sucking the flesh from them.*

The Texas redbeds

By Permian times, the continents were moving into closer contact than in the Carboniferous, as the northern and southern supercontinents of Laurasia and Gondwana began to assemble into the single great landmass called Pangaea. A vast icecap had formed over Antarctica, southern Africa, South America, and India in the late Carboniferous, to disappear in the early Permian, when the climate of Euramerica grew hotter and more arid, and the great

swamplands, lakes, and floodplains dried out. Their flora, and particularly the giant club mosses and horsetails, went into fast decline, to be replaced by conifers descending from the uplands, and by other gymnosperms such as cycads, ginkgoes, and seed ferns.

The Gondwana icecap disappeared in the early Permian period. A new flora then spread southward, dominated by the seed fern *Glossopteris*, which gave its name to this distinctly different area of vegetation, one of several "floral provinces" that were now developing. But conditions for reptile

OPPOSITE *Texas, in the early Permian, some 270 My ago, was largely arid, but with monsoonal rains and watercourses. Although large amphibians, such as Eryops were still present, the dominant land animals were reptiles. Three carnivorous pelycosaurs, Dimetrodon, come to slake their thirst.*

DIPLOCAULUS, A SWIMMING BOOMERANG

One of the most bizarre of the early amphibians was *Diplocaulus* from the early Permian of the midwestern United States. In adults, the back corners of the skull were tweaked out into long prongs, making the head look like a boomerang. What was the function of these unusual structures?

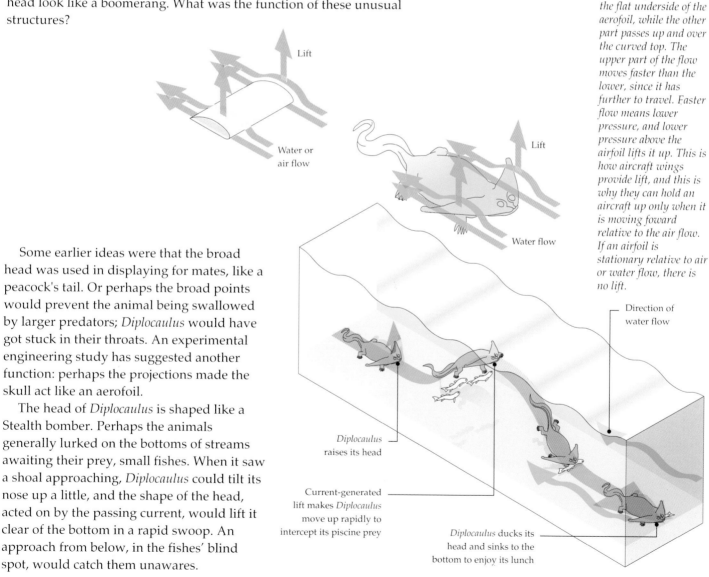

An airfoil (or aerofoil) acts by splitting a smooth flow of air or water. Part of the flow continues uninterrupted beneath the flat underside of the aerofoil, while the other part passes up and over the curved top. The upper part of the flow moves faster than the lower, since it has further to travel. Faster flow means lower pressure, and lower pressure above the airfoil lifts it up. This is how aircraft wings provide lift, and this is why they can hold an aircraft up only when it is moving foward relative to the air flow. If an airfoil is stationary relative to air or water flow, there is no lift.

Lift

Water or air flow

Lift

Water flow

Direction of water flow

Diplocaulus raises its head

Current-generated lift makes *Diplocaulus* move up rapidly to intercept its piscine prey

Diplocaulus ducks its head and sinks to the bottom to enjoy its lunch

Some earlier ideas were that the broad head was used in displaying for mates, like a peacock's tail. Or perhaps the broad points would prevent the animal being swallowed by larger predators; *Diplocaulus* would have got stuck in their throats. An experimental engineering study has suggested another function: perhaps the projections made the skull act like an aerofoil.

The head of *Diplocaulus* is shaped like a Stealth bomber. Perhaps the animals generally lurked on the bottoms of streams awaiting their prey, small fishes. When it saw a shoal approaching, *Diplocaulus* could tilt its nose up a little, and the shape of the head, acted on by the passing current, would lift it clear of the bottom in a rapid swoop. An approach from below, in the fishes' blind spot, would catch them unawares.

evolution, meanwhile, may still have been rather too cold in the south, just as they were growing too dry for the amphibians elsewhere; and most reptile finds come from the Northern Hemisphere.

The richest early Permian deposits are the redbeds of northern Texas and Oklahoma, lying between the Red River and the Salt Fork of the Brazos River. The area was hotly contested by Native Americans and white settlers until about 1880. A Swiss botanist named Jacob Boll had found some fossil plants and scrappy bones, and he was employed as a collector by Edward Drinker Cope (1840–97), famous for his part in the great dinosaur bone wars of the late nineteenth century.

In the winter of 1877–8 Boll collected the beginnings of a fauna, from Archer County, now assigned to the Wichita Group, and Cope based a number of new species on these specimens. Boll died in the field in 1880, and Cope next employed a traveling frontier preacher named W. F. Cummins as his collector. He worked in the redbeds for a number of years, and turned up a younger fauna in western Baylor County (Clear Fork Group). Later the area was worked by a string of eminent paleontologists who included Charles H. Sternberg, E. C. Case, Samuel W. Williston, and Alfred S. Romer.

One of the best-known fossil sites in the Texas redbeds is the Geraldine Bonebed in Archer County, found by Romer in 1932, which has produced superb skeletons of amphibians and reptiles. The fossils appear in a sequence of red and gray mudstones with occasional sandstone lenses. The gray mudstones are deposits from ponds, the red mudstones ancient tropical soils, and the sandstones were laid down in stream channels that cut across the pond muds and soils from time to time.

Abundant plant remains found with the bones confirm the seasonal tropical conditions. The giant horsetail *Calamites* was still able to live in the ponds, while ferns, seed ferns, and conifers lived in the flood basin and on the margins of oxbow lakes. Other conifers and seed ferns lived on higher ground.

Animal fossils include some ostracods, insects, a freshwater shark, a rhipidistian, and a lungfish, all of which lived in or above the pond waters. Some of the amphibians too may have died while in the water, but most of the tetrapods came from the surrounding dry land. Several skeletons were excavated of the massive temnospondyl and key predator *Eryops* and the anthracosaur *Archeria*, both animals measuring about $6^1/2$ ft (2 m) long. Also found were fragments of the advanced reptile-form *Diadectes*, which was 12 ft (3.7 m) long when full grown. *Archeria* was a slender aquatic amphibian with a deep tail fin and a skull like a crocodile's; it must have fed on fishes. The other amphibians, *Eryops* and *Diadectes*, were much more adapted to the land, and *Diadectes* was one of the earliest vertebrate herbivores.

The reptiles from the Geraldine Bonebed include a small short-snouted anapsid, *Bolosaurus*, and several superb skeletons of the spectacular sail-backed pelycosaurs *Edaphosaurus* and *Dimetrodon*. We shall return to this important and fascinating group.

The Geraldine animals are a mixture of nearly complete skeletons and assorted fragments. The skeletons seem to be aligned, and are found mixed up with similarly aligned logs and plant stems, which strongly suggests that they were all caught in the grip of a powerful current. Martin Sander of the University of Bonn concluded that the terrestrial animals were driven into the water by a forest fire. They suffocated in the flames, and their carcasses were swept away, with burnt pieces of wood, until the river dumped the whole mass.

The pelycosaurs

The early synapsids, or mammal-like reptiles, of the late Carboniferous and early Permian are classed as pelycosaurs. Often they are also called the "sail-backed reptiles," which is a misleading term. Certainly the Texan *Dimetrodon* and *Edaphosaurus* sport dramatic sails on their backs, and have stolen a little of the dinosaurs' thunder; but the majority of pelycosaurs, such as *Ophiacodon*, *Haptodus*, and the herbivore *Cotylorhynchus*, which was 10 ft (3 m) long, did not have sails. Pelycosaurs were unquestionably the key early Permian group, generally supplying 70 percent of all species found.

The pelycosaurs are divided into six families, with animals ranging in size from 2 ft (60 cm) to 13 ft (4 m) total length. In the Texas redbeds, *Dimetrodon* is a carnivore, with a high-sided skull, powerful jaws, and long slashing teeth. *Edaphosaurus* is a herbivore, equipped with equally powerful jaws, but smaller leaf-shaped teeth at the sides of the jaws, and masses of rounded crushing teeth in the middle of the palate. It is plain that herbivores arose later than carnivores; all the Carboniferous reptiles ate insects or larger prey, and *Edaphosaurus* was one of the earliest reptile herbivores.

The sails of *Edaphosaurus* and *Dimetrodon* are not typical in pelycosaurs, and may draw more attention than they should, but some interesting detective work has gone into interpreting them. In both cases the sail is made by the elongated neural spines of the vertebrae of the trunk, together with the hindmost neck vertebrae and those in the hip region. These spines form a kind of railing which is uniformly graded to give a smooth rise and fall at the back and front of the sail when it is viewed from the side. The spines have grooves near the base that probably conducted blood vessels to a covering of skin.

Here was a thin flat surface, rich in blood vessels. An elegant theory proposes that the function of this sail was temperature control. As primitive reptiles, the pelycosaurs were unable to keep their blood at a constant temperature. They were "poikilothermic" – literally "variably hot," their blood heat more or less following the temperature of their surroundings.

As we have seen, the climates of the Texas redbeds were tropical, which means cold nights and hot days. A cold reptile tends to be torpid; warmth makes it more active. The idea is that in order to get moving faster after the nightly power-down, the sail-backed reptiles could have crept out of their sleeping quarters early in the day and stood broadside-on to the Sun to absorb heat. By midday, when air temperatures became too high and they might run the risk of overheating, they could either face the sun or stand in the shade, so that the sail would radiate heat. The temperature-control model seems to make sense, especially since both the herbivorous *Edaphosaurus* and its main predator *Dimetrodon* possessed the sail and could use its boost to prey or to escape. But what about the rest of the pelycosaurs, and all the other sailless reptiles? They seem to have managed well enough.

The first aquatic amniote

It took 80 My of reptilian evolution before a member of that group became adapted to life in the water. This first aquatic effort was also rather small-scale and limited. *Mesosaurus* was a lightly built animal, about 40 in (1 m) long, that had a number of unmistakable adaptations to a life in the sea: a long steep-sided tail that it beat from side to side when swimming; broad paddle-like hands and feet, probably used for steering; weak limb girdles; thickened ribs; a long neck; and a crocodilish skull with long jaws and bristling with long sharp teeth. These it clearly used for grasping fishes and holding them while water drained from the mouth.

Mesosaurus is known only from one geological horizon both in Brazil and on the west coast of southern Africa, which were neighbours in Gondwana in early Permian times, when the Atlantic Ocean did not exist.

The first marine reptile, Mesosaurus, *swam in shallow seas and fresh waters on the western side of Africa and in eastern Brazil. It was a small reptile, fully adapted to underwater life, with a long, deep, flat-sided tail and paddle-like limbs.* Mesosaurus *caught small fishes in a delicate cage made from needle-like interlocking teeth.*

EVOLUTION OF THE MAMMAL JAW

One of the most striking transitions in the evolution of life took place when reptiles turned into mammals; the jaw joint switched position. The reptile jaw joint is formed between the quadrate (in the skull) and the articular bone (in the lower jaw), while in mammals it lies between the squamosal and dentary bones. How could such a dramatic switch have taken place?

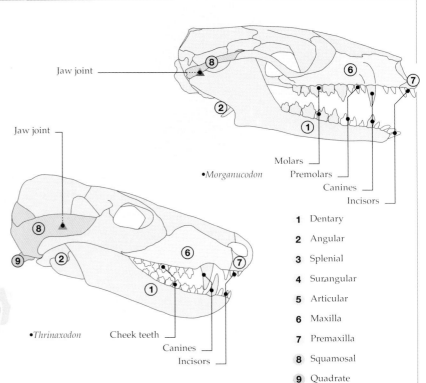

•*Morganucodon*

Molars
Premolars
Canines
Incisors

1 Dentary
2 Angular
3 Splenial
4 Surangular
5 Articular
6 Maxilla
7 Premaxilla
8 Squamosal
9 Quadrate

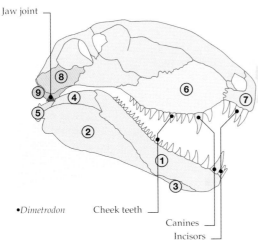

•*Dimetrodon*   Cheek teeth
Canines
Incisors

•*Thrinaxodon*   Cheek teeth
Canines
Incisors

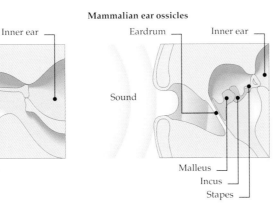

Reptilian ear ossicles
Eardrum — Inner ear
Sound
Stapes

Mammalian ear ossicles
Eardrum — Inner ear
Sound
Malleus
Incus
Stapes

The reptile lower jaw contains several bones (dentary, splenial, angular, surangular, prearticular), that of mammals only one (the dentary). Hence, five of the reptilian bones seem to have been lost, and, astonishingly, the whole process is shown in the fossil record of the mammal-like reptiles, a group that existed from the mid-Carboniferous until the Jurassic. The crucial stages may be seen by comparing a Permian form, Dimetrodon, with a

Triassic mammal-like reptile, Thrinaxodon, and one of the first mammals, Morganucodon, from the early Jurassic. Between the Permian and the Triassic, the dentary had come to dominate the mammal-like reptile lower jaw, and the other bones were restricted to an internal posterior position. They went from there to form parts of the inner ear in mammals; our auditory ossicles once formed the reptilian jaw joint.

It clearly represents an advanced stage in reptile body-building for life at sea, but earlier mesosaurs have not been found, and its relationships are obscure.

The South African therapsids

Pelycosaurs were mostly an early Permian group. At the same time some new kinds of mammal-like reptiles begin to show up in the very uppermost levels of the Texas redbed sequence. But the new wave of reptile evolution in the late Permian happened in what is now European Russia and South Africa, and with its wealth of fossil therapsid remains we shall focus on South Africa.

The Karroo Basin of South Africa is an enormous area of sandstones and mudstones laid down by lakes and rivers. There were lush subtropical forests here, in the Permian, populated by the seed fern *Glossopteris* and by horsetails and ginkgoes. Reptiles came to light in this now semi-arid region in the 1840s, when Andrew Bain, a Scottish engineer, uncovered a "charnel house" of bones near Fort Beaufort. In 1845 a shipment reached

London, where Sir Richard Owen described the new and striking animals – *Dicynodon*, toothless but for a pair of fangs, and other unfamiliar creatures. After this, several eminent Victorian paleontologists collected Karroo specimens, and there has been large-scale collecting in the twentieth century. Robert Broom (1866–1951), another Scotsman, was involved in making the largest collections and in naming dozens of new species. He belonged to the old school, and was never seen in the field wearing anything less formal than his black tail coat, stiff wing collar, and top hat. It was Broom who showed the crucial importance of the South African material in bridging the evolutionary space between the North American pelycosaurs and the later mammal-like reptiles, and eventually the mammals themselves.

The "advanced" mammal-like reptiles of the Karroo are all therapsids, the pelycosaurs' successors in the synapsid line. They display a number of "mammalian" features, among them a larger opening behind each eye socket, fewer palatal teeth, and an enlarged dentary, the bone that housed the teeth in the lower jaw. (In these early mammal-like reptiles, the dentary was one of several bones that made up the lower jaw; it has grown to become the only one left in modern mammals. Experiments with teeth have been a key to mammal development. No other animal has such a variety, with so many specialized functions.) The late Permian therapsids fall into four main groups: dinocephalians, dicynodonts, gorgonopsians, and therocephalians.

Dinocephalians include both herbivorous

In the tropical forests of southern Africa about 250 My ago, two male Moschops *clash their massive, thick-framed skulls. These large herbivores were about 16 ft (5 m) long. Violent head-bashing contests were probably fought to establish dominance within the herd, in the same way as present-day mammals, such as wild rams, prepare for the mating season.*

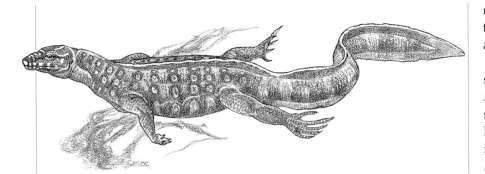

A representative of a short-lived aquatic group of diapsid reptiles from the late Permian of Africa and Madagascar, Hovasaurus *swam in freshwaters, using its narrow, deep-sided tail for propulsion and its lizard-like limbs for steering. It was clearly adept on land, too, and may have hunted small prey in both habitats.*

and carnivorous forms. *Titanosuchus* is an obvious meat-eater, with its well-developed canine teeth, ideal for grasping prey and tearing flesh. It has very short legs and a heavy skull, and is clearly not equipped for hot pursuit. The herbivore *Moschops* is a bulky animal, 16 ft (5 m) long, with a massive barrel-like body, heavy limbs, and a superficial resemblance to an outsize Paleozoic pit-bull terrier. In proportion to its bulk, the legs seem to belong to an animal half the size. The shoulders bulge with bone and muscle, supporting a skull whose roof is immensely strong – the bone thickness was 4 in (10 cm).

The dicynodonts were easily the dominant therapsids of their time, with a span of species ranging in length from 1 ft (30 cm) to 10 ft (3 m). Dicynodonts had either no teeth at all, or simply a pair of tusks. They were herbivores, and could cut up their plant food successfully by using horn-covered beaks, just as turtles do. Their jaw joints performed a chewing cycle in which the lower jaw ran forward a little, closed on the food, tore backward very firmly, and then relaxed. This was powered by enormous muscles that ran from the boatlike lower jaw to anchor on a massive bone base at the back of the skull.

The gorgonopsians were the saber-toothed carnivores of their day. Typical examples like *Lycaenops* were generally about 40 in (1 m) long, and more actively built than the carnivorous dinocephalians, with longer limbs and a smaller skull. They probably fed on the large and thick-skinned herbivorous dinocephalians and dicynodonts, using canine teeth to pierce the hide of their prey.

The therocephalians were a mixed group of smaller carnivores that may have fed on small dicynodonts, as well as on some of the lizard-like reptiles of the Karroo. These smaller reptiles are probably less well known than the

mammal-like reptiles, partly at least because they are not so easily spotted by prospectors as they comb the hot Karroo landscapes.

The other modest-sized animals from the southern African late Permian are diapsids. *Hovasaurus* was an aquatic animal, lizard-like in shape, whose hind limbs have broad paddle-like feet for steering. The tail is deep, just as it is in *Mesosaurus* and many other aquatic reptiles. *Hovasaurus* swam in freshwaters, probably feeding on fish. It dived with the help of stones in the gut region which it must have swallowed in order to counteract its natural floating tendency. Crocodiles do this today. Perhaps we should count this practice of shipping ballast as the first use of tools.

A spectacular diapsid from the late Permian of Madagascar, Germany, and England is *Coelurosauravus*. The head is short and triangular in side view, with a remarkable frilled back margin, and the legs are relatively short. What distinguishes this small, lightly built animal is the long projecting ribs that in life must have supported a gliding membrane, a feature later to evolve independently in some lizards. The flying insects were no longer alone in the air. *Coelurosauravus* could glide from tree to tree to catch them.

Crash

The end of the Permian was also the end of the Paleozoic, the era of ancient life. All the periods in the geological time scale receive their names in recognition of obvious changes in the fossil record; but the shift from the Permian period to the Triassic introduces a new era, the Mesozoic, middle life, so named following a suggestion made by John Phillips in 1840. (Adam Sedgwick had named the Paleozoic in 1838.)

At the end of the Permian there was a massive upheaval in the land faunas. The captorhinids, gorgonopsians, dinocephalians, weigeltisaurs, and 81 percent of amphibian families died out. The therocephalians and procolophonids, as well as the dicynodonts, were severely depleted. Altogether nearly 75 percent of all amphibian and reptile families disappeared. Similar drastic extinctions seem to have happened in the ocean: one-half of all marine families, four-fifths of all genera. By all estimations, the end-Permian extinctions amount to the most catastrophic event, or close series of events, that life has ever suffered – whether in terms of the sheer numbers of species lost, or of the traumatic effects on subsequent evolution. What happened, and why, at the end of the Permian?

Fossils are not a recent discovery. All through the span of human history, shells and bones have been weathered out of the ground, dug out of fields and quarries, exposed by earthquakes or landslides. In many parts of the world, the landscape itself records how rock strata have been stood on end, folded miles into the air, or chiseled flat and then covered again by horizontal layers whose color and texture are far too different to come from the same source as the rocks they cover. Long before the sciences of paleontology and geology were ever invented and named, it was obvious to some observers and scholars that animals now unknown had once existed, and that enormous forces must have acted to raise mountains, carve river valleys, or leave beds of seashells hundreds of miles inland. How could this have happened, in a world that was supposed to be only a few thousand years old?

One answer was provided by the concept now known as "catastrophism." It proposed that only a series of tremendously violent events could have produced these effects in so short a time. One of these events had been Noah's Flood, but there might have been others. Earth had been stocked with life since the beginning, but a disaster so huge that it could drain seas and level mountains was bound to have wiped out animal populations, leaving the remaining fauna to repopulate the planet. Literal interpretations of the Bible reinforced this concept, and although it was not universally shared it remained a dominant paradigm in the Christian West for several centuries.

Perhaps the concept of an Earth so young would be more easily believed in a country covered by great plains or tropical forests. James Hutton was born in 1726 in Scotland, much of whose geology is not hidden by grass, trees, or even heather. His studies of landscape convinced him that the world was so ancient that "we find no vestige of a beginning – no prospect of an end." In 1788 he took his friend, the clergyman and mathematician John Playfair, to see Siccar Point, on the coast of Berwickshire, where a gently sloping layer of Old Red Sandstone lies on top of upended Silurian mudstones and sandstones.

Playfair described the experience of realizing what must have happened in order to create what modern geology calls an "angular unconformity"; to him, it was a vision of "the abyss of time." Hutton's *Theory of the Earth*, published in 1795, insisted that a landscape was also a timescape, and that if only science would discard the blindfold concept of a practically newborn Earth there was no need to invoke fantastic forces in the past to explain events that had only required forces acting through a longer time.

Hutton's new theory is known as "uniformitarianism." Its opponents included the great French paleontologist Georges Cuvier, a brilliant reconstructor of fossil remains and pioneer in zoological classification, but unable to make two vital connections: between extinction and evolution, and between geological change and time. Resistance lasted for decades. It was broken down by the work of Charles Lyell, who published the first volume of his classic *Principles of Geology* in 1830. Its subtitle sums up the idea of uniformitarianism as "*an Attempt to Explain the Former Changes of the Earth's Surface by Causes Now in Operation.*"

A small, gliding, diapsid reptile, Coelurosauravus, *achieved remarkable success during the late Permian of Madagascar, Germany, and England.* Coelurosauravus *was the earliest vertebrate to take to the air and so became the first vertebrate to prey on airborne insects.* Coelurosauravus *could only glide, supported by a membrane of skin stretched over its expanded ribs (similar to the gliding structure of* Draco, *a modern gliding lizard). Flapping flight came later in evolution.*

The same idea of slow gradual change was absorbed by the early theorists of evolution, who believed that the fossil record would eventually show that today's species had evolved along slow consistent pathways into the present, with natural selection working in a style just as uniform and consistent as geological processes.

Later research has tilted the balance again, toward a new kind of "catastrophe" scenario. Plate tectonics has expanded our ideas about the possibility of change; Earth's ecology looks far less durable now than it did in the early nineteenth century, before pollution, the greenhouse effect, and holes in the ozone layer. We know much more about what *has* happened, both to our own planet and to our neighbors in the solar system, and we know that not all change has been slow. Seas can run dry, glaciers retreat and expand, climates transform themselves, oceans turn cold – faster than life can adjust. Asteroids can strike, and so can comets.

The universe itself is a more variable place than the people of the nineteenth century knew. We also know much more about the ups and downs of life, as recorded in the geological record. It has been obvious to geologists for well over 150 years that major changes must have taken place from one period or era to the next – this is why they have their different names.

Paleontologists have discovered two major patterns in life that make it difficult to support a totally uniformitarian view of its development. First, the origin of species (speciation) does not appear in the fossil record as a curve of almost imperceptible changes but as a series of jumps, some of them quite big ones. Some theorists argue that these jumps are merely gaps in the record, which time and work will someday fill; others suggest that the jumps are real, and that evolution tends to consist of long periods of relative standstill interrupted by spurts of rapid change. These spurts may take place when a species is isolated from the mainstream of development, in conditions that compel or encourage change; the original shift is unlikely to appear in the fossil record until the new species has spread far enough and grown numerous enough for chance fossils to survive.

These two schools of thought are known as phyletic gradualism and punctuated equilibrium – the second term first used in a paper published in 1972 by Niles Eldredge and Stephen Jay Gould.

The second pattern is much less controversial. The fossil record shows that several of the changes from one period or era to the next have resulted from mass extinction events severe enough to destroy not only species but whole families and orders. This story has already covered the mass extinction event at the end of the Ordovician period, when over one-fifth of families died. Later on, this present chapter will deal with the late Triassic extinctions that cleared the way for the dinosaurs, and Chapter Four – *Dinosaur Summer* – will cover the best-known of all mass extinctions, the one that put an end to the dinosaurs and enabled the mammals to take off, 65 My ago at the end of the Cretaceous. None of these events, however, can rival the vast scale of the extinction in the late Permian that wiped out so many species that their loss meant the close of the era of ancient life, the Paleozoic.

Both on land and in the sea, the end-Permian extinctions were at least twice as severe as any other, and possibly five or even ten times greater. It is estimated that only 5 percent of species survived, compared with values of about 50 percent for the worst of the other events. Yet this harshest of all extinction events has also turned out to be one of the hardest to study. Problems with dating the rocks make the time scale controversial, and we lack a good sample of fossil-bearing rock sections that span the crucial interval.

None of the great extinctions has a single proven cause, and this is not surprising when scientific detectives are investigating "crimes" that happened hundreds of millions of years ago. The bodies in the morgue are mineralized bones and the scene of the crime has been lost or transformed. Nearly all the evidence is circumstantial. We can round up a number of suspects – changes of sea level, atmosphere, or climate, volcanic episodes, extraterrestrial impacts – but any of these may turn out to have been present during periods when no mass biocide was being committed. None of them is a clinching piece of evidence, found at the scene of the crimes and nowhere else.

GLOBAL PATTERNS

The great crash 245 My ago may have been engineered by the unusual state of the world at the time. In the late Permian the continents of earlier times were moving together and fusing into the great landmass Pangaea. This had the simple geometric effect of reducing the total area of shallow coastal seas; as the two continents fused, the intervening seaway was lost and, with it, much of the life of the region.

On land, too, the continental fusion created profound effects. Continental interiors formed huge deserts (see below), presumably devoid of life, as a result of their distance from the sea. These deserts showed temperatures fluctuating between extremes of hot and cold. Mountainous regions, many of them formed by the continents pushing into each other and crumpling along the join, experienced cold temperatures, and there may have been glacial activity. There was also a marked increase in seasonality worldwide. All of these effects could have led to a long-term decline in the diversity of land.

These physical effects have been determined from studies of paleoclimatic evidence in late Permian rocks, and from theoretical modeling. All of these aspects of climatic deterioration could explain extinction, but it is still hard to imagine effects such as these inducing extinctions of such magnitude as occurred during the great crash.

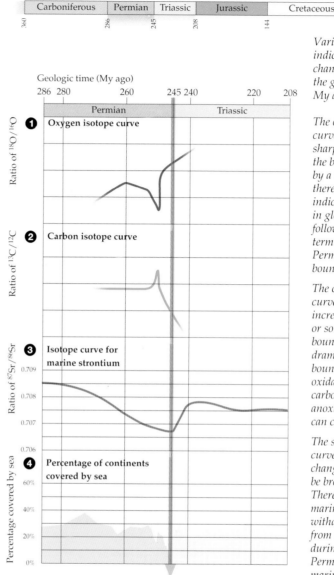

Carboniferous	Permian	Triassic	Jurassic	Cretaceous		

Various environmental indicators show clear changes at the time of the great extinction 245 My ago.

The oxygen isotope curve shows a matching sharp decrease before the boundary, followed by a steep increase thereafter, which could indicate a dramatic rise in global temperature, followed by a longer-term drop across the Permo-Triassic boundary.

The carbon isotope curve shows a sharp increase a million years or so before the boundary, and then a dramatic fall across the boundary, indicating oxidation of organic carbon, followed by anoxic events, which can cause extinction.

The strontium isotope curve and the sea-level change curve appear to be broadly correlated. There was a substantial marine regression, or withdrawal of the sea from continental areas, during the late Permian, followed by a marine transgression, or flooding of the continents ,in the early Triassic. High strontium ratios early in the Permian drop sharply during the late Permian, and rise thereafter. This may relate to changes in plate tectonic activity, and in the nature of erosion of terrestrial rocks.

LEFT *Map of the late Permian world, showing all the continents fused into one as Pangaea. There were four distinctive floras, remnants of the time before this fusion when the world was divided into four major continents.*

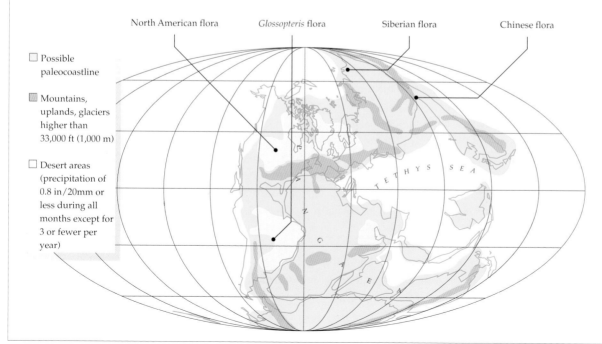

Extinctions in the sea

The richest records of animal life in the sea come from rocks laid down on the continental shelf, especially those that lie a little offshore, away from the influence of breaking waves that disturb the conditions of preservation. Some of the finest fossil evidence has come from warm waters where limestones are commonly laid down. These tend to be packed with fossils large and small, many in their life positions.

It has been possible to draw up remarkably detailed pictures of such marine habitats for the Carboniferous and Permian periods. Foremost are the complex and extensive coral reefs, with their associated bryozoans, arthropods, crinoids, starfish, sea urchins, brachiopods, gastropods, and bivalves on the surface, burrowing worms, bivalves, and arthropods beneath, and nautiloids and fishes swimming above. Two other shallow marine communities were hardgrounds – areas of half-consolidated limestone muds occupied by attached filter feeders such as bryozoans and crinoids, and heavily bored by sponges, bivalves, worms, and others – and non-calcareous muddy bottoms, with poorer communities of bivalves, arthropods, worms, and the like. All these communities were in ruins at the start of the Triassic. The typical Paleozoic communities, dominated by attached filter feeders – articulate brachiopods, bryozoans, crinoids – were never to be seen again. New kinds of communities of highly mobile mollusks, arthropods, and echinoderms replaced them, and still dominate today.

In a survey of all marine animals, J. John Sepkoski Jr of the University of Chicago calculated that 54 percent of families died out during the last 5 My or so of the Permian. He gave a figure of 78 to 84 percent for the proportion of marine genera destroyed, and this has been scaled up to give a drop in species diversity of 96 percent. It is simpler to count the survivors: 46 percent of families, only 16 to 22 percent of genera, and as few as 4 percent of species. Each family of animals usually contains ten or more genera, and each genus contains several species. The scale of survival changes because it is harder to kill off the larger group: one species left, and its genus is preserved, and therefore the whole family. The 46 percent of surviving families were nearly all heavily reduced, and some of them barely squeezed through into the Triassic.

The severity of the extinction varied greatly. At the family level, the seas lost 98 percent of crinoids (sea lilies), 78 percent of articulate brachiopods, 76 percent of bryozoans, 71 percent of cephalopods, and 50 percent of the microscopic planktonic Foraminifera. Several major groups disappeared altogether, including the blastoids (stalked echinoderms), eurypterids, rugose and tabulate corals, already hard hit in the late Devonian crisis, and the remainder of the trilobites. Overall, the typical Paleozoic invertebrate animals lost 79 percent of their families, compared with a 27 percent loss among the gastropods, sponges, and bivalves, the "modern" faunas that replaced them.

Extinction patterns

Attention is now focusing on the patterns of extinction for particular groups and in particular localities. Detailed studies have already shown that, although some of the late Permian extinctions may have been sudden, most of them happened as long-term declines. The dominant Paleozoic coral groups, the rugose and tabulate families, spent the whole of the Permian in decline. They were replaced in the middle Triassic by their descendants the scleractinian corals, the main modern group; but this leaves an interval of millions of years when it seems that no new coral reefs were born. Another persistent colonial group, the bryozoans, were declining only in the last 10 My of the Permian. Among the microscopic plankton, the extinction of Foraminifera lasted all through the Permian but with a rapid drop at the Permo-Triassic (P-Tr) boundary.

Slow changes appear to suggest a gradual alteration in the conditions of life, and a "uniformitarian" pattern; sharper falls and rises inside shallow ones may indicate that mass extinction events happen when a trend already under way is pushed into a faster lane by some sudden event or combination of events. While opponents of catastrophism have objected that it is not scientific to explain major events in terms of imagined causes no longer in operation, this objection vanishes if science can show that there are causes of potential catastrophe, past, present, and future, built into the working machinery of our Earth and solar system.

There may be biological features that assist or resist extinction. For example, it has been suggested that the survival of bivalves, brachiopods, and some other groups may have been affected by the way their species reproduced. These forms that produce larvae able to live and feed as members of the plankton, floating near the surface of the oceans, before eventually maturing to settle as bottom-dwelling forms, seem to have been more likely to die out than those without a plankton-feeding stage. The ecological groups that declined most were members of the zooplankton, attached filter feeders, and high-level marine carnivores.

Another suggestion for a selective agent in the end-Permian extinctions is nearness to the equator. Certainly 75 percent of families of tropical articulate brachiopods died out, compared with 56 percent of non-tropical families. Professor Steven Stanley of Johns Hopkins University has pointed out that by the final stage of the Permian, three major groups, the fusulinid Foraminifera, rugose corals, and bryozoans, were confined to the region of the Tethys Sea, before the first two became extinct and the bryozoans lost 65 percent of their families.

Stanley is a strong advocate of climate change as an agent of extinction. He suggests that global cooling would have been particularly damaging to tropical species that had nowhere else to go, and that equatorial Tethys might have been a last refuge before it too was overtaken by cold. Although there is definite evidence for a long glacial episode in Gondwana starting in the Carboniferous and lasting well into the Permian, there is no strong evidence for large-scale cooling across the later Permian and early Triassic.

Extinctions on land

Either the same force threatened life on land as in the sea, or some other but equally lethal force was at work there in the later Permian. Michael Benton has documented a worldwide loss of 27 out of a total of 37 families of amphibians and reptiles during the last 5 My

of this period. These include six families of amphibians, both temnospondyls and reptiliomorphs, as well as captorhinids, pareiasaurs, probably the weigeltisaurs and younginiforms, and at least fifteen families of mammal-like reptiles, including the last of the gorgonopsians and dinocephalians, and most of the dicynodonts and therocephalians. In all, the extinction wiped out 73 percent of tetrapod families on land, higher than the 54 percent in the sea, and suggesting an extinction rate of 98 to 99 percent of tetrapod species. A drawback of these figures is that they are based on a total diversity of only 37 terrestrial families worldwide, compared with about 500 marine families, so the margin of error is likely to be greater.

The transformed world of the earliest Triassic offers further evidence of the disastrous situation of land animals at the end

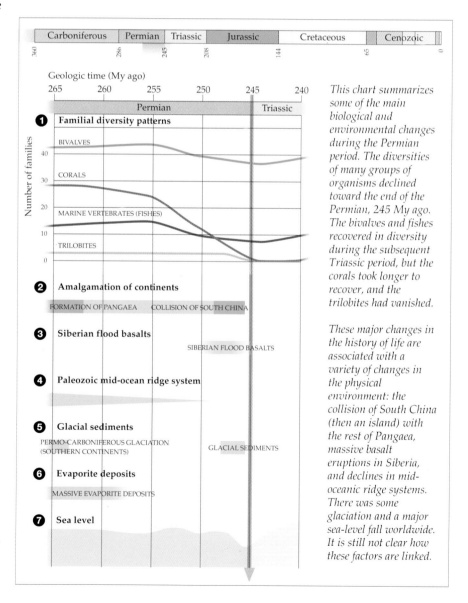

This chart summarizes some of the main biological and environmental changes during the Permian period. The diversities of many groups of organisms declined toward the end of the Permian, 245 My ago. The bivalves and fishes recovered in diversity during the subsequent Triassic period, but the corals took longer to recover, and the trilobites had vanished.

These major changes in the history of life are associated with a variety of changes in the physical environment: the collision of South China (then an island) with the rest of Pangaea, massive basalt eruptions in Siberia, and declines in mid-oceanic ridge systems. There was some glaciation and a major sea-level fall worldwide. It is still not clear how these factors are linked.

of the Permian. The first known fauna comes from South Africa, and is dominated by the lumbering dicynodont *Lystrosaurus*. This one animal accounts for all but 5 percent of the specimens collected, and the rest are a handful of smaller mammal-like reptiles, relics of a major liquidation of species. Similar evidence of monotonous herds of *Lystrosaurus* comes from China, western Russia, and India.

Extinctions among land plants occurring about this time appear to have followed a pattern of very long-term change. Typical late Paleozoic (or more correctly paleophytic) floras consisted of broad-leaved seed ferns, cordaites, and pecopterid ferns in equatorial regions. Cordaites dominated in high northern latitudes, while the key seed fern *Glossopteris* and its relatives headed the list in high southern latitudes. These floras were replaced piecemeal by mesophytic floras consisting of conifers, ginkgoes, cycads, bennettitaleans, and new groups of spore-bearing plants – club mosses, ferns, horsetails, psilophytes – at different times in different parts of the world.

There were extinctions among the Permian plants, and overall diversity was halved during this interval, but the decline took place in fits and starts and was linked to a variety of environmental factors in different parts of the world, and to the presence or absence of physical barriers to migration. The losses cannot be tied to a single episode of "catastrophic" extinction, nor can they be correlated with any events affecting marine animals or land vertebrates.

This comparative resilience among plants has been the general finding of research into extinctions. They are unaffected by most of the events that have wiped out animals, probably because they have fail-safe techniques, means of long-term ecological survival such as resistant seeds that can lie dormant for years, vegetative reproduction that can generate new growth from a small part of a fallen individual, and rhizomes and root systems protected underground even after destruction of the parts exposed to the elements.

If there are problems in dating the extinctions in the sea and corroborating their worldwide timing, these problems are ten times worse for the land plants and animals. Although there is no doubt that terrestrial floras and faunas experienced different and complex patterns of extinction, the deposits are very hard to date on an absolute scale, and just as hard to correlate with what we know about the marine sections.

Physical changes on the Earth

There is evidence for a variety of changes in the physical makeup of the Earth, the oceans, and the atmosphere during the late Permian, and many of them figure in scenarios aimed at explaining the extinction events. Changes took place in sea level, climate, stable isotope values for key elements in the oceans, and the magnetic polarity of the Earth. Large-scale volcanic eruptions occurred.

Sea levels fell throughout much of the Permian, from the Sakmarian stage onward, at a rate that accelerated later in the period and reached a maximum close to the Permo-Triassic boundary. According to some estimates, so steep a fall would have exposed 70 percent or more of continental shelf areas. The regression was caused by major plate tectonic processes associated with the formation during the late Permian of the single great landmass of Pangaea, when all the continents fused together. They fused at the margins of their continental shelves, of course, not along their coastlines, which are variable features placed somewhere along the gradual slope from the edge of a shelf to the higher uplands. The outermost shelf usually marks the edge of a tectonic plate, the sharp boundary between continental and oceanic crust.

The fusion of continental shelves inside the patchwork of jostling plates pushed upward and drained massive areas that had once lain underwater and fostered a variety of marine life. It seems possible that the temporary "log jam" in the new supercontinent was also connected with a pause in the output of new crustal material from its sources beneath mid-ocean ridges. When these ridges began to subside, more water drained from shallow seas and continental margins.

A glacial episode at the end of the Permian has been suggested on the basis of evidence from Siberia and eastern Australia. Pangaea rotated during the late Permian and drifted northward, placing Siberia close to the North Pole, while Australia and Antarctica remained near the South Pole.

When landmasses lie near the poles, a sharp temperature gradient is set up, and ice sheets can develop from the continental interiors. Once ice starts to build up, sunlight is reflected rather than absorbed (this is called the albedo effect), and temperatures cool further and so expand the ice sheet. When ice builds up, it lowers global sea levels by removing water from circulation. Some evidence for glacial conditions is that sea levels appear to have dipped and risen every 2.5 My during the mid- and late Permian, possibly as a result of cyclically active ice ages.

The formation of a supercontinent stretching almost unbroken between the poles and twinned with a single world ocean, Panthalassa, was bound to have a drastic impact on climates. The end of the southern glaciation during the Sakmarian stage (277 to 268 My ago) led to rising temperatures and drier conditions worldwide. Sure indicators are the massive deposits of salts formed at low levels in mid-Permian times by shrinking seas, and preserved in the form of the sedimentary rocks known as evaporites.

Other sedimentary evidence points to highly seasonal climates, typical of inland continental regions far from the moderating effects of the ocean; these regions were extensive inside a single landmass. It has been estimated that mean temperatures had a range of 90°F (50°C) from summer to winter. Severe monsoons seem to have been another feature of these continental interiors. Coal deposits in China, India, and Russia suggest that these areas, which lay in the Permian temperate belt, had humid climates.

Major changes in ocean chemistry also took place in the late Permian, and we can trace them by using a mass spectrometer to measure the proportions of the stable isotopes of various key elements contained in sediments of that age.

Many elements occur in nature in at least two stable isotopic forms, each with a different atomic weight and identified by a different mass number: the lower the number, the lighter the isotope. In carbon, the stable isotopes are ^{12}C and ^{13}C; one form has 12 nucleons, the other and much rarer form has

13. The Earth contains a fixed supply of both of them, and science has agreed on an international standard to define the ratio of the lighter to the heavier isotope that a substance should contain in conditions of ordinary random distribution. Biochemical processes in plants and animals tend to take up the lighter isotope in preference to the heavier, and this unbalances the standard ratio, so if a substance contains a higher ratio of the lighter, "organic" isotope, ^{12}C, it is evidence of biological activity. A higher than standard ratio of the heavier, "inorganic" isotope, ^{13}C, in a substance suggests that somewhere in the time it came from the processes of life must have absorbed a massive amount of ^{12}C and removed it from circulation.

Inorganic ^{13}C can also be taken out of circulation by being locked into various forms of calcite (calcium carbonate: $CaCO_3$), either in the skeletons of marine organisms or in limestones. Measurements of the proportions of the two isotopes have been made on brachiopod shells or limestones taken from Spitsbergen, Greenland, Texas, Austria, northwest Europe, the Transcaucasus, Pakistan, Iran, and south China. They record a sharp rise in the ratio of ^{13}C to ^{12}C that comes in the latest Permian and is followed by a steep drop across the P-Tr boundary. The rise is caused by the removal of organic carbon buried in the coal levels that built up during much of the Permian just as they had done in the late Carboniferous. The drop indicates either the disappearance of a vast amount of ^{13}C, or the reappearance of ^{12}C. How can these alternatives be explained?

Tony Hallam of Birmingham University and Paul Wignall of Leeds University believe that falling sea levels in the late Permian exposed huge areas of land to chemical weathering. In particular, they think that the organic carbon previously buried in coal deposits reacted with atmospheric oxygen to produce a great increase in the volume of carbon dioxide, and of course to reduce the volume of free oxygen. (The same "greenhouse effect" has been methodically created in modern times by industrial societies dependent for their energy on burning fossil fuels.)

Oxygen isotopes measured in the same samples also show an abrupt peak in the late

Permian followed by the same sharp decline just before the P-Tr boundary and lasting well into the Triassic. It has been estimated that the level of atmospheric oxygen fell from a normal value of about 30 percent to as little as 15 percent. The loss of half of their oxygen supply might have been fatal to those land animals which had evolved active high-energy lifestyles.

With so much less oxygen left in the air, the supply to the sea was inevitably reduced, and this is recorded in deposits such as black shales and pyrites, which can only be formed in the absence of oxygen. Wignall estimates that oxygen in the sea may have declined to less than one-fifth of its normal level as the waters rose again in the early Triassic.

In this scenario, the agent of extinction was slow suffocation while the land fell silent and the sea stagnated. One advantage of this explanation is that it would account – as some others do not – for the universal power of an agent that could kill throughout the world, and kill in the sea as easily as on the land.

Two large-scale volcanic episodes may also have affected the global climate during the late Permian. Volcanoes in Siberia poured out lavas over an area of some 600,000 sq miles (1.5 million sq km), while eruptions in south China showered a widespread layer of ash all over the area. Both of these events came just before the Permo-Triassic boundary. If the ash they erupted reached a high enough level in the Earth's atmosphere, it may have remained there long enough to reduce global temperatures, and so possibly to trigger glaciation.

Causes of extinction

Several well-documented environmental changes have been linked to the late Permian extinctions. Some geologists trace them to salinity. The deposition of huge masses of salt made the oceans considerably less saline during the Permian, and it has been suggested that this wiped out all the marine groups that could not tolerate less brackish conditions. However, there is no evidence that ocean salinity changed sharply enough during the period to have these effects, and in any case the greatest thicknesses of salt were laid down in the mid-Permian, a good

10 or 15 My before the end of the period.

Steven Stanley, of Johns Hopkins University, Baltimore, has made a more convincing case for global cooling as the cause of extinction. He believes that the increasingly arid and warm climates of much of the Permian were reversed at the end by a short phase of global cooling with icecaps at both poles. As temperatures fell worldwide, the temperate zones were pushed toward the equator, and the tropical belt was simply squeezed out of existence. Temperate-adapted forms could still find homes, but tropical specialists had nowhere to go. This hypothesis could account for the extinction of corals and other warm-adapted marine animals, as well as of the large cold-blooded reptiles on land, and some of the plants. But the glacial episode is poorly dated, there is as yet no evidence that it coincided with the extinctions, and the theory runs into a problem of causal consistency. Like causes should produce like effects.

Why have other glacial episodes, some of them longer or sharper than the facts seem to indicate at the end of the Permian, failed to make an equally lethal impact?

The most popular theories for the late Permian extinctions relate to the major physical changes taking place: the fusion of continents to form Pangaea, the lower sea levels, and the transformed climate. With fusion came a dramatic loss of the global area of continental shelves, and Jim Valentine, now of the University of California, Berkeley, and others have theorized a direct connection. Modern ecology has shown that the number of species that any area can support is related to the size of that area. Why not suppose that the reduction in shelf area produced extinctions in proportion?

Here a number of problems arise. First, the modern age happens to be a time of exceptionally low sea levels. We have no widespread examples of life conditions on shallow sea floors, few practical parallels to examine. Second, in large faunal provinces of modern seas, diversity does not seem to be governed by the area available. Local geography, food supply, community history, temperature, sediment types, and other localized factors combine into patterns that are too complex to produce a simple species–area effect.

Third, David Jablonski of the University of Chicago and Karl Flessa of the University of Arizona have argued that all but a few families of present-day marine invertebrates would be able to survive a fall in sea levels that exposed all continental shelves. They found that 87 percent of the marine mollusk groups they looked at had representatives that lived on offshore oceanic islands. Regressions there would tend to increase the total shelf area, because such islands are roughly conical in shape. Their shelf populations would flourish.

Valentine's geographic theory did not see the end-Permian extinction as a shotgun effect working at the level of species, but rather as a selective impact related to reductions in the number of faunal provinces right across the globe. When Pangaea was formed from the mid-Permian to the late Triassic, whole marine faunal provinces disappeared, such as those around the Chinese and southeast Asian block and the Siberian block, those in central Europe and over the present polar regions, and those over Arabia and parts of India and Madagascar. It does seem reasonable to expect the loss of these shallow seas to correlate with reductions in marine invertebrate diversity, but the fossil record shows no very close connection. Lastly, the question of causal consistency arises again. Sea levels have fallen in other periods, and particularly in the middle Oligocene, without large-scale extinctions.

The fusion of continents and the retreat of seas cannot have been ignored by life on land, and in particular it has been linked by Bob Bakker, then of Johns Hopkins University, with the extinctions of terrestrial vertebrates. Just as Pangaea's formation removed some marine faunal provinces, so it would also have reduced the number of independent lowland provinces. On a larger scale, the diversity of habitat types must have been globally reduced by the amalgamation of formerly separate continents into a single massive interior region, now a scene of climatic extremes, and a single huge coastal area. This, and the global cooling suggested for the latest Permian, could explain the losses among terrestrial vertebrates.

Studies of the extinction event at the end of the Cretaceous that wiped out the dinosaurs and other groups have lately focused

attention on extraterrestrial causes for mass extinction. Chapter Four will refer to the strong evidence, now broadly recognized, of a major impact or series of smaller impacts of asteroids or possibly comets about 65 My ago. High concentrations of the rare element iridium have been found in many boundary layers for that period, and seem likely to record the explosive impact of one or more large asteroids. Searches for iridium spikes at Permo-Triassic boundaries in Armenia, Russia, Italy, and Austria produced equally negative results. Iridium levels seem to be generally low at the Permo-Triassic boundary, and any slight peaks tend to resemble those that occur in some volcanic rocks of strictly terrestrial origin.

No evidence confirms an extraterrestrial impact as the cause of the greatest of extinctions. Cometary impacts may not create an iridium layer, and an asteroid need not contain enough of the element to leave traces after a hit, but the surface of the modern Earth reveals no trace of a crater big enough and old enough to have damaged the balance of life so profoundly.

So, do geophysical changes, or indeed any of the other changes in sea level, climate, and ocean chemistry, offer a single culprit to be found guilty beyond a reasonable doubt? The answer is no. The scene of the crime

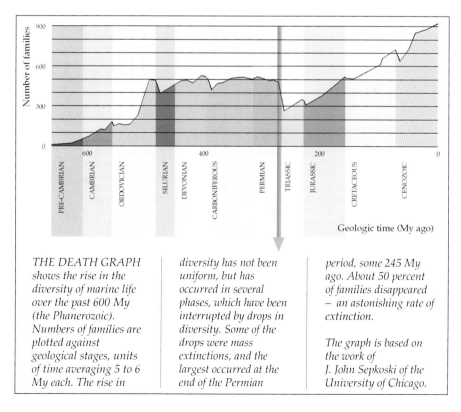

THE DEATH GRAPH shows the rise in the diversity of marine life over the past 600 My (the Phanerozoic). Numbers of families are plotted against geological stages, units of time averaging 5 to 6 My each. The rise in *diversity has not been uniform, but has occurred in several phases, which have been interrupted by drops in diversity. Some of the drops were mass extinctions, and the largest occurred at the end of the Permian* *period, some 245 My ago. About 50 percent of families disappeared – an astonishing rate of extinction.*

The graph is based on the work of J. John Sepkoski of the University of Chicago.

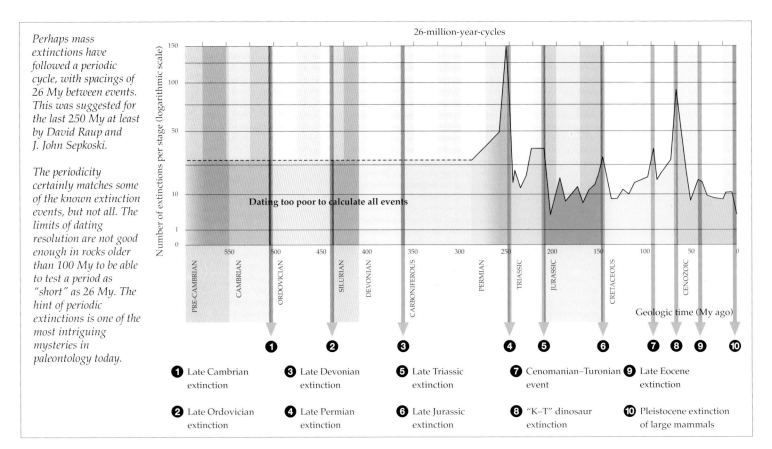

Perhaps mass extinctions have followed a periodic cycle, with spacings of 26 My between events. This was suggested for the last 250 My at least by David Raup and J. John Sepkoski.

The periodicity certainly matches some of the known extinction events, but not all. The limits of dating resolution are not good enough in rocks older than 100 My to be able to test a period as "short" as 26 My. The hint of periodic extinctions is one of the most intriguing mysteries in paleontology today.

26-million-year-cycles

Number of extinctions per stage (logarithmic scale)

Dating too poor to calculate all events

PRE-CAMBRIAN · CAMBRIAN · ORDOVICIAN · SILURIAN · DEVONIAN · CARBONIFEROUS · PERMIAN · TRIASSIC · JURASSIC · CRETACEOUS · CENOZOIC

Geologic time (My ago)

❶ Late Cambrian extinction
❷ Late Ordovician extinction
❸ Late Devonian extinction
❹ Late Permian extinction
❺ Late Triassic extinction
❻ Late Jurassic extinction
❼ Cenomanian–Turonian event
❽ "K–T" dinosaur extinction
❾ Late Eocene extinction
❿ Pleistocene extinction of large mammals

contains no obvious weapon, and the location and timing of the events themselves are still in shadow. What we can say for sure is far too vague to make a verdict: that the mid- to late Permian saw 20 My of tremendous global changes, and that the most plausible causes of extinction are interactions between continental fusion and climatic change.

Some scientists would prefer there to be a uniformity of causes, even of extinction – declining sea levels, cooling climates, catastrophic impacts. David Raup of the University of Chicago admits that he would prefer to find a single cause because a series of unique causes would be hard to prove: "If all extinction events are different, the deciphering of any one of them will be next to impossible." This seems uncomfortably close to the position of the judge who finds it unthinkable to allow an appeal against an unjust verdict because once the possibility has been admitted, verdicts become too difficult to reach. Raup favors the theory of catastrophic collisions, acting at a single time and place – the most certain of causes.

Other scientists have absorbed some of the ideas of chaos theory developed by

mathematicians and physicists since the 1960s, a way of understanding the behavior of dynamic systems whose apparently random behavior may contain underlying patterns, even though they never work quite the same way twice. A famous image comes from the proposition that in a truly "chaotic" system such as the behavior of weather, a small cause acting at the right time and place – say, the flutter of a butterfly's wings – might be the distant trigger of a hurricane.

If the evolution and extinction of living creatures turned out to be governed by such causes, able to make sudden turbulent changes in what seem like orderly systems, then any one, or any combination, of the forces described above might be capable of putting life through occasional "hurricane" episodes, before the weather calmed down again. Science has recognized this possibility by finding more and more potential "butterflies," knowing that it may not be the strongest and most obvious force that produces the most significant effects. In the study of extinctions, it has to be accepted that certainty may simply not be available, but causes still have to be sought. If a butterfly can make a difference, so can a better theory!

Mass extinctions and the history of the Earth

Twenty or more global mass extinctions have been identified, some more convincing than others, during the history of life. They can be rated according to their casualty figures, and Category 1 contains only the late Permian event, which removed more than 50 percent of families and more than 95 percent of species.

Category 2 contains those mass extinctions during which 20 to 25 percent of families and up to 50 percent of species died out throughout the world. In chronological order they are:

1 The late Ordovician event, 438 My ago, when brachiopods and trilobites took the heaviest losses.

2 The late Devonian event, 367 My ago, when many brachiopod, ammonoid, gastropod, trilobite, tabulate coral, stromatoporoid, and fish groups died out.

3 The Triassic–Jurassic boundary event, 208 My ago, when some groups of brachiopods, mollusks, arthropods, and some land vertebrates died out. This was preceded by another late Triassic event about 225 My ago, when most of the dominant land vertebrates disappeared and there were big losses among echinoderms, bryozoans, conodonts, and fishes.

4 The Cretaceous–Tertiary boundary event, 65 My ago, when the dinosaurs, pterosaurs, marine reptiles, ammonites, and other marine groups died out.

Other undisputed phases or points of mass extinction took place in the late Cambrian, at the Jurassic–Cretaceous boundary, in the late Cretaceous, and in the late Eocene. Other outbreaks, such as the one that came between the Ediacaran faunas and the start of the Cambrian, are less well defined, and many others are linked only to certain groups, or to certain parts of the world. Statistics for the number of groups lost, and the duration of the key events, tend to be controversial. It is sometimes possible to argue that one or another of these events was not so much an epidemic of dying as a failure of new species to develop at the usual rate. Suggestions of poisoning by trace elements or by outbreaks of disease may offer easy explanations but are impossible to prove, and ought not to account for a spectrum of deaths in many species, and both on land and in the sea.

Impact explanations may be offered even for events where a pattern of change and decline can be measured long before the suspect crater appeared. Proposed extraterrestrial causes such as cosmic radiation, possibly generated by supernovas, are not supported by any evidence.

The Earth and its tenants have come to seem more vulnerable than they once appeared. Persuasive evidence of repeated mass extinctions has compelled science to recognize that they could happen again, and to investigate their history and possible causes. Here, paleontology is being reinforced by many other fields of study in the common quest for clues to the present in the distant past. The same pattern can be seen in physics, as we look for the traces of events that happened long before our Sun was born. In both cases, the exotic theories of one generation may turn out to be the elementary assumptions of the next.

Periodic cycles of mass extinction

One of the most dramatic recent suggestions about mass extinctions is that they show periodicity – that is, they have coincided with some of the beats of a regular rhythm in time, spaced across equal intervals. This remarkable suggestion was made in 1983 by David Raup and J. John Sepkoski Jr of the University of Chicago, when they came to analyze their extensive data base on extinctions among marine animals. Even after various statistical tests to eliminate possible biases in the way the data were collected, their graphs of extinctions among both families and genera indicate that over the last 250 My at least, these have tended to peak at intervals of roughly 26 My.

It is hard to imagine what planetary mechanism could work at such regular tempo, so if this observation is correct it would most likely demand an extraterrestrial forcing agent able to influence the Earth, perhaps through large-scale impacts, every 26 My. There is no independent evidence for such an agent, but a number of possible candidates have been suggested, all of them acting as triggers for comet showers. Astronomers have believed for some time that the solar system is surrounded by a shell of billions of comets orbiting at a distance of thousands of billions of miles. A distant tenth planet with the right orbit could deflect showers of comets from their position in what is known as the Oort Cloud. So could a distant companion of the Sun, too dim for telescopes to observe. No trace of "Planet X" or of the companion star sensationally baptized "Nemesis" has ever been observed.

So far, the Raup/Sepkoski hypothesis of periodicity has been heavily criticized by some scientists, eagerly accepted by others. There is no question that their data yielded this result, and that it took them by surprise. There is also no doubt that the ten or so mass extinctions over the past 250 My cannot be equated with each other. Some may barely exist, while other mass extinctions, as least as valid, do not fit into the timetable. In addition, the quality of absolute dating is not yet good enough to confirm (or to deny) the accuracy of our timing on a scale of tens of millions of years. Current paleontological work is focusing on the nature of the global data bases and on the individual events, in an effort to achieve a sharper focus and a more reliable chronology.

If mass extinctions really have occurred in a cyclical rhythm, a pattern so vital to the history of life would have far-reaching implications. In particular, it would tip the balance of evolutionary theory heavily against the notion of "fitness" of species and toward the idea that contingency – bad luck, not bad genes, as David Raup has put it – has been the ultimate driving force. Until periodicity has been proved or disproved, the scales will continue to hang fairly level: at one extreme, the belief that inbuilt biological and genetic features dictate the course of life; at the other, the belief that changing conditions and planetary events can and do intervene to shuffle the pack. Are the rules of life on Earth an eternal, constant standard, or can they be changed and broken? On this level, the argument between uniformitarian and catastrophist views merges with the broader human problem of how to behave on a smallish planet. There is now an intense and worldwide effort devoted to the study of mass extinctions, involving geologists, paleontologists, geophysicists, geochemists, and many others. It is remarkable how much, and yet how little, we know.

Triassic transition

Paleozoic, Mesozoic, Cenozoic – Old Life, Middle, and New: When nineteenth-century geologists realized the scale of the story in the rocks, they saw it as a drama in three acts. We know now that there was a prologue that lasted more than 3 billion years, and in 1932 it received the slightly apologetic title of Cryptozoic, "hidden life." No one knows when this began, so many paleontologists prefer to leave "life" out of the prologue's title and refer to the whole of Earth's history before 590 My ago as the Precambrian.

These are vast tracts of time. In a book of 100,000 words, if life began around 4 billion years ago, each word covers 40,000 years, each letter a period longer than the whole of recorded human history. This is a broad focus, to say the least – especially when the subject is not a single storyline with a few main twists, but the bundle of lifelines represented by the billions of species that have spun their threads since the start of the Cambrian, plus their innumerable unknown precursors.

Act Two, the Mesozoic, Scene One, the Triassic (245 to 208 My ago), sounds like a precise enough division. But plenty of "ancient life" survived the Paleozoic; and whole orders of creatures that enter the record in the Mesozoic must surely have come onstage during the final scene of Act One.

After the wholesale liquidations of the late Permian – clearly not a brief event, but a lengthy episode orchestrated by diverse causes – recovery was slow. It took at least 10 My for faunas to return to previous levels, and some marine faunas, reefs in particular, took upward of 20 My to flourish again. Where new patterns did begin to emerge, a further extinction episode during the late

Triassic, about 225 My ago, stopped some in their tracks and gave others new momentum. The supercontinent Pangaea grew to its maximum size when the twelve main tectonic blocks huddled together. A single great ocean stuck a tongue into its eastern side to form the Tethys Seaway. In the massive interior of Pangaea, increasingly warm and arid climates prevailed. Shrunken inland seas were no barrier to animal species now able to explore.

Vertebrate land faunas of the earliest Triassic are surprisingly well documented. In the Karroo Basin of South Africa they contrast sharply with the prolific scene of the latest Permian, which ranged from tiny scuttlers and lithe carnivores to lumbering herbivore hulks. Now there was just one genus of medium-sized dicynodont, *Lystrosaurus*, a snub-nosed, thickset animal.

Several hundred identifiable skulls have been collected from the Karroo's *Lystrosaurus* Zone over the past 150 years, and more than 95 percent are of *Lystrosaurus*. The remainder consist of a small dicynodont, a therocephalian, a cynodont mammal-like reptile, a procolophonid, a few small diapsids, and the thecodontian carnivore *Proterosuchus*. Thecodontians ("socket-tooth") were a rising force, part of the archosaur group that would dominate the later Triassic. Many of them had a crocodilian look, and this one's name means "early crocodile," though it was not till the late Triassic that the first true crocodiles appeared. In general, though, this animal community was uncannily uniform; everywhere you looked you would have seen identical piglike dicynodonts feeding on low plants among the ponds and watercourses of the Karroo. The only other cases of such restricted, species-poor communities are human cities and farms.

Elements of the *Lystrosaurus* Zone fauna, and in particular the dicynodont whose name it bears, have also been found in the earliest Triassic of Antarctica, China, Russia, and India. This general distribution enhances its importance; for it seems to have been virtually a worldwide community, derived from the few survivors of the late Permian upheavals. Nowhere do large animals appear, and *Lystrosaurus* evidently had no predators.

The early Triassic of Madagascar contains the oldest representative of today's amphibian groups. The primitive frog *Triadobatrachus*

shows the low, wide, round-fronted temnospondyl skull type, as well as the beginnings of specialized frog features: long legs, reinforced pelvis, reduction of ribs, and shortening of the back. From here onward, the frogs (order Anura) become more and more common in the fossil record; having arrived at their highly successful leaping lifestyle, it seems that nothing has forced them to change.

Dicynodonts and cynodonts

The mammal-like reptiles were virtually wiped out at the end of the Permian, with the loss of all but one or two families. Some therocephalians survived, but in minor roles. *Lystrosaurus* led a significant recovery among the dicynodonts. Within the first 5 My of the Triassic the South African species *Kannemeyeria* had evolved a frame as massive as its late Permian forebears. This heavy herbivore, 10 ft (3 m) long, was the first in a radiation of similarly large dicynodonts that show up in the middle and the early late Triassic of North and South America, India, China, Russia, and Australia.

The most important mammal-like reptile lineage of the Triassic was one that arose right at the end of the Permian. The early cynodonts were dog-sized animals such as *Procynosuchus* from the late Permian and *Thrinaxodon* from the early Triassic of South Africa. These therapsids were much more mammal-like than relatives such as the therocephalians and dicynodonts. In particular there are key changes in the skull. A secondary hard palate in the roof of the mouth separated it from the nasal passage, allowing the animal to eat and breathe at the same time; the teeth were differentiated into mammalian incisors, canines, and cheek teeth; the lower jaw was dominated by the dentary bone, which in earlier animals had been little more than a thin bedding plane for teeth, laid on top of a mosaic of larger components; and there was a wide arch of bone in the cheek region – the zygomatic arch – which flared outward to make room for stronger muscles controlling the bite of the lower jaw.

During the Triassic, the cynodonts radiated into all sorts of carnivorous and herbivorous versions that achieved worldwide success.

They show ever more mammal-like features, until they reach a point in the late Triassic where some forms are hard to classify either as reptiles or as mammals. The skull is even more mammalian, with the eye socket and single (synapsid) temporal opening fused and the dentary bone almost completing its takeover of the jaw. The skeleton shows evidence of an advanced upright posture, with both pairs of limbs tucked under the body instead of sprawling. Also, most of the Triassic cynodonts seem to have been endothermic, which means that they generated their own body heat, independent of their surroundings – they were "warm-blooded." There is evidence in the snout region for nerves and blood vessels serving sensory whiskers; whiskers mean the presence of hairs on the body; hair means insulation of a warm-blooded body.

The transition stage between reptile and mammal was marked by two further transformations in the skull. In reptiles, the jaw hinges between the articular bone in the lower jaw and the quadrate bone located low down at the back of the skull. In mammals, the joint is formed between the dentary bone in the lower jaw and the squamosal bone of the skull. For a while the transitional group, the advanced cynodonts, had both kinds of jaw joint, both of them functioning. The fossil record is good enough to display the whole sequence by which the dentary bone moved up inside the widening zygomatic arch until each end of the dentary's rough U-shape made light contact with the squamosal bone in the back corner of the arch. In a related sequence the "reptilian" jaw joint between articular and quadrate bones grew smaller, and these bones drifted closer to the new hinge point till the two were virtually in contact.

Now came an extraordinary piece of evolutionary make-do-and-mend. Reptiles went on using the original articular-quadrate hinge; mammals did not need two sets of hinges, and the outer skull has lost them. In both animals, the ear lies near the hinge. Reptile hearing uses a single ossicle (little bone) called the stapes, a thin rod that picks up vibrations from the eardrum and transmits them to the inner ear inside the brain case. Mammalian hearing uses three ossicles to make a delicate instrument in the middle ear that transmits sound from the eardrum to a second membrane, the "oval window," which connects with the inner ear. These three bones are, starting at the eardrum, the malleus, incus, and stapes – Latin for hammer, anvil, and stirrup. We have already met the stapes. We first met the malleus and the incus as the articular and quadrate bones. They have shrunk and migrated, after all these hinges and structures were brought close together in the transitional reptile/mammals. It seems that there were twin forces at work to speed the shift; the jaw needed a better joint, and the nearby ear needed better machinery. The reptilian jaw made a third contribution when the angular bone traveled to become the mammalian ectotympanic, a C-shaped ossicle that holds the eardrum taut.

On this high note, the storyline shifts back to the reptiles in their Mesozoic heyday. No mammal bigger than a cat will challenge them for 150 My.

The middle Triassic seas of Germany

A number of Triassic reptiles led fully aquatic lives – a great change from the Permian, when the water produced very few experiments along the lines of *Hovasaurus*.

What looks at first sight like a sudden burst of marine expansion is unlikely to have been triggered by the after-effects of the late Permian extinction. In fact the quite early appearance of many fully aquatic reptiles strongly suggests that they must have arisen in the Permian but have not yet been found there. Highland and high seas species of any kind are badly underrepresented in the fossil record, the first because they are especially vulnerable to the forces of decomposition and erosion, the second because nowhere in the world do ocean crusts predate the mid-Jurassic. Anything from an earlier time has long since been squeezed to the edge of tectonic plates by the pressure of new material spread from mid-ocean ridges, and then subducted beneath.

In any case, there is no observable change in potential food supplies that would stimulate marine evolution in the early Triassic. If anything the sources of both fishes and invertebrates became rather more restricted than they had been in the late

Permian. Triassic shallow warm seas were populated by coiled swimming nautiloids (the ceratite ammonoids), as well as brachiopods, bivalves, gastropods, and sea urchins, all of them re-radiating after the Permian crash. Many of the bivalves and echinoderms had actually developed new abilities to burrow into the sediment and so avoid predation. New forms of marine life were the hexacorals, which began to build reefs by middle Triassic times, and oysters, a variety of bivalves that lived in great colonies fixed to rocks in shallow seas.

Bony ray-finned fishes became faster and more lightweight in the Triassic. Relatives of the Devonian, Carboniferous, and Permian *Cheirolepis* family, the paleonisciforms, still lived on, but in reduced numbers. The main groups were those that are often called holosteans, with nearly symmetrical tails and a more complex mouth system than *Cheirolepis*. When the mouth opened, the jaws were projected a little further forward – a feature seen fully developed in the later teleost fishes.

The Muschelkalk ("clam limestone")

deposits of central Europe and particularly Germany have produced a generous array of middle Triassic marine fossils, which include some spectacular reptiles. Varieties such as *Tanystropheus* were only partially adapted to aquatic life. This bizarre animal grew up to 23 ft (7 m) long. The body was modest, the limbs generally lizard-like, the tail quite ordinary, and the head standard issue for a fish- or meat-eater. It kept its uniqueness in its slender neck, whose nine to twelve vertebrae were normal enough, except that each one was up to 1 ft (30 cm) long, and the neck was more than half the creature's length. With so few joints, this was a stiff and anything-but-serpentine extension. Young *Tanystropheus* had relatively commonplace necks, and may have fed on insects; but the body's rate of growth was outstripped by the neck's, which kept on sprouting. (This is an extreme form of the well-known process called allometry – differential rates of growth.) The skeletons have been found in coastal marine sediments, so it is assumed that adult *Tanystropheus* swam, or stood on rocks, and used their long necks to fish. They belonged

One of the most preposterous reptiles of all time, Tanystropheus *was a common prolacertiform that lived in the seas of the mid-Triassic Germanic Basin. An adult's neck was composed of twelve vertebrae, each up to 12 in (30 cm) long.* Tanystropheus *may have used its neck as a kind of fishing rod, swooping down at surface-swimming fishes and pulling them from the water. Juveniles had more standard neck lengths and may have fed on insects.*

to a Permo-Triassic group of land animals called prolacertiforms, close relatives of the rhynchosaurs and archosaurs.

The other Muschelkalk reptiles are more fully adapted to life at sea. The nothosaurs were streamlined long-necked animals with four paddle-like limbs. They swam by beating their tails from side to side and steered with their paddles through Muschelkalk seas that must have been nothosaur playgrounds, for individual slabs are found littered with their carcasses. They died out in the late Triassic.

The Muschelkalk sea also sported the heavy coastal flounderers called placodonts. These retained a rather land-bound kind of body – narrow tail, legs without paddles, heavy bulky frame – but their heads were spectacularly adapted for a new kind of marine diet. Placodonts were malacivores, mollusk-eaters, and they exploited the new oysters of the Triassic using two different sets of teeth. Spatulate incisors jutted forward at the front of both upper and lower jaws, while broad platelike teeth were set further back in the palate and lower jaw. The incisors would prise the oysters off the rocks; the millstone rear teeth crushed the shells; then the fleshy parts were swallowed and the splintered shells ejected. The placodonts, like the nothosaurs, died out in the late Triassic; they left no descendants.

The final Triassic marine group, and the most highly adapted of all for the aquatic life, were the ichthyosaurs. Unlike the nothosaurs, they could never have gone ashore, even to lay eggs. Ichthyosaurs are a classic example of convergent evolution; they share with sharks and dolphins (fish and mammal respectively) a shape that casts them as strong, fast

The small "nothosaur" Pachypleurosaurus, *from the mid-Triassic of Switzerland, was an early member of the group that gave rise to the placodonts and the plesiosaurs. This efficient swimmer was immensely common in the Muschelkalk seas, where it captured small fishes in jaws lined with thin, needle-like teeth.*

swimmers, and their feeding abilities were probably identical. The oldest ichthyosaurs, from the early Triassic of Japan, Spitsbergen, and Canada (the last reported only in 1992), give few clues to the ancestry of the group. These early forms already have the streamlined fishlike body, long snout, and paddles of their later descendants. *Mixosaurus* from the Muschelkalk was 3–6$^{1}/_{2}$ ft (1–2 m) long and had a long asymmetrical tail, with a fin only on the lower half. The long narrow jaws lined with sharp conical teeth show that it fed on nimble unarmored fishes like the new holosteans.

By late Triassic times, some massive ichthyosaurs had evolved in North America. *Shonisaurus*, from Nevada, reached a length of 50 ft (15 m). It had a narrow bullet-like head, a deep body, and hugely elongated paddles; and it must have swum in schools, because one deposit contains dozens of specimens mostly in the same alignment, as if stranded all together. These particular ichthyosaur families did not survive the end of the Triassic, although the ichthyosaurs as a whole went on to even greater success in the Jurassic.

The relationships of the nothosaurs, placodonts, and ichthyosaurs remain to be unraveled. The view now is that all three arose from diapsid ancestors still unknown.

The rise of the ruling reptiles

Scrappy remains from Russia show that the archosaurs, the ruling reptiles, arose at the very end of the Permian, but the first reasonably well-known archosaur is *Proterosuchus*, already encountered among the Karroo *Lystrosaurus* fauna, and also found in those of China and Russia. *Proterosuchus* was 5 ft (1.5 m) long, shaped roughly like a large lizard, and certainly a carnivore. The early archosaurs are also often referred to as thecodontians. They include the ancestors of the crocodilians on the one hand, and the pterosaurs and dinosaurs (and therefore the birds) on the other. The pathways are not always agreed, however, and "archosaur" is a baggy sort of category that can contain all of these groups without having to define the point where one becomes another, or whether

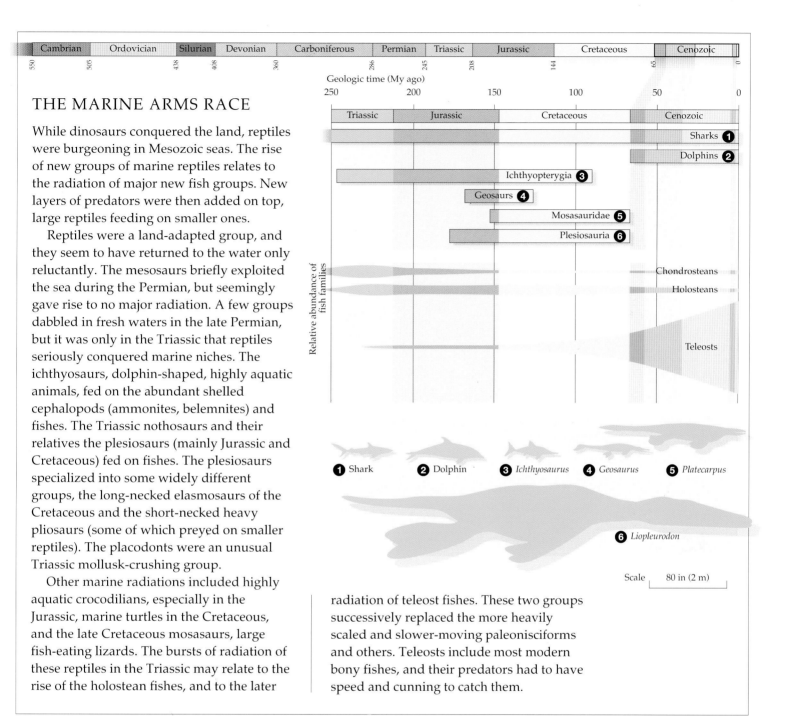

Cambrian	Ordovician	Silurian	Devonian	Carboniferous	Permian	Triassic	Jurassic	Cretaceous	Cenozoic

550 · 505 · 438 · 408 · 360 · 286 · 245 · 208 · 144 · 65 · 0

Geologic time (My ago)

250 · 200 · 150 · 100 · 50 · 0

Triassic	Jurassic	Cretaceous	Cenozoic

Sharks ❶

Dolphins ❷

Ichthyopterygia ❸

Geosaurs ❹

Mosasauridae ❺

Plesiosauria ❻

Relative abundance of fish families

Chondrosteans
Holosteans
Teleosts

❶ Shark ❷ Dolphin ❸ *Ichthyosaurus* ❹ *Geosaurus* ❺ *Platecarpus*

❻ *Liopleurodon*

Scale | 80 in (2 m)

THE MARINE ARMS RACE

While dinosaurs conquered the land, reptiles were burgeoning in Mesozoic seas. The rise of new groups of marine reptiles relates to the radiation of major new fish groups. New layers of predators were then added on top, large reptiles feeding on smaller ones.

Reptiles were a land-adapted group, and they seem to have returned to the water only reluctantly. The mesosaurs briefly exploited the sea during the Permian, but seemingly gave rise to no major radiation. A few groups dabbled in fresh waters in the late Permian, but it was only in the Triassic that reptiles seriously conquered marine niches. The ichthyosaurs, dolphin-shaped, highly aquatic animals, fed on the abundant shelled cephalopods (ammonites, belemnites) and fishes. The Triassic nothosaurs and their relatives the plesiosaurs (mainly Jurassic and Cretaceous) fed on fishes. The plesiosaurs specialized into some widely different groups, the long-necked elasmosaurs of the Cretaceous and the short-necked heavy pliosaurs (some of which preyed on smaller reptiles). The placodonts were an unusual Triassic mollusk-crushing group.

Other marine radiations included highly aquatic crocodilians, especially in the Jurassic, marine turtles in the Cretaceous, and the late Cretaceous mosasaurs, large fish-eating lizards. The bursts of radiation of these reptiles in the Triassic may relate to the rise of the holostean fishes, and to the later radiation of teleost fishes. These two groups successively replaced the more heavily scaled and slower-moving paleonisciforms and others. Teleosts include most modern bony fishes, and their predators had to have speed and cunning to catch them.

a given thecodont is an uncle or a father to a later species. *Proterosuchus* is identified as an archosaur/thecodont because of features such as the antorbital fenestra, which is an additional opening in the skull between the eye socket and the nostril, and the possession of teeth that are flattened from side to side instead of having a rounded cross-section. The extra skull opening may have housed a gland that discarded surplus salt from the blood – a useful gift in desert climates. *Proterosuchus* was a sprawler, its limbs held outward like a lizard's or salamander's, and it probably managed little more than a leisurely waddle in pursuit of prey, but that was all it needed in the early Triassic.

The first really large carnivores appear among the erythrosuchids, "crimson crocodiles," a thecodont family that arose toward the end of the early Triassic and spread to South Africa, China, and Russia in particular. *Vjushkovia* from Russia was up to 10 ft (3 m) long, its massive skull equipped with jaws strong enough to crush the bones of almost any other animal it came across, including the large dicynodonts. A key advance was some changes in the limbs toward a more upright stance that looked

forward to the continuing rise and future domination of its descendants.

More progress appears in the little thecodont *Euparkeria*, found side by side with the erythrosuchid *Erythrosuchus* in the late early Triassic *Cynognathus* Zone of the Karroo. *Euparkeria* is a remarkable archosaur because it had the option of walking on its hind legs and was the first known tetrapod to do so. It was 20 in (50 cm) in length, with a heavy skull, shortish limbs, and a very long tail that it may have used for balancing when it ventured on to its hind limbs – probably only when it wanted to peer about or to run fast. Twin rows of bony plates along the backbone show that *Euparkeria* was also developing armor plating, another feature to be riotously exploited in later groups.

Changes in posture

During the middle Triassic, the archosaurs embarked on an ambitious and vivid evolutionary binge. A posse of lineages took over the major guilds – carnivores, fish-eaters, even herbivores – until the end of the period, and their offspring bestrode the Mesozoic for 150 My.

The first key advances seem to have involved posture. Whereas *Proterosuchus* was a primitive sprawler, and *Vjushkovia* could keep its belly off the ground, archosaurs after the middle Triassic became fully upright, or erect in posture. This does not mean that they were bipeds. Certainly it is impossible to have a bipedal sprawler (it would fall over); but an elephant or a sauropod dinosaur like *Diplodocus* or *Brachiosaurus* is just as erect as a human being or a bird. Erect posture is the necessary step before bipedalism. It is also the necessary prelude to large size – a trademark of the dinosaurs, and to a lesser extent of the mammals.

Sprawling tetrapods are limited in their maximum size because of biomechanical requirements. When a sprawler stands up straight with its trunk off the ground, its weight is transmitted through its legs to the ground. The force of gravity pulls the weight straight down, and tends to the bellyflop resting position, but when the animal stands up the forces have to run horizontally from the hips and shoulders and then down through the legs and arms, both pairs of limbs doing the equivalent of a press-up exercise. Great stresses act upon the knee and ankle joints, which have to resist the tendency to bend. An animal whose limbs tuck vertically under its body can transmit its weight straight downward through those limbs. There is no leverage to bend the knee and ankle joints, no pressure to collapse.

A striking feature of reptilian evolution during the middle Triassic is the number of times that different animals achieved the erect gait by separate pathways; it was a mechanism whose time had come. Close study of the various evolutionary pathways, and of the limb skeletons, shows that it arose at least once in the cynodont mammal-like reptiles, several times along the archosaurs, and perhaps as many as ten times in all. Was this because it would allow the descendants of those innovating lineages to become large and two-legged? The answer is no. These features are consequences, not cause. Nature has no intentions. In the deserts and steaming jungles of the middle Triassic, what counted was to be a better whateverosaur, whether at catching your lunch or escaping from your predators.

Any adaptation that enabled either predator or herbivore to become marginally faster or marginally larger was likely to give a selective advantage. Erect posture was rewarded by these immediate benefits. With our advantage of hindsight, we can see what superb opportunities it offered. Once it existed, erect posture could act to enable the dinosaurs and (eventually) the mammals to be big; it could enable many dinosaurs, the birds, and sooner or later certain mammals to be bipedal.

Ankles and archosaurs

For the archosaurs to switch to an erect gait required a number of major changes in the limb skeleton that the fossils clearly illustrate. In particular, all of the joints – ankle, knee, hip – had to change their orientation. The legs had sloped outward during a stride; now they moved right under the body and simply swung backward and forward parallel to the midline of the animal. The changes happened independently at least four times among the archosaurs, and the details of posture development plot out the main lines of their evolution.

At the start of the middle Triassic, 240 My ago, the post-*Euparkeria* archosaurs split into two main lines: the Ornithosuchia (ornithosuchids, pterosaurs, dinosaurs, and later the birds) and the Crocodylotarsi (phytosaurs, pseudosuchians, and the crocodiles and crocodile ancestors). Not all of these developed the fully upright gait.

The ornithosuchians are certainly characterized by a superficially mammalian type of erect posture, and by the so-called "advanced mesotarsal" ankle. To achieve erect posture, they pulled the leg partly under the body and evolved a ball-and-socket type of hip joint, like our own. The femur, or thigh bone, in tetrapod vertebrates fits into a bowl-like depression on the side of the hip bones, the acetabulum. In a sprawler like *Proterosuchus*, the femur has a rounded head that fits directly into the acetabulum and causes the femur to slope out as well as down. In dinosaurs and pterosaurs, the acetabulum still faces mainly sideways, and the femur has a spherical extension protruding inward at its top end. This fits into the socket of the acetabulum, allowing the femur to run directly downward with the limb. The advanced mesotarsal ankle of these forms is a straightforward hinge joint in which the two main ankle bones, the astragalus and the calcaneum, form a simple roller that is firmly attached to the tibia and fibula, the bones of the calf.

The Crocodylotarsi are a more diverse

POSTURE AND GAIT

During the Triassic period, 245 to 208 My ago, major changes were taking place in the posture of several groups of reptiles. They were shifting from the standard "sprawling" mode to an "erect" posture. Does this explain the later success of the dinosaurs?

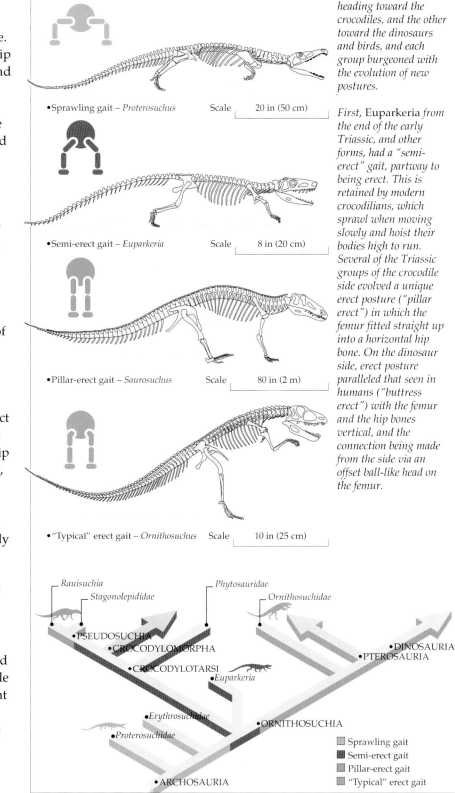

• Sprawling gait – *Proterosuchus*   Scale   20 in (50 cm)

• Semi-erect gait – *Euparkeria*   Scale   8 in (20 cm)

• Pillar-erect gait – *Saurosuchus*   Scale   80 in (2 m)

• "Typical" erect gait – *Ornithosuchus*   Scale   10 in (25 cm)

The archosaurs were a major group of Mesozoic reptiles during the Triassic. The cladogram below shows how some older groups retained the primitive sprawling gait. But the archosaurs then split into two groups, one heading toward the crocodiles, and the other toward the dinosaurs and birds, and each group burgeoned with the evolution of new postures.

First, Euparkeria from the end of the early Triassic, and other forms, had a "semi-erect" gait, partway to being erect. This is retained by modern crocodilians, which sprawl when moving slowly and hoist their bodies high to run. Several of the Triassic groups of the crocodile side evolved a unique erect posture ("pillar erect") in which the femur fitted straight up into a horizontal hip bone. On the dinosaur side, erect posture paralleled that seen in humans ("buttress erect") with the femur and the hip bones vertical, and the connection being made from the side via an offset ball-like head on the femur.

Rauisuchia
Stagonolepididae
Phytosauridae
Ornithosuchidae
•PSEUDOSUCHIA
•CROCODYLOMORPHA
•CROCODYLOTARSI
•DINOSAURIA
•PTEROSAURIA
Euparkeria
•*Erythrosuchidae*
•ORNITHOSUCHIA
•*Proterosuchidae*

◼ Sprawling gait
◼ Semi-erect gait
◼ Pillar-erect gait
◼ "Typical" erect gait

•ARCHOSAURIA

Crocodile-like phytosaurs, Parasuchus, *stalk their prey beside a late Triassic river in northern India. The phytosaurs generally hunted the thick-scaled fishes of these rivers, but occasionally tackled the larger, slow-moving, herbivorous rhynchosaurs that fed on low-level plants at the waterside.*

group, some still mainly sprawling, others with erect posture. The crurotarsal (or crocodilian) type of ankle allows these archosaurs to walk fully erect, or to switch to a semi-sprawling posture when they wish. In fact modern crocodiles still have these options. When they are well fed and sluggish they haul themselves around in a semi-sprawling mode, belly scraping the ground; when they are slightly hungrier the gait becomes semi-erect, with the limbs tucked partly under and belly well clear of the ground. In cases of urgency they can run at some speed with a fully erect gait, and in absolute emergencies they can even break into a flat-out gallop. Rare film has been taken of crocodiles bounding like hares, moving the forelimbs and hind limbs in pairs and bending the backbone up and down to lengthen the stride.

The phytosaurs are generally seen as the original crocodiles. Phytosaurs are known so far only from the late Triassic, but they must have arisen earlier.

Parasuchus, the Indian phytosaur, is one of the most completely known forms. It was 8 ft (2.5 m) long and looked very like a crocodile: long snout lined with pointed teeth; head broad at the back with nostrils placed high up for lurking under the water; long armored

body; short limbs; and deep-sided tail for swimming. But the similarities to crocodiles are superficial – for example, the snouts of phytosaurs and crocodiles are made from different skull bones, and whereas in phytosaurs the nostrils are raised on a bump between the eyes, in crocodilians they lie at the tip of the snout. *Parasuchus*, like the other phytosaurs, evidently hunted freshwater fishes that they caught in their long snouts and trapped with their many pointed teeth. Two specimens have been found with stomach contents that indicate other elements in the diet: some smaller land reptiles that must have been snatched from the riverbank, perhaps when they came down to drink.

The ornithosuchids and pseudosuchians, also late Triassic thecodontian forms, had fully crurotarsal ankles that seem to be more advanced than those of the phytosaurs. The ornithosuchids were moderate-sized carnivores that preyed on smaller mammal-like reptiles. They had a partially upright posture, achieved in a similar style to the dinosaurs and pterosaurs.

There were two groups of pseudosuchians: the heavily armored "ruling reptiles," which were the first herbivorous archosaurs, and the rauisuchians. Both groups are known mainly from South America, with rarer representatives in North America, Europe, and India. The aetosaurs had flexible armor on their backs and bellies and around the tail. The armor was made up of numerous heavy rectangular plates held together by articulating pegs and by the skin. They had short snub noses with which they probably snuffled for edible tubers and roots. The armor gave protection against the rauisuchians, a large group of carnivores, some of them bipedal, which produced some species up to 20–23 ft (6–7 m) long. One of these was *Saurosuchus*, from Argentina, which had a fully upright quadruped gait, and looks like the most formidable carnivore of its time.

The later pseudosuchians had achieved erect posture in a different way from the ornithosuchians. Theirs was a "pillar-like" construction in which the primitive straight femur fitted up into a near-horizontal downward-facing acetabulum. It had nearly the same effect as the ball-and-socket joint with sideways-facing acetabulum of the dinosaurs, pterosaurs, ornithosuchids, and mammals, but it may not have allowed as much flexibility of the limbs for long striding, let alone a full gallop.

The Santa Maria Formation of Brazil

Halfway through the Triassic comes the Carnian stage (230 to 225 My ago), which is classified as "early late Triassic" because "early," "middle," and "late" in geological periods are terms that identify the sequence of the main episodes, not their duration. The momentum of the post-Permian recovery had been all the greater because of the sheer volume of ecological space left vacant, and a number of Carnian faunas record the variety of the recovery, as well as providing the prelude to a sweeping revolution in land vertebrates. The Santa Maria Formation of Brazil is a well-studied example, but remarkably similar equivalents are known from Argentina, India, Morocco, Scotland, and the United States. These are the pre-dinosaur faunas.

The Santa Maria Formation of the Rio Grande do Sul consists of a lower level, 230 ft (70 m) thick, of river-borne sandstones and conglomerates with lenses of finer-grained sediment. Fossil plants, and especially the seed fern *Dicroidium*, have been found near the base. The upper unit, 360 ft (110 m) thick, consists of interbedded siltstones and sandstones, with remnants of fossil tropical soils represented by limestone concretions. The rocks throughout are red, and display much evidence of originally hot and arid conditions. Many of the bones were sun-cracked in the Triassic before being covered by sediment. Since 1929, expeditions from German, Brazilian, and US institutions have built up oodles of bones from the Santa Maria, and documented twenty-four species.

The main herbivores in Santa Maria times were medium-sized cynodonts such as *Traversodon* and *Massetognathus*, about 20 in (50 cm) long and equipped with broad grinding cheek teeth. These have the very mammalian capability of occlusion, which means direct tooth-to-tooth contact between upper and lower teeth. Reptilian herbivores generally have teeth that cut past each other like a pair of scissors but do not occlude, which limits their ability to crush plant food. The herbivorous cynodonts were able to rotate their jaws a little, and this caused the cheek teeth to cut across each other, and so reduce the food to small digestible pieces. Now they could extract more value from their food, as well as avoiding excessive gas.

The most populous of Santa Maria herbivores were the rhynchosaurs, a group related to animals like *Tanystropheus*, and to the archosaurs. Specimens of the rhynchosaur *Scaphonyx* – first described and named early this century – accounted for 70 percent of finds in the Santa Maria fossil fields.

Scaphonyx was 3–6^1/$_2$ ft (1–2 m) long and roughly pig-shaped. Its skull was triangular as seen from above, and deep and wickedly grinning in side view. As in the placodonts, the breadth of skull at the back provided ample space for the jaw adductor muscles to attach and operate, and the depth of the lower jaw confirms that the rhynchosaurs had a powerful bite, easily strong enough to take your fingers clean off. Viewed from below, the upper teeth are arranged in multiple rows on a pair of triangular tooth plates, one on either side, and a groove runs down the middle of each plate. The crest of the lower jaw forms sharp ridges that slot precisely along the grooves in each palate plate. When

the jaws shut firmly the action was like the blade of a penknife folding up into its handle, a graphic analogy made in the nineteenth century by Thomas Henry Huxley.

Some scientists argued at first that the rhynchosaurs were oyster-catchers. But teeth like these are not designed for crushing; each has a delicate enamel point that is readily worn away. If a rhynchosaur took a fancy to a bivalve and managed to coax it into its mouth, the shell would pop out like a bullet as the pressure came to bear, and some of the tooth crowns would fall out. Rhynchosaurs probably fed on the seed fern *Dicroidium* and on other low bushy plants. The batteries of teeth served to clutch and cut the tough stems. The food was gathered together by the curious hooked bones at the front of the head, and pulled in with a muscular tongue. Rhynchosaurs also had powerful back legs, and they could dig up roots and tubers by kicking backward and using strong toe claws.

The rarest elements of the Santa Maria were some medium-sized bipedal inhabitants. *Staurikosaurus* was about $6^{1}/_{2}$ ft (2 m) long, quite a lot of it the long tail that stuck straight out behind when it was running. The relatively short skull had the sharp teeth of a carnivore, and the slender limbs identify a sprinter, clearly capable of preying on any of the cynodonts and other small reptiles. Ecologically, there is little to single out *Staurikosaurus*; after all, it represented less than 1 percent of the Santa Maria fauna. And yet with hindsight it stands as a key component, because it is an ornithosuchian, and the oldest dinosaur. It has the advanced mesotarsal ankle, an erect hind limb, a perforated acetabulum, and a forelimb less than half the length of the hind limb – all defining characters of the Dinosauria.

Mass extinction and the radiation of the dinosaurs

By the end of the Carnian stage of the late Triassic, about 225 My ago, a radical upheaval had overtaken the faunas of the land. It wiped out the rhynchosaurs and dicynodonts, and most of the cynodonts and thecodontians, the paramount animals in Santa Maria times, but

The early late Triassic scene, 225 My ago, in the Santa Maria Formation of southern Brazil. The small dinosaur Herrerasaurus, *one of the oldest dinosaurs, is harassed by small meat-eating cynodonts,* Belesodon. *These cynodonts were probably warm-blooded, and probably had hair; they lie close to the origin of true mammals. In the background loom a bulky rhynchosaur,* Scaphonyx, *and a dicynodont,* Dinodontosaurus *(right), grumbling and burping through the indigestible seed ferns and horsetails of the lakeshore areas.*

This scene is a precursor to the Age of the Dinosaurs; 5 My later, the rhynchosaurs would be extinct, together with many others, and the dinosaurs would take over. The cynodonts and mammals carried on largely unaffected, but did not achieve large size until after the extinction of the dinosaurs some 160 My later.

Three small dinosaurs, Coelophysis, *sniff warily for danger. These late Triassic dinosaurs are known in a great deal of detail, thanks to the astonishing find of dozens of skeletons piled on top of each other in a death assemblage in New Mexico in 1947.* Coelophysis *was an active predator that fed on a range of small animals, seizing them with strong hands and tearing them apart with its jaws and strong, birdlike feet.*

it offered new horizons to the survivors, and especially to the dinosaurs. They seized their opportunity, and spread pell-mell to fill the world again with carnivores and herbivores small, medium, large, and – in the case of the dinosaurs – gigantic.

The facts of this mass extinction event are hotly disputed at present. It happens that there were also major extinctions in the seas: massive turnovers in reef communities, large-scale losses of whole families of fishes, crinoids, sea urchins, bryozoans, scallops, and others. However, marine life was more profoundly affected by the end-Triassic extinction event some 15 to 20 My later. That episode had little impact on dinosaur evolution; the pattern was set already, and only a few stragglers expired – some thecodontians and cynodonts. It seems likely that there were two late Triassic mass extinctions, and on much the same scale, but with the heavier blows falling first on the land, and second in the sea.

What was the new pattern? What replaced the faunas lost in the end-Carnian liquidation? There were the dinosaurs, of course, the "terrible reptiles," and these fall into two initial groups, the ceratosaurs and the prosauropods, the first group carnivores, the second herbivores.

The best-known ceratosaur is *Coelophysis*, illustrated by the dozens of skeletons excavated in 1947 from a quarry called Ghost Ranch in New Mexico. Scattered remains of this dinosaur had turned up in New Mexico and Connecticut, but the Ghost Ranch collections made by teams from the American

Museum of Natural History seem to have represented more than a chance accumulation, and may even be the skeletons of a herd. There were young and old animals, males and females, and all preserved in good condition, piled in the same heap. They appear to have been washed along by a river and dumped on a sandbar downstream from the spot where they were drowned.

Coelophysis was a slender bipedal dinosaur with strong hands and a long snout lined with sharp serrated teeth. The limbs show all the adaptations for speed seen in *Staurikosaurus*, and it is likely that the New Mexico dinosaur lived on smaller lizard-like animals. The Ghost Ranch specimens are notorious, however, because the stomach contents of some of the adults contain the skeletal remains of juveniles of their own species, which makes *Coelophysis* a cannibal. Its relatives are also known from post-extinction faunas in Europe and Africa. The ceratosaurs lived on through the Jurassic period, and they were the first of the theropods, the general title given to several different lines of carnivorous dinosaurs.

The first herbivorous dinosaurs, the prosauropods, are known from late Triassic faunas worldwide. They range in size from the 6^1/$_2$–10 ft (2–3 m) length of *Anchisaurus* from North America and South Africa and *Thecodontosaurus* from Britain, to the 40 ft (12 m) melanorosaurs of South Africa and South America. The best-known prosauropod is *Plateosaurus*, from France and Germany, 20–26 ft (6–8 m) long, and well studied from some German locations.

The Stubensandstein of Germany

Sandstone quarries around Stuttgart in southwest Germany proved to be a fossil storehouse when they were mined for building stone in the nineteenth century. Some of the first Triassic reptiles to be identified anywhere were the phytosaur remains reported there in 1828, and dozens of fine amphibian and reptile skeletons were collected. Two fish-eating groups were the broad-skulled amphibian temnospondyls, which also appear in the Triassic of England and North America, and the phytosaurs.

Procompsognathus and *Halticosaurus* are theropod dinosaurs, possibly related to *Coelophysis*, but known from only one or two skeletons each. The third dinosaur, and much the commonest animal in these Stubensandstein deposits, is *Plateosaurus*, named in 1837 by Hermann von Meyer, who was later to name *Archaeopteryx*.

Plateosaurus has since been found in over fifty localities in central Europe and is known from dozens of specimens. The most extensively excavated quarry is at Trössingen in southern Baden-Württemberg, not far from the Swiss border. Here Friedrich von Huene and others excavated a series of yellow and red sandstone sites in the 1920s and 1930s, and recovered thirty-five skeletons, together with fragments of seventy more. At first Huene got it into his head that the Trössingen deposits represented a mass grave in which a herd of *Plateosaurus* had come to grief while migrating across a desert plain in search of food. This image of a heroic last hope has been refuted by later examinations of the site.

David Weishampel of Johns Hopkins University found that the skeletons occurred in water-laid mudstones and sandstones, deposited by rivers. The remains have been disturbed by scavengers as well as by water movement, and the mode of death and deposition seems to have been the classic one for dinosaur preservation. Flash floods killed the animals and washed the carcasses downstream before depositing them at a bend in the river. When the same thing reoccurred over the course of several years it might produce a mass deposit of carcasses.

Plateosaurus was a prosauropod that could grow up to 26 ft (8 m) long, the biggest dinosaur yet, and only made to look relatively small by the colossal sauropods of later periods. It had a long tail, long hind limbs, and a small head at the end of a fairly long neck. The teeth were leaf-shaped and the jaws not powerful, so *Plateosaurus* was certainly a plant-eater. It could rear up on its hind legs to feed from the tops of trees, and may have used its sizable thumb-claw to pull in succulent branches.

Herbivores had a long menu to choose from, including conifers, seed ferns such as *Caytonia*, tall cycads or the now extinct cycadeoids, the thriving ginkgo varieties, or the wealth of ground ferns. Grasses still lay a long way ahead. *Plateosaurus* had a much tougher diet, and had to use gizzard stones to grind up plant stems and foliage that its teeth could cut but not chew. Digestion must have been hard labor, what with so much gastrolith percussion, the slosh of digestive juices in tun-sized bellies, and the gargle of gases escaping at one end or the other.

The prosauropods were a highly successful group that lived on into the Jurassic period. Somewhere along the line they bequeathed their small heads, long tails, and solid four-footed form to their sauropod descendants. No wonder that one of these still sticks to its first official name of *Brontosaurus* – "thunder-lizard."

The other animals from the Stubensandstein include the oldest turtle, *Proganochelys*, a whole new reptile order appearing out of nowhere, but already in full armor, with the characteristic carapace above and flat plastron below. *Proganochelys* still had some teeth; modern turtles have none. No halfway forms have yet emerged to give a clue to turtle ancestry, but a form as specialized as this must have a lengthy, if hidden, evolutionary history.

Other small animals are bound to have been present in the Stubensandstein region, but have not been found among the fossils there. At around the same time there lived the first mammals – small shrewlike animals. Northern Italy provides three species of the first known vertebrate fliers, the pterosaurs, which lived along the northwest shores of the Tethys Sea. The oldest, *Eudimorphodon*, was a fish-eater with a wingspan of 40 in (1 m) and the long tail that is featured in all of the Triassic pterosaurs.

Pterosaur wings are membranes rigged along the arm and wrist bones and then along the enormously lengthened bone of the fourth finger. The membrane is reinforced by an elaborate system of thousands of parallel fibers of an unknown material, possibly keratin, sandwiched between layers of skin, and serving not only to hold the wing rigid in flight but also perhaps to alter the conformation of the wing for flying maneuvers, if they could be made to bend by muscular control. The design is quite different from those of the birds and bats, and it carried the pterosaurs all the way through the Jurassic and then the Cretaceous, some with weirdly

sculptured beaks, others with wingspans wider than a small aircraft.

Most pterosaurs are known from aquatic environments, but as usual this may say more about the conditions that favor preservation than about the full range of their habitats.

Dinosaur origins: one group or several?

Late Triassic dinosaurs included ceratosaurs such as *Coelophysis*, prosauropods such as *Plateosaurus*, and a third group, the ornithischians, known from scrappy remains in North and South America. These were destined for a spectacular future.

Until recently it was fashionable to see the dinosaurs as a polyphyletic group: not a natural assemblage of related creatures, but a mob converging from various directions, each with different ancestors, linked only by superficial similarities and the name of Dinosauria, invented by Sir Richard Owen in 1842. The Victorians had believed that the dinosaurs were a single phylogenetic unit.

The current view is that most of the animal groups in the traditional arrangement have similar features because they really are related to each other, not because these are "convergent" features that happened to develop independently in quite separate lineages. A number of cladograms of dinosaurs and other archosaurs constructed independently by various European and North American paleontologists in 1984–5 confirmed that the Dinosauria made up a valid clade, defined by all sorts of complex modification to the limbs, relating to erect gait, as well as by other features too complex in their detail to be merely convergent.

The dinosaurs arose at the end of the middle Triassic, existed through Carnian (Santa Maria-type) faunas at very low density, and then exploded into a range of forms in the last 10 to 15 My of the Triassic.

New beginnings

The dinosaurs were not the only strikingly new group to appear among the late Triassic faunas. They appear alongside the first turtles, the first sphenodontids ("wedge-tooth": ancestors of living tuataras of New Zealand), the first crocodilians, pterosaurs, and mammals. It can be argued that this was a key turning-point, when the foundations of modern land faunas were laid. The Paleozoic reptiles had disappeared, with the extinction of the mammal-like reptiles, and most of today's leading groups were making their debut. (The amphibians were out of step – frogs had come earlier, and the oldest of them, the salamander, belongs to the mid-Jurassic. As for the birds, none of them flew until some 40 My later.)

Not even the most biased mammal observer can argue about the identity of the most powerful and innovative newcomers in the late Triassic, however. Of all the animals that survived to radiate into the ecospace created by the mass extinction of the end-Carnian phase, one group seized the time. Chance made the opportunity, but the dinosaurs took it. They rose to a position of worldwide dominance in every major role in as little as 4 or 5 My. After that, for well over 100 My, *they* were the modern fauna; every other creature lived along the byways.

OPPOSITE *The first large dinosaur,* Plateosaurus, *was a plant-eater that lived 215 My ago in Germany, during the late Triassic. Typically, this dinosaur grew to between 16^{1}/$_{2}$ and 20 ft (5 and 6 m), and some achieved 26 ft (8 m). Plateosaurus was also immensely abundant; hundreds of remains have been found at over fifty separate localities in central Europe, and some accumulations may contain dozens of skeletons. As it feeds in the trees, two kuehneosaurs glide away from their treetop perch. They extend their "wings," which were made from skin stretched over hinged, elongate ribs, in order to glide long distances, perhaps in pursuit of insect prey or to escape the unheeding jaws of large plant-eaters.*

One of the best-known living fossils, the tuatara, Sphenodon punctatum, *lives today on only a few small islands lying off the main north and south islands of New Zealand. Before the Maoris reached New Zealand, the tuatara was widespread on the mainland. The sphenodontids formerly had a global extent, and they were abundant and diverse in the late Triassic, 225 My ago, at the time of the origin of the dinosaurs. The fossil examples would have looked nearly identical to the living tuatara; the length of the specimen is 20 in (50 cm).*

The early Jurassic seas extended over much of Europe, flooding lands that had supported abundant dinosaurs not long before. Shoals of dolphin-like ichthyosaurs swam and leapt in shallow waters, in pursuit of fishes and belemnites (swimming mollusks superficially like cuttlefish). Ichthyosaurs were reptiles which clearly had land-going ancestors, but they were entirely marine and gave birth to live young, just as dolphins do today. Plesiosaurs were long-necked marine reptiles that also preyed on fishes; they came ashore from time to time, like present-day marine turtles.

Chapter Four

DINOSAUR SUMMER

Michael Benton

Thigh bones taller than the tallest human being, skulls over 8 ft (2.5 m) long, jaws with batteries of three or four hundred teeth – if ever an animal group was equipped to survive even its own extinction, it is the dinosaurs. In continental fossil deposits from the late Triassic to the latest Cretaceous, theirs are the most common and most widespread remains. Their footprints stroll or stampede over ancient surfaces all the way from Queensland, Australia, to Wyoming, USA, and the Isle of Wight, in England. And although it tends to be the giants whose bones survive, it is clear that the dinosaurs covered every size, down to lengths of about 2 ft (60 cm), and weights of $4^1/2$–$6^1/2$ lb (2–3 kg). Small, large, or truly colossal, rearing on two legs or lumbering on four, eaters of meat, insects, or plants, the Mesozoic lands belonged to them.

In that world, the mammals were tiny creatures no bigger than a mouse or shrew. They have been compared to vermin – the rats, or possibly the cockroaches, of their day. The dinosaurs have captured the imagination of us later mammals ever since we began to realize that they once existed, and how long they existed. Paleontologists are bound to ask themselves what gave the dinosaurs their momentum and what qualities kept them supreme throughout the Jurassic period (208 to 114 My ago) and all through the succeeding Cretaceous (until about 65 My ago).

Some of the dinosaurs' advantages are clear. They began their radiation in the late Triassic, a time of hot dry climates more favorable to reptiles than to mammals. They did not have to out-compete their predecessors, but moved into fairly empty ecospace. Their upright gait bestowed mechanical advantages that permitted quick advances in speed and size. (But there were other upright-walking animals that either died out or else proliferated in the space that dinosaurs stayed clear of.) By rising during an era of maximum linkage between all the continents assembled into the supercontinent of Pangaea, the dinosaurs could exploit every habitat that suited them, wherever it was. Likewise, no other group could rise and spread elsewhere, behind impassable mountain barriers or across secluding seas or oceans; there were no such enclaves large enough to nurture major rival groups.

Science is full of cautionary episodes with a single basic plot: worn-out old theories turn into blinkered assumptions (paradigms) that ignore or even persecute new views that fit the facts more closely. Copernican astronomy and Darwinian evolution are obvious examples. So is Alfred Wegener's theory of "continental drift"; seeing that the continents must once have fitted together, he argued that they must have moved, but he could not devise a mechanism that would enable them to do so. Instead of looking for the mechanism that would explain the facts, some geologists chose to reject the theory and ignore the facts. Plate tectonics now offers us the mechanism, and the movement of continents is taken for granted.

Of course the history of ancient times affords us too few facts, so it has to be wide open to theories. They swarm around the various extinction events, shedding both heat and light, and the more sweeping the theory

the greater is its power of attraction. The latest theory that tries to account for the dinosaurs' long dominance argues that in order to shut out the mammals for so long they must have shared the mammals' great advantage: they must have been warm-blooded. This idea is voiced with tremendous enthusiasm and scholarship by the American paleontologist Robert Bakker, in his book *The Dinosaur Heresies* (1986). Warm blood would have given the dinosaurs their competitive edge: a more efficient heart, more energy and speed, a better brain, the ability to survive in cold conditions (and not to be immobilized at night).

There are counter-arguments here, too lengthy to go into. Bakker's book is as lively as the non-stop hyperactive animals in his spectacular illustrations, and by arguing from a single underlying cause he is able to give complex events a dynamic form that is simple and dramatic. On the other hand, in the story of life there is good reason for believing that causation is not so straightforward. Life is an open system, a cluster of interlocking, interacting forces possibly so crammed with key events at every level from electron-microscopic to planetary that they could never produce the same result twice. Systems like this are sometimes called "chaotic," which does not mean that trends occur at

JURASSIC WORLD
During the Jurassic, Pangaea began to break up. Rifting between North Africa and the eastern coast of North America led to the opening of the North Atlantic Ocean. Southern continents began to rotate away from each other, but Gondwanaland broke up only later.

Climatic evidence in Jurassic rocks shows a simple set of climatic belts, much as today, but with no evidence for polar ice. Hence, the tropical

and subtropical belts are much wider today, and the polar regions were merely temperate in climate. Fossil forests and other plant accumulations

are found worldwide, as are the dinosaurs.

There is little evidence for distinctive floral and faunal zones during the

Jurassic. Identical genera of plants and dinosaurs are found essentially worldwide, from Australia to Siberia, and Africa to North America.

- Jurassic rich fossil forest sites
- Jurassic rich plant fossil sites
- Jurassic rich dinosaur fossil sites
- Jurassic rich other reptile fossil sites
- Wet
- Seasonally humid
- Arid desert

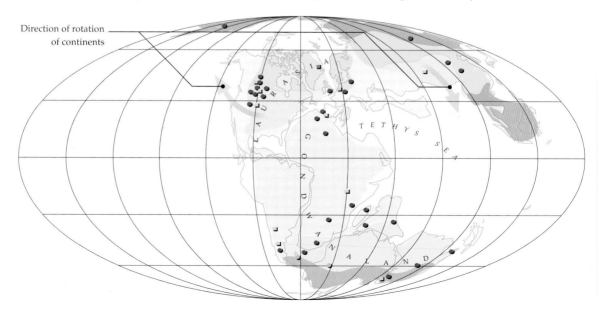

random, but only that their pattern of causes may be too variable to predict.

Here we are dealing with a major trend: the long dominion of the dinosaurs. The warm-blooded dinosaur is a powerful idea, but in the absence of basic information that fossils do not give, it cannot be either proved or disproved. The notion of a single cause is too attractive to ignore. If it oversimplifies an intricate story, it is also too good to be true. Dinosaurs have left us no soft tissues. We don't know what kind of hearts they had, what organs they developed, or how they worked. What we do have is circumstantial evidence, and as usual this is open to very different interpretations.

It is likely that the dinosaurs had opportunity, climate, and geography on their side. What other factors worked for them? The most obvious is sheer size; the larger sauropods achieved weights of possibly 70–80 tons (tonnes), several times more than the largest-ever land mammal, *Indricotherium*, the giant rhino of the Oligocene epoch (about 35 My ago). It was the upright gait that enabled them to grow; no sprawling posture could bear such weights even while standing still. Managing their vast bulk was partly an engineering problem – *Diplodocus* has been compared to a walking cantilever bridge – and partly mechanical, a matter of force and leverage. The dinosaurs needed powerful muscles, solidly anchored, most obviously for walking or running and for eating. Some dinosaur skulls are marvelously sculpted to make room for jaw muscles that had to deal with cartloads of tough plant materials, or disable or kill their prey with a single bite.

For the biggest plant-eaters, their size must have made them almost immune to attack during their maturity, although their young, sick, or elderly would still have offered rich pickings. Likewise, the biggest flesh-eaters grew as large as it was possible to go and still be a predator; no point in being a 50-ton (tonne) carnivore if you were too slow to catch very much, but had to keep hunting in order to fuel your bulk. Size may also have had advantages for the armed carrion-eaters, enabling them to loot what the more agile killers had caught.

Bulk meant warmth. An animal that doubles its volume (a cubic measure) comes nowhere near doubling its surface area (a square measure). Since heat loss is a function of the surface area exposed, the greater the bulk, the slower the change in internal temperature. It has been calculated that in warm conditions a medium- to large-sized dinosaur could have maintained a regular temperature without having the equipment for endothermy, internal heat production. Four-fifths of a mammal's food consumption is invested in keeping it warm. A reptile has much lower running costs – about one-tenth of the food needed by an endotherm of the same size. Throughout the Jurassic and for much of the Cretaceous, the world was warm even in regions relatively close to the poles.

These economies of scale would have offered real advantages to the larger dinosaurs, but what about their smaller relations? How did they keep up their temperatures, and therefore their energy levels? The answer is that we do not know, and dinosaurs are such a varied group that no possibility should be utterly ruled out. But plenty of reptile groups survived the Mesozoic and many survive today. Ectothermy, allowing body heat to vary with the outside environment's, was and still is a perfectly viable strategy.

The Earth was changing radically during the long summer of the dinosaurs, moving away from the largely arid conditions of the Permian and the Triassic. In the Jurassic, most modern groups of land animals diversified, but major changes also affected Jurassic seas, with the evolution of modern fish groups, and the rise of the ammonites and of large marine reptiles, all of them flesh-eaters.

Break-up of Pangaea

The supercontinent Pangaea, which had come together during the Permian and existed throughout the Triassic, began to split apart in Jurassic times. The first rift in this vast landmass appeared between Europe and Africa, at the western end of the Tethys Sea that separated Asia from India, Antarctica, and Australia. Rifting spread westward, across the region of Spain and down the east coast of neighboring North America, running down from Nova Scotia to Florida. Further west, it separated North and South America. Southward, it divided South America from

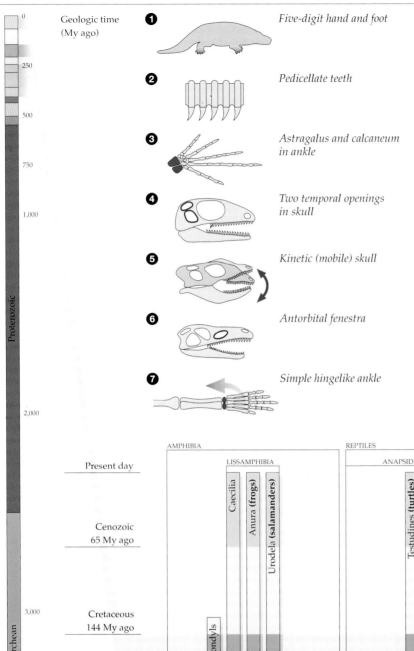

1 *Five-digit hand and foot*

2 *Pedicellate teeth*

3 *Astragalus and calcaneum in ankle*

4 *Two temporal openings in skull*

5 *Kinetic (mobile) skull*

6 *Antorbital fenestra*

7 *Simple hingelike ankle*

TETRAPOD PHYLOGENY

The tetrapods, four-footed land-living animals, arose from the sarcopterygian fishes during the Devonian. The earliest amphibians dominated on land during the Carboniferous. One line gave rise eventually to the modern amphibians, the other to the reptiles.

The oldest reptiles are Carboniferous in age, but the group as a whole split into three main lineages. The anapsids were important early on, and gave rise to the turtles. The synapsids, or mammal-like reptiles, dominated in Permian and Triassic times, and gave rise to the mammals proper in the late Triassic.

Most of the age of reptiles was dominated by diapsids. These include lizards, snakes, and the tuatara on the one hand (the lepidosaurs), and crocodilians and birds on the other (the archosaurs). Extinct relatives of the latter are the pterosaurs and dinosaurs, which jointly dominated on land during the Jurassic and Cretaceous.

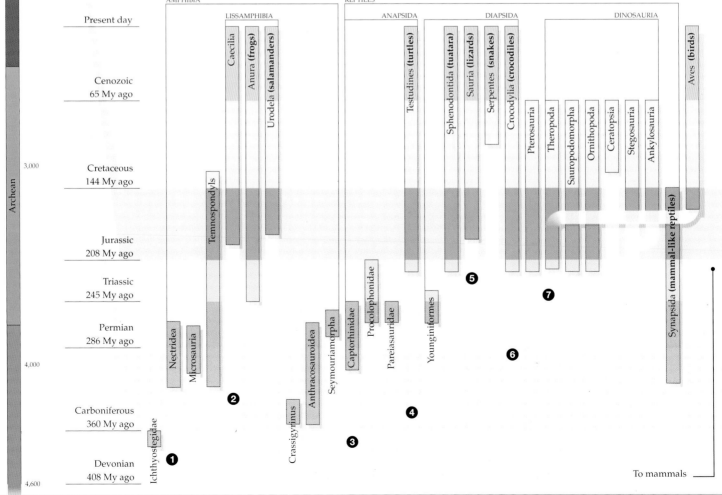

THE MORRISON FORMATION

The year 1877 was a crucial one for tetrapod paleontology. A chance find in the Jurassic foothills of the USA's Rocky Mountains, near Morrison, Colorado, launched a decade that revealed the unique variety, colossal size, and outstanding preservation of dinosaur fossils that transformed the world's vision of an era far less richly represented in the rocks of Europe.

The first person to stumble on the bone hoard as it weathered out of the ground was the English missionary and schoolteacher Arthur Lakes. He wrote excited letters to one of the two great rivals for supremacy in the world of American dinosaurs, Othniel Charles Marsh (1831–99). When Marsh failed to answer, Lakes consigned him ten crates of bones, but it was not till Lakes also sent some finds to Edward Drinker Cope (1840–97) that jealousy spurred Marsh to action.

Marsh's own assistants were soon digging up 1 ft (30 cm) long vertebrae and 8 ft (2.5 m) thigh bones from the Morrison Formation. First Cope's and then Marsh's team raced in to dig at sites bang next door to each other, and each discovered several more dinosaurs. Later that summer, two railway engineers came across some enormous bones at Como Bluff, Wyoming, later identified as belonging to *Brontosaurus*. They wired Marsh, who took credit for the discovery – an important blow in the struggle between the two men that was to last for the next twenty years.

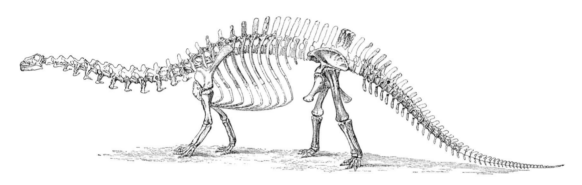

TOP *Edward Drinker Cope (1840–97).*

ABOVE *Othniel Charles Marsh (1831–99).*

LEFT Brontosaurus excelsus – *the reconstructed skeleton as drawn by Marsh.*

SOLNHOFEN, BAVARIA

In 1861 the Solnhofen quarries in Bavaria yielded their most astonishing discovery, a creature that seemed to combine features of both birds and reptiles, soon named as *Archaeopteryx*, "ancient wing."

The new animal was dragged straight into the quarrel between supporters and enemies of Darwin's new theory of evolution. Richard Owen (1804–92), a famous anatomist and Superintendent of Natural History at the British Museum, swept aside the reptile features of *Archaeopteryx* and bluntly insisted that it was "unequivocally a bird."

In the competitive world of Victorian science, Owen fought a running battle with his former protégé Thomas Huxley (1825–95), now Darwin's most outspoken supporter. In a famous lecture in 1868, Huxley showed that *Archaeopteryx* was "descent with modification" made flesh, a bird transforming its reptilian origins.

ABOVE *Sir Richard Owen (1804–92).*

LEFT *Thomas Huxley (1825–95) as portrayed in a contemporary cartoon.*

Africa; these continents appear to have been neighbors in Gondwana for a period of more than 400 My.

This series of rifts brought wide-ranging effects. Thick beds of salt were laid down in North Africa, Texas, and the Caribbean, and off the coast of Brazil. In these areas, the rifting seems to have produced depressions between the continents, where sediments from rivers could be deposited but the sea was as yet unable to break in. The salt beds represent millions of years during which water evaporated and accumulated in cycles. The rifting down the east coast of North America opened a series of basins parallel to the main rifting direction (on a modern map, northeast to southwest) from Nova Scotia to Virginia, and the basins filled with sediments washed out from the land to form the Newark Supergroup. This geological formation spans the late Triassic and early Jurassic and contains a detailed record of land life of the times – plants, insects, fishes, and reptiles.

By late Jurassic times, some 160 My ago, the Tethys Sea extended almost all round the world, just north of the then equator. It split Asia from the Gondwana continents, and also now separated North and South America, forming a forerunner of the Caribbean Sea. The animal life of the Tethys Sea was distinct from that of other parts of the world, and consisted of tropical reefs and a number of specific groups of shellfish. This equatorial Tethyan ocean contrasted with the cooler realms of northern Europe.

Jurassic plants

The separate northern and equatorial provinces are reflected in the distribution of plants. Distinct realms developed because now there were sea barriers between north and south, and because there was a greater range of temperatures from the poles to the equator. The temperature gradients were not so steep as they are today, and there is no evidence for polar ice during the Jurassic, but floras that were distant from the Jurassic equator were more like temperate-zone plants.

Many of the typical Permian and Triassic plants failed to survive into the Jurassic. Lycopods – club mosses – had already declined during the Permian, but clung on at low levels. The seed ferns had flourished ever since the late Devonian, sometimes as the most prominent plants (*Glossopteris* in the Permian). Now they were fading, and they barely survived to the start of the Cretaceous. Ferns, too, were less spectacularly exuberant, though big tree ferns lasted into the Cretaceous, and the family survives today in more modest forms.

The key Jurassic trees were the various groups of gymnosperms, the plants that bear exposed seeds, usually in a cone of some sort, and tend to be fertilized by wind-blown pollen. Most of these groups arose from the seed ferns in the late Carboniferous and the Permian. They were the cycads and cycadeoids, and the ginkgoes. The bigger cycad species grew up to 60 ft (18 m) tall and looked like bulky palm-trees, with a crown of fronds and a pattern of armor all the way up the trunk, formed from bases that remained in place after the frond was shed. They carried their seed and pollen cones on separate plants, and if the pollen landed close enough to the female ovules, the sperms still had to swim. The cycads survive, though sparsely, in tropical and subtropical areas.

The single ginkgo species that survives today is also called the maidenhair tree. It looks very like its ancestors, and like the cycads produces its spores and ovules on different trees. Unlike the cycads, it has a modern treelike shape, with a narrower trunk and recognizable if unusually shaped leaves, borne on thin branches. Apart from the pines, modern conifer families had all arisen by late Triassic and Jurassic times. The famous Petrified Forest of Arizona consists of the trunks of Triassic monkey puzzle trees, not very different from their modern descendants.

Conifers are the most advanced of the gymnosperms, and the pollen grains grow a tube that delivers the male sperms straight to the ovule – a far less chancy technique than random drift and water-dependent swimming.

Jurassic landscapes had a curious halfway look. They were dominated by cycads, and their cousins the cycadeoids, with the air of giant pineapples crowned by waving fronds and covered in the season by colorful "flowers" that were not true flowers. Forests of ginkgoes and especially of conifers gave the scene something of a modern appearance; but true flowering plants, hardwood trees, and particularly grasses were still absent. There is evidence that beetles may have become involved in the pollination of some gymnosperms, but otherwise few of these plants were trying to attract insects, and it looks as if cycad cones were designed to keep them out. Well over half the known insect species were equipped for drilling, nibbling, and sucking at plants, and it was the angiosperms, the flowering plants, that later specialized in contracts by which board and sometimes lodging were exchanged for pollination services.

Rise of the dinosaurs

All the early dinosaurs were carnivores such as *Staurikosaurus*, and it seems to have been these meat-eaters that radiated first, in the form of the ceratosaur bipeds like *Coelophysis*. By the end of the Triassic the dinosaurs had also diversified into plant-eating forms like the prosauropods, and *Plateosaurus* was already a hefty species. All of these animals were saurischians ("lizard-hips"), but the other main dinosaur group, the ornithischians ("bird-hips"), was already on the rise before the Jurassic began. All the ornithischians were plant-eaters. They began as agile bipeds, but appear during the Jurassic in a wide variety of forms that include the massive armored ankylosaurs and stegosaurs. Both these latter groups were quadrupeds specializing in static defense, their size varying from 8 to 25 ft (2.5 to 7.5 m) long, and their weight ranging up to 2 tons (tonnes). All these ornithischian dinosaurs were dwarfed by the huge sauropods that will be described later in the present chapter.

Recent cladistic analyses of the dinosaurs have been carried out by Jacques Gauthier of the California Academy of Sciences, Paul Sereno of the University of Chicago, and David Norman of Cambridge University, among many others. Their studies of the main lines of evolution may differ in details but they settle a number of arguments. For example, they agree that the dinosaurs are a monophyletic group descending from a single common ancestor, that their closest relations are the pterosaurs, and that birds are the direct descendants of the theropods, a group of carnivorous dinosaurs. These investigations also confirm the division of dinosaurs into the two main groups of saurischians (theropods plus prosauropods plus sauropods) and ornithischians (bipedal ornithopods plus all armored and plated forms). The British paleontologist Harry Seeley proposed the names in 1887. They are based on two different layouts of the hip bones when seen in side view. Later evolution modified the designs, but the original pattern is distinctive.

Tetrapods have three hip bones that fuse together to form the pelvic girdle. Seen in side view, the ilium appears at the top, solidly fixed to the backbone by fusing with one or several vertebrae. The other two bones are the pubis in front and the ischium behind. One end of the pubis and one end of the ischium make up the lower edge of a socket whose upper edge is made by part of the ilium. This socket is the acetabulum, where the hind leg of the animal joins the rest of the skeleton. Great strains fall on this structure, whether the legs are propelling a biped sprinter or propping up the rear tonnage of a sauropod. The difference between the saurischian and the ornithischian hip lies in the arrangement of the pubis and the ischium at the point where they meet the anchoring bone, the ilium. The standard arrangement is for the pubis to point forward and downward and the ischium backward and downward; this is the lizard-hip design. In ornithischians, instead of pointing forward the pubis is folded back and down from where it joins the acetabulum, so that it runs parallel to the backward-sloping ischium.

That is the simple version. Two main complications arise. In the later ornithischians, part of the pubis continues to slope backward, but it also develops a further extension at the top, forming a kind of arch that slopes upward from the rear, attaches itself to the ilium, and then slopes forward and downward again, so that at first sight it looks like the saurischian arrangement. The second complication is that the birds did not arise from the bird-hip ornithischians but from lizard-hip theropods that folded back the pubis at least 50 My after the ornithischians did it, and in a different way.

These designs had far-reaching effects on

the dinosaurs that used them. Saurischians carried their guts in front of the forward-sloping pubis. This set no problems for the meat-eaters, which did not need elaborate digestive systems to process their food, and so were able to retain a bipedal posture without being tipped forward; a heavy tail could counterbalance the weight of the body organs. For the plant-eaters, there was no choice. They needed a much larger and heavier digestive system, and this forced them forward to walk on four legs. The plant-eating ornithischians could continue to operate as bipeds because the backward-sloping pubis made room for carrying the guts around the center of gravity, slung between the hind legs.

Alan Charig has shown that these different pelvic structures also involved different ways of moving. Saurischian dinosaurs moved their hind legs forward by contracting strong muscles fixed at one end to the jut of the pubis and at the other to the thigh bone. For a big quadruped saurischian there was only a narrow angle between pubis and thigh bone, and therefore a small arc of movement for the pillar-like legs, but they carried so much weight that a longer step would have caused dangerous stresses. This was the case with the ponderous sauropods. Even the heaviest of their relations, the meat-eating theropods, weighed much less. Large or small, they also needed to move faster in pursuit of prey, and they could do that because they were bipeds; when they stood up, the angle between pubis and thigh bone grew wider because the pubis pivoted upward, and this made room for longer muscles that would power a longer stride. It seems that the plant-eating ornithischians folded back the pubis right from the start, perhaps to relocate the guts, and they moved one end of the same muscle so that it attached to the front of the ilium. This arrangement was mechanically awkward for a bipedal bird-hip, and this accounts for their developing a new front extension on the pubis, and shifting the same muscle to anchor there. Dr Charig's is a superbly logical account of 150 My of evolutionary research and development.

One of the lessons of this story is that evolution does not move in straight lines. The bird-hips made far more experiments with pelvic design, and they are extinct; the more conservative lizard-hips made one vital set of modifications, and they survive as birds. Other bird-hip innovations were the predentary bone at the front of the lower jaw, supporting the lower half of horny beaks, and the systems of bony rods or tendons running more or less parallel or as angular struts along the spine, to reinforce the back or tail. This was particularly useful for bipeds that used their tails as balancing poles when walking or running, and did not have to invest more muscular energy to keep them horizontal.

The heavyweight meat-eaters

Most of the ceratosaurs were small to medium-sized members of the theropod suborder of lizard-hips, but some Jurassic ones outgrew them. *Dilophosaurus* and *Ceratosaurus* itself, from that dinosaur Valhalla the late Jurassic Morrison Formation of Utah and Colorado, reached lengths of 20 ft (6 m). *Ceratosaurus* had a pair of crests along the top of its snout which may have been used in pre-mating contests. The early Jurassic *Dilophosaurus* had even more conspicuous crests, each shaped like half an upturned dinner plate, and standing parallel along the top of the snout from the nostrils to behind the eyes. The ceratosaurs did not outlast the Jurassic.

A second, less clearly defined group of theropods are the megalosaurs, known mainly from the mid-Jurassic. *Megalosaurus* was the first dinosaur to be named, on the basis of a piece of jawbone and other elements. It was baptized in 1824 by Dean William Buckland, an Oxford theologian and geologist. Bones of *Megalosaurus* had turned up much earlier, and one of the first had appeared in 1676 in a book on the natural history of Oxfordshire. It was not recognized for what it was. The seventeenth century had no conception of dinosaurs; the notion was hardly imaginable in an age when science was based broadly on biblical revelation, and Archbishop Ussher had dated the creation of the world to Sunday, 23 October, 4004 BC.

The Megalosauridae were a prominent carnivore family in the mid-Jurassic of Europe and China, and probably elsewhere, but sediments of this age are rare worldwide. *Megalosaurus* bones and teeth have been found in plenty in the Cotswold region of

England, and reputed megalosaurs have been noted from the late Jurassic and even the Cretaceous of Europe, North Africa, China, and elsewhere. It reached a total length of 30 ft (9 m), and with its deep jaws and crunching bite it probably fed on the plated plant-eaters, stegosaurs, crocodiles, and other large reptiles of its day.

The third main theropod group of the Jurassic was the carnosaurs, "flesh-lizards." *Megalosaurus* may be one of these, but there is no doubt about the identity of *Allosaurus*, from the late Jurassic of North America. It was big enough to tackle most of the plant-eating dinosaurs that shared its territory, such as *Stegosaurus*, the ornithopod *Dryosaurus*, and even massive sauropods like *Apatosaurus* and *Diplodocus*. No rival predator outweighed the 1½ tons (1.5 tonnes) of *Allosaurus*, which had three main toes on its foot, leaving a huge birdlike footprint, as well as a shorter fourth toe at the back, and three more claws on each powerful hand. It may have hunted the large sauropods in packs.

During mid-Jurassic times in China, some 175 My ago, stegosaurs and sauropods dominated the scene. Here a herd of the spiny plant-eating stegosaurs Tuojiangosaurus *trot past in search of waterside plants. Behind them are the long-necked plant-eating sauropods* Shunosaurus *and a flock of tiny pterosaurs.*

Middle Jurassic dinosaurs from China

More exciting new finds are coming from China, which has produced by far the greatest number of new dinosaur species over the past twenty years. Some of the most important have come from the mid-Jurassic, a time that is rather poorly known in most other parts of the world.

Dong Zhi-Ming and his colleagues from the Institute of Vertebrate Paleontology in Beijing have dug widely in the Xiashaximiao Formation of Sichuan Province, where they have turned up a rich fauna of dinosaurs large and small. At the bottom of the food chain is *Xiaosaurus*, a small bipedal ornithopod ("bird-foot").

So far *Xiaosaurus* is known only from a jaw and a leg. It was about 40 in (1 m) long, with a short snout, and jaws lined with roughly diamond-shaped teeth whose scalloped edges form rounded projections quite different from the sharp serrations on theropods' teeth. It used these knobbly edges for cutting up leaves, and it may have been able to chew in a crude way, instead of just gnashing its jaws and having much of their contents fall out each time the jaws parted. Chewing was a major advance over typical reptiles, reducing waste and improving digestion; it was crucial in the later evolution of ornithopods.

Tuojiangosaurus was one of the first stegosaurs, "plated reptiles." The skeleton is known in some detail, and corresponds closely to its later and larger relative, *Stegosaurus*. Stegosaurs walked on all fours all of the time, but their bipedal ancestry is clear, and *Tuojiangosaurus* has much longer hind limbs than forelimbs. The head is small and the jaws long, evidently not adapted for heavy chewing of tough vegetation. Like the rest of its family, *Tuojiangosaurus* was no genius; for all of its weight and length – $1^1/_2$ tons (1.5 tonnes), 13–20 ft (4–6 m) – it had a brain the size of a walnut.

The key feature of this animal, and the one that gives the group its name, is the series of plates and spines down the middle of the back and along the tail. Their possible function is discussed in the next section.

Other mid-Jurassic Chinese dinosaurs include a possible megalosaur, *Gasosaurus*, known from early complete remains and big enough to feed on anything from *Xiaosaurus* to *Tuojiangosaurus*. Two of the earliest sauropods were *Shunosaurus* and *Datousaurus*, 30 ft (9 m) and 45 ft (14 m) long. This group was to dominate late Jurassic landscapes.

The mystery of the stegosaurs' plates

We recognize stegosaurs by the series of plates and spines down the middle of the back and tail. *Tuojiangosaurus* and the other mid-Jurassic stegosaurs had triangular plate-spines. So did the late Jurassic *Kentrosaurus*, from Tanzania, a smaller animal, only 8 ft (2.5 m) long. Its remains were dug out of the famous Tendaguru Hill site early in the twentieth century by major German expeditions led by Werner Janensch. He sent the skeletons to Berlin, where they can be seen today – no small undertaking, since Tendaguru lay many miles inland and there were no roads. Each bone was carefully packed and carried on the head of a local porter to be shipped from Lindi. *Stegosaurus* itself, from the Morrison Formation of Utah and Colorado, had broad diamond-shaped plates down its back, and businesslike spikes on its tail.

This is the standard stegosaur image, though it happens that *Stegosaurus* is the only one to have had such broad plates. Most of the group died out at the end of the Jurassic, but a few appear in the Cretaceous – one of them, *Dravidosaurus*, right at the end. It seems to have been marooned when India broke loose and drifted northeastward from Gondwana, sheltering from the advanced predators that elsewhere killed off the rest of its relations.

The function of the plates has long been debated. They may have had a deterrent effect against predators, but they were not useful armor, because they stood upright and did not protect any vital organs. They may also have been shown off in displays between males seeking to establish territory and impress females before mating. Certainly the vicious spikes on the end of the flexible tail could have been used to considerable advantage in any contest, and could do real damage to predators; but display alone seems a very expensive explanation for growing those larger plates on the back.

Could it be that the plates had a function in temperature control, like the dorsal sail of the pelycosaurs? If so, the process worked in reverse. Wind-tunnel experiments on engineering models of stegosaur spines show that they make efficient heat-dissipating fins. An overheated stegosaur could stand face-on to a breeze and shed heat fast through the spines. It is known that the spines and plates were covered with skin in life, and thick with blood vessels – features necessary for dumping heat, though not logical for covering armor. Yet even for the purpose of controlling body temperature, this array of extra bones, fixed in the skin by tendons, still seems like an expensive luxury. All the other dinosaurs living alongside *Tuojiangosaurus* & Co. had no such arrangement, yet they must have avoided overheating somehow. Maybe the stegosaurs' digestive apparatus produced more heat than most because their teeth were not specialized for chewing and they gobbled their plant food whole, leaving their stomachs so much work to do that they heated up like fermentation vats.

This illustrates the fun, and the frustration, of the speculative exercises that are forced upon paleontologists in the absence of concrete proofs. The approach employed in those wind-tunnel studies was a major advance over simple armchair speculation in biomechanics. Models were made, and theories tested, using the laws of engineering and physics. Yet all we produce in the end is a plausible likeness to a modern machine. There is nothing like the stegosaurs' plate devices in living animals, so straight biological parallels cannot possibly be drawn.

What we have is a proposed function for the plates, with a certain level of probability based on their engineering design, but no way of proving it. Of course it is worthwhile to recall that, in theory, bumble bees are too fat to fly with such small wings, and dolphins should never achieve the speeds they do within the terms of our mechanical understanding. With our limited design abilities, we cannot match evolution.

Life in the Jurassic seas

Typical animals of the Jurassic seabed would be familiar today. Hexacorals formed reefs, just as they still do, and abundant bivalves burrowed in soft sediments and built up oyster reefs. Gastropods such as whelks and limpets moved slowly about, seeking prey, together with starfishes and sea urchins. Sea lilies, stalked crinoids, grew in abundance, especially in the early Jurassic, while giant specimens colonized every available hard surface, including pieces of driftwood. Sponges and bryozoans, sea pens, and sea fans grew, and even the crab–lobster group put in an appearance, with the first lobster-like animals coming on the scene, although these important marine predators did not really take off until the Cretaceous.

Some significant changes were happening among the pelagic animals, swimming in open seas. Two new groups of mollusks, the ammonites and belemnites, arose in the Jurassic. Both had close relatives that had virtually died out in the end-Triassic mass extinction, a few squeezing through to the Jurassic to spread into these two major groups. Both groups are cephalopods, distantly related to the modern octopus, cuttlefish, and *Nautilus*.

The belemnites had an internal calcite skeleton, shaped in the main like a rifle bullet, and often of smaller size. This skeleton was the typical mollusk shell that usually lies outside the animal but in this case had migrated inside. The bullet-like portion was made from a single calcite crystal and acted as a counterweight to keep the animal level while it was swimming. Its fossils litter the marine deposits of the Jurassic and Cretaceous, and are used for dating the rocks in some places. They were frequently found long

before the dawn of paleontology in the late eighteenth century, and were called "devil's thunderbolts." When a complete belemnite is found, the bullet turns out to be only one element of a more complex internal shell, a delicate structure divided into chambers and tapering to fit neatly on to the blunter end of the bullet. No doubt the belemnite did what cuttlefish do with their chambered shells, pumping liquid in or out to maintain buoyancy at a given level. Some near-perfect fossils show that the shell was covered with a muscular fleshy coat that followed the shape, and that the head end looked very much like a cuttlefish: round head, sharp jaws, large eyes, and numerous tentacles equipped with various hooks for seizing prey.

The ammonites were more plentiful and varied than the belemnites. They had coiled shells, chambered in the usual cephalopod style, and used for buoyancy control. The animal lived in the final outside chamber, using the standard kit of eyes, jaws, and tentacles, and apparently able to withdraw them behind a pair of plates that could close up to seal the shell like a pair of doors. Ammonites ranged in size from less than $2/5$ in (1 cm) in diameter to a full $6^1/2$ ft (2 m); the real giants look like tractor tires and come from the latest Jurassic.

Ammonite species are distinguished by a broad range of intricate patterns – ridges, bosses, spines, keels, and so on – on the outside of the shell, and by the involved shapes of the margins of the plates that separated the internal chambers, visible as "sutures" on the outer wall of the shell. It takes a dedicated ammonite fancier to identify the hundreds of short-lived species produced throughout the Jurassic and Cretaceous by this fast-evolving group. They are used extensively in dating the marine rocks of these periods, with a level of precision often as good as a quarter of a million years – pinpoint aiming, in rocks that may be over 150 My old.

Major developments also took place in the Jurassic among the bony fishes and the sharks. The holostean bony fishes were still important, but now a new group, the teleosts, started to spread. Today the teleosts dominate the seas, with 20,000 species classified into 40 orders. They range from salmon to seahorses, tuna to tench, eels to anglerfishes, and they appear to owe at least a part of their success to

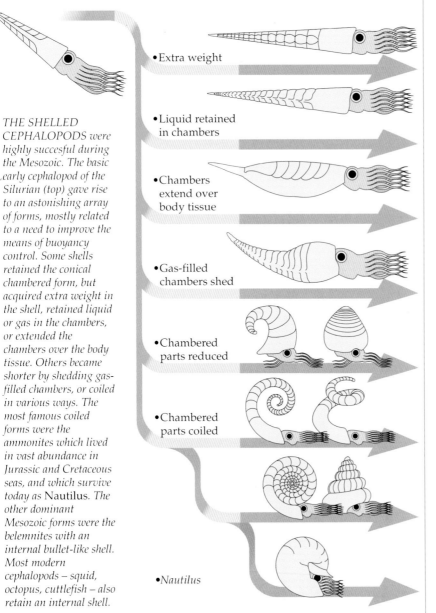

THE SHELLED CEPHALOPODS *were highly succesful during the Mesozoic. The basic early cephalopod of the Silurian (top) gave rise to an astonishing array of forms, mostly related to a need to improve the means of buoyancy control. Some shells retained the conical chambered form, but acquired extra weight in the shell, retained liquid or gas in the chambers, or extended the chambers over the body tissue. Others became shorter by shedding gas-filled chambers, or coiled in various ways. The most famous coiled forms were the ammonites which lived in vast abundance in Jurassic and Cretaceous seas, and which survive today as* Nautilus. *The other dominant Mesozoic forms were the belemnites with an internal bullet-like shell. Most modern cephalopods – squid, octopus, cuttlefish – also retain an internal shell.*

- Extra weight
- Liquid retained in chambers
- Chambers extend over body tissue
- Gas-filled chambers shed
- Chambered parts reduced
- Chambered parts coiled
- *Nautilus*

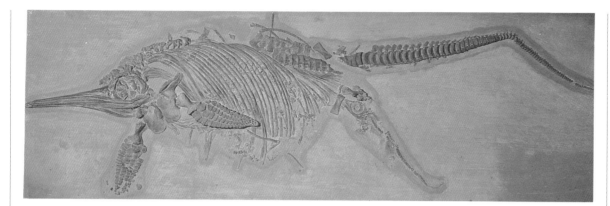

An icon of the Jurassic, and one of the most astounding fossils of all time. Here a female ichthyosaur, about 10 ft (3 m) long, is fossilized apparently in the process of giving birth. Three tiny skeletons may be counted inside her rib cage, and a fourth lies outside her body. This is one of fifty or more mothers with embryo young that have been found in the extraordinary early Jurassic limestone quarries of Holzmaden, southern Germany.

improvements in their jaw apparatus, compared with that of the holosteans.

The jawbones of a teleost are jointed in such a way that when it opens its mouth the whole mouth shifts forward and forms a short tube. When the mouth closes, the mobile upper and lower jaws snap rapidly back into place. The advantages of this system are that it allows teleosts to manipulate food items and it provides the fish with an effective suction device when the jaws first open. Once it has taken in the food, a teleost is also more likely to retain it, because the collapse of the tube produces a further surge of inward suction. Holosteans lack this system, so food often falls out again before the jaws can fully close. The suction system has also enabled teleosts to exploit wholly new ways of feeding. They can pick up pieces of food from the seabed, and some have specialized, for example, in nibbling corals. Other teleosts will hold their mouths wide open and wait for prey to come in range; simply closing the mouth then sucks in the helpless victim.

Jurassic teleosts include the long-snouted aspidorhynchids and the small leptolepids. These already possessed the fully symmetrical tails of modern teleosts, and lighter scales than the ancestral holosteans. They were fast-moving and elusive fishes, and their evolution led to sweeping changes among their predators, such as the sharks. These cartilaginous fishes (chondrichthyans) had diverted from the bony fishes (osteichthyans) sometime in the Silurian, about 250 My before. One of the curious effects of cladistic systems of defining animal relationships is that they place the herring closer to the apes than to the sharks, since both apes and herring are direct descendants of the early bony fishes.

Hybodont sharks stayed active throughout the Jurassic, and the varied designs of their teeth indicate that prey types included both fishes and crustaceans. But the key event in these times was the origin of all the modern sharks and rays, classed together as the neoselachians. One crucial advance on earlier sharks is the greater efficiency of the mouth, set back below the snout and able to open wider than a hybodont's. In large neoselachians the mouth, with its rows and rows of teeth, is a lethal weapon for gouging flesh – in fact, as well as in cinema mythology.

The Jurassic neoselachians were large inshore hunters that probably specialized on the new bony fishes and on squid. Later species became fast offshore hunters. Some groups specialized in krill, which they strained out of the water rather as the large modern whalebone (baleen) whales do. Others, and notably the skates and rays, have concentrated mainly on mollusks, which they crush between pavements of tough knobbly teeth.

Jurassic marine reptiles

Reptiles responded to the diversity of new food sources in the seas by spreading and developing. The ichthyosaurs changed least. Jurassic species show subtle differences in the skull and in the shape of the tail, compared with their Triassic ancestors, but they followed the same lifestyle. These dolphin-like reptiles ranged in length from 3 to 16 ft (1 to 5 m), and hunted belemnites and fishes, as we know from their stomach contents and coprolites (fossilized feces).

Remarkable specimens of ichthyosaurs appear in the Posidonienschiefer of Holzmaden and other early Jurassic localities in southwestern Germany, where slate deposits form a vital fossil *Lagerstätte*. These

THE BIOMECHANICS OF VERY LARGE DINOSAURS

Dinosaurs are famous for having been very large. Of course, many were of human dimensions, or even smaller, but most were indeed large. The sauropods were all larger than any living land animal. How could they survive at such size, and what did they do? Had they reached the limits of size?

Size limits for land-living animals may be tested by some simple calculations. (These calculations do not refer to aquatic animals, since physical limitations are different in water.) The crux of the engineering approach is to understand the volume/area effect.

As the volume (or weight) of a land animal increases, the diameter of its legs increases in proportion to volume (a cubic measure), and not in proportion to body area (a squared measure).

This is because the legs have to support the weight of the animal. Hence, legs get relatively fatter the larger the animal.

A graph may be plotted (below right) in which the circumference of the leg bones corresponds to body weight, and indeed this technique may be used to estimate weights of unknown animals (like dinosaurs) from bone dimensions.

The block diagrams (above right) show how leg diameter scales to body volume. At a body weight of about 140

tons (tonnes), the legs are so wide that they touch, and the animal can no longer walk. This is the upper limit to body weight.

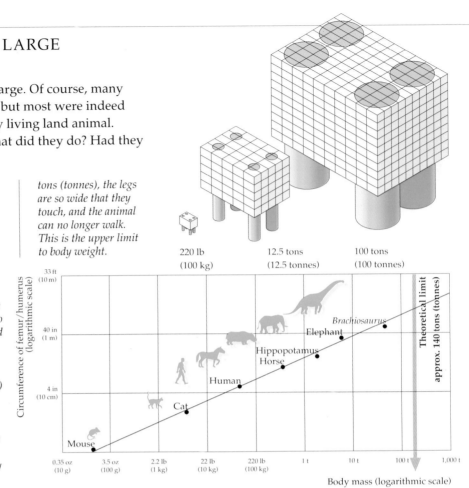

fossils preserve the ghost of the body outline in black carbonaceous deposits that tell us the shape of the parts that lacked bones, such as the upper fin of the tail and the dorsal fin in the middle of the back. There are Holzmaden specimens that preserve embryo ichthyosaurs inside their mothers, two or three young at a time. Some clearly died while giving birth, since the small ichthyosaurs are halfway out of the rib-cage region. This confirms that the ichthyosaurs were so modified for marine life that they no longer came ashore to lay reptile eggs, and the Holzmaden region may even have been a long-term birthing ground. The young were born quite large, and must have been able to fend for themselves from birth, as is broadly the case for dolphins and whales.

Jurassic seas were also populated by crocodiles. There are seagoing crocodiles today, but none is so modified for a marine existence as the late Jurassic geosaurs such as *Metriorhynchus*, 6¹/₂–10 ft (2–3 m) long. This crocodile has a tailfin, paddles, and no armor – a move toward greater streamlining. The geosaurs were a short-lived but important fish-eating group.

The most spectacular Jurassic marine group was the plesiosaurs, which divide roughly into the long-necked plesiosaurs and the short-necked pliosaurs. Both lines arose in the early Jurassic, but they are well known from European marine sediments of the late Jurassic. The plesiosaur *Cryptoclidus* was 10–13 ft (3–4 m) long and had over 30 vertebrae in its neck – a great increase over the standard 7 or so, but still a long way from the 70 or more in the neck of some Cretaceous forms. These reptiles swam rapidly by beating their large paddles rather like underwater wings, and they fed on shoals of teleost fishes by darting their smallish heads about on their mobile necks, and seizing fishes in their long needle-like teeth.

The pliosaur *Liopleurodon* was much larger: up to 40 ft (12 m) long. It had a short neck and a massive head, and it is probable that it fed on other marine reptiles such as the ichthyosaurs, plesiosaurs, and geosaurs. The pliosaurs amount to a new level in the food chain, occupied by a "top" carnivore that preyed on other large carnivores, which in turn fed on fishes, which fed on smaller

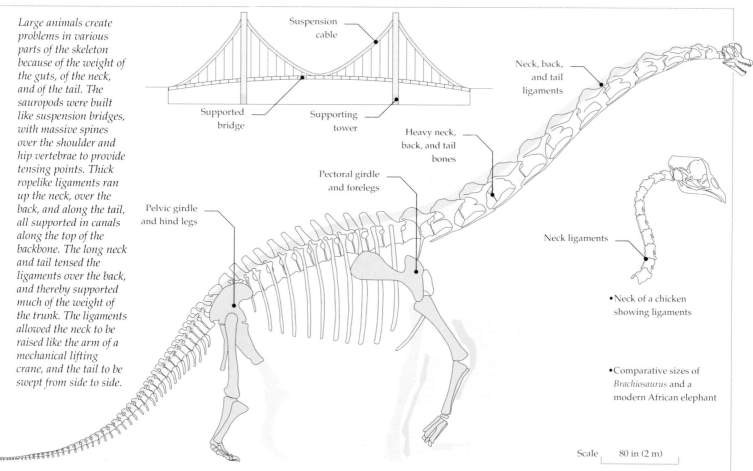

Large animals create problems in various parts of the skeleton because of the weight of the guts, of the neck, and of the tail. The sauropods were built like suspension bridges, with massive spines over the shoulder and hip vertebrae to provide tensing points. Thick ropelike ligaments ran up the neck, over the back, and along the tail, all supported in canals along the top of the backbone. The long neck and tail tensed the ligaments over the back, and thereby supported much of the weight of the trunk. The ligaments allowed the neck to be raised like the arm of a mechanical lifting crane, and the tail to be swept from side to side.

Suspension cable

Supported bridge

Supporting tower

Neck, back, and tail ligaments

Heavy neck, back, and tail bones

Pectoral girdle and forelegs

Pelvic girdle and hind legs

Neck ligaments

• Neck of a chicken showing ligaments

• Comparative sizes of *Brachiosaurus* and a modern African elephant

Scale   80 in (2 m)

creatures, and so on down to the microscopic plankton. They peopled the seas worldwide, and there is no convincing modern rival for the pliosaurs, with the possible exception of killer whales, which feed on seals and other large marine predators.

The giant sauropods

The sauropods ("reptile-feet") included the largest dinosaurs, and therefore the largest land animals, of all time. *Diplodocus* reached lengths of up to 90 ft (27 m). *Brachiosaurus* towered 33 ft (10 m) above the ground. Even larger dimensions have been claimed for species with challenging names like *Supersaurus, Ultrasaurus, Seismosaurus* – suggested lengths of 130 ft (40 m), heights of 50 ft (15 m), and weights of 100 tons (tonnes) and more. None of these last three is complete, and the measurements are scaled up from the available remains – a huge thigh bone, a massive shoulderblade. They are likely to be exaggerated. Dinosaurologists are like anglers in this regard, except that when

dinosaurologists spread their arms out maximally wide, they're probably saying: "And that was only the jawbone!"

The sauropods descended from prosauropod ancestors sometime in the early Jurassic, and no clear line divides the two groups. The earliest widespread species were the cetiosaurids, known from the middle Jurassic of England, where they shared a habitat with *Megalosaurus*, and of China (*Shunosaurus, Datousaurus*), but not confined to the Northern Hemisphere. Sauropods had their heyday in the late Jurassic, when they radiated into three main lines, the diplodocids, brachiosaurids, and camarasaurids, many of them extinct by the start of the Cretaceous. A fourth group, the titanosaurids, rose to prominence in the late Cretaceous, mainly on southern continents.

Truckloads of flesh arrayed around girders of bone, ten times the size of a typical elephant – how big could the sauropods get? Theoretical calculations have shown that some of the wilder estimates of length and weight noted above may approach the uppermost limit. The calculations were based

Herds of herbivorous **Brachiosaurus** *inhabited both East Africa and midwestern North America. These behemoths have close relatives in all other parts of the world, suggesting that they may have undertaken long-distance migrations in search of food. With a long neck and raised forequarters reminiscent of the giraffe,* Brachiosaurus *probably browsed on the leaves of tall trees.*

on the simple and provable fact that the diameter of an animal's legs is proportional to the animal's weight, so that lightweight animals like gazelles can have very thin legs, while heavy animals like elephants must have proportionally much thicker legs. The diameter, a linear measure, is proportional to the weight, which is a cubic measure: increase an animal's length by a little, and you increase its weight quite a lot; increase the weight by a lot, and you increase the diameter of the legs in the same proportion. At a weight of 50 tons (tonnes), typical for many sauropods, the leg diameter is already large, and the animals probably could not run fast without the risk of breaking their legs. At estimated weights of 100 tons (tonnes), a sauropod would have had massive pillar-like legs that would keep its top speed to a slow stroll. The limit is reached at about 140 tons (tonnes), when the diameter of the legs is so great that the four legs are locked vertical under the body, unable to move at all.

Dinosaur body weights are estimated from meticulous scale models. The model is suspended in a vessel full of water, and the amount of water displaced gives a measure of volume. The volume of a dinosaur is directly proportional to its weight, of course, but the trick is in the conversion factor. Do 2 pints (1 liter) of dinosaur weigh 2 lb (1 kg) or less, and how much less? The conversion factors used range from 1.0 down to 0.7, a typical value for a live bird, because birds have many hollow bones that are filled with air, and this reduces their overall weight–volume ratio. Sauropod dinosaurs too had hollow bones, at

least in the backbone, and these may have been filled with air sacs. What with the problematic conversion factor and the inevitable variations in 3-D models of dinosaurs, the accounts of weight vary hugely. For instance, estimates for one of the biggest dinosaurs, *Brachiosaurus*, found in Utah, Colorado, and Tanzania, range from 15 all the way to 78 tons (tonnes).

The sauropods all had long necks and tails, and small heads with a small number of peglike or spoonlike teeth. As herbivores, they must have fed almost constantly in order to fuel their huge bodies. Their biology has been a controversial topic, with theories that span a variety of lifestyles on land or in the water, and see them as anything from sluggish to highly active creatures. Elmer Riggs, one of the early sauropod specialists, speculated in 1904 that *Brachiosaurus* was a terrestrial animal and noted the similarity of its limbs to those of elephants and rhinos. His view went very much against the prevailing image that viewed the sauropods as slow-moving creatures that squelched about in deep water most of the time, in order to support their huge weight.

For most of the rest of the twentieth century opinion swung to the aquatic theory. Some paleontologists argued that *Brachiosaurus* stood in water 33 ft (10 m) deep, with just the top of its head projecting. (The nostrils are very oddly sited, above the level of the eyes.) Others considered that the sauropods stood in shallower water and gathered in soft waterside plants with their long flexible necks. The deep-water idea had to be dropped when Kenneth Kermack, of University College London, showed that a sauropod could not have breathed with its chest immersed in water more than 40 in (1 m) or so deep, because a tremendous suction force would have been required in order to draw oxygen down to the lungs, acting against the external water pressure, whose effect was to keep them compressed. Kermack referred to the work of the Austrian physiologist Robert Stigler, who experimented with long snorkels in a swimming pool, attempting to inhale air. He found that he had difficulty breathing through a 40 in (1 m) snorkel. When he tried breathing through a 6^{1}/$_{2}$ft (2 m) snorkel, the effort did fatal damage to Stigler's heart, and he died shortly afterward.

Bob Bakker, then of Yale University, challenged the amphibious model of sauropods in 1971, and offered the alternative proposal that they were fully land-going "giraffes" that could use their long necks to reach for foliage high in trees. He envisaged *Brachiosaurus* feeding 40 ft (12 m) above ground level, and the shorter sauropods, *Diplodocus* and *Apatosaurus*, rearing up on their hind legs to feed at heights of 50–60 ft (15–18 m), as high as a six-story building. Bakker's dynamic view of the dinosaur design and metabolism also assigned to the Cretaceous *Tyrannosaurus* speeds of 45 mph (72 kph), and argued that *Triceratops* must have been capable of a gallop as fast as a charging rhino, about 35 mph (56 kph). His version of *Brontosaurus* could manage the "running walk" of an elephant. Most paleontologists have rebelled at these images, which seemed too dashing even for his fairly sedate sauropods, but there is broad agreement nowadays that the sauropods were largely terrestrial animals.

Solnhofen's pterosaurs and birds

Following their appearance in the late Triassic, a number of pterosaurs are known from the early and middle Jurassic of England, India, and Arizona, as well as in the Holzmaden Posidonienschiefer already described. The strongest evidence for major success among these aerial archosaurs,

however, comes from a range of late Jurassic deposits, and particularly from one of the most famous of all fossil *Lagerstätten*, the Solnhofen limestones of Bavaria.

The Solnhofen region, 150 My ago, was a quiet warm-water lagoon lying behind coral reefs on the northern shores of Tethys. It enjoyed a tropical climate, and around the shores were forests tenanted by small theropod dinosaurs, lizards, pterosaurs, and – most remarkable – birds. Outside the reefs, sharks, rays, teleost fishes, turtles, and crocodiles teemed in the Tethys waters, feeding on each other and on the rich fauna of corals, worms, mollusks, lobsters, starfishes, and sea urchins beneath. The stagnant waters of the lagoon itself sheltered very few animals, either on the surface or on the seabed; but occasional tropical storms forced surges of fine sediments in across the reefs, with whatever dead animals they contained. They sank to the bottom again, and were buried beneath the particles that came with them and fell as a soft lime mud.

These anoxic conditions sometimes allowed soft parts to survive, and the fossils include jellyfish. The same well-sifted sediments also entombed carcasses that drifted out from shore, or fell from the skies. These carcasses are now preserved in the most memorable detail. Occasional trace fossils show the tracks left by crayfish or horseshoe crabs swept helplessly into the lethal environment of the lagoon bed and reeling toward graves where they now lie fossilized. With more than 600 species immaculately preserved in limestone so fine that once it was used for lithographic printing, the Solnhofen *Lagerstätte* is one of the world's most precious treasuries of paleontological knowledge.

The small dinosaur *Compsognathus* is known from a single beautifully detailed Solnhofen skeleton. At 2 ft (60 cm) long, it is the smallest known adult dinosaur, with long birdlike hind limbs, strong fingers, and a stomach that contains evidence of its efficiency in pursuit of agile prey, in the shape of the skeleton of its final meal, the lizard *Bavarisaurus*.

Pterosaur fossils are remarkably common here, with hundreds of specimens that belong to eight different genera. *Rhamphorhynchus* and *Scaphognathus* are small animals, about

Archaeopteryx, the world's oldest bird, climbs a treelike cycad on the shores of a tropical, lagoonal sea in southern Germany, around 150 My ago. Archaeopteryx had strong, clawed hands and feet for climbing trees, and its ability to fly probably arose as a means of moving rapidly between treetops. Its feet and legs show features of a dinosaurian ancestry and enabled it to run fast. Perhaps Archaeopteryx pursued insects on the run and used its wings to aid their capture.

OPPOSITE *Adult pterosaurs, members of the genus* Pterodactylus *and contemporaries of* Archaeopteryx *in southern Germany, return to their clifftop nest with fishes for their ravenous young. Baby pterosaurs hatched from eggs and were probably unable to fly at first. They had tiny undeveloped wings, but large heads, and large demanding mouths, just like birds. A few weeks of feeding on highly proteinaceous fishes allowed the baby pterosaurs to grow strong enough to fly.*

the size of a small seagull. They resemble the late Triassic pterosaur forms because they retain a long tail, and soft-tissue impressions record a variety of leaf- or delta-shaped vertical tail vanes that could act in flight as rudders.

Other fine impressions show that *Rhamphorhynchus* had a throat pouch for storing fishes, which were the standard diet of the Solnhofen pterosaurs, probably caught by swooping to snatch them from the surface waters of the sea. However, the broad jaws of *Anurognathus* suggest that it caught insects on the wing – an aeronautic feat – and the Solnhofen fauna obligingly offer some possible victims: cicadas, dragonflies, wood wasps, and many others. None of the rhamphorhynchoids, the long-tailed pterosaurs, survived the Jurassic, but in episodes untold by the fossil record they had already given rise to their successors. *Pterodactylus*, also found in East Africa, England, and France, is the first of a new group of pterosaurs, the pterodactyloids, which had very short tails, larger heads, and longer necks than the rhamphorhynchoids.

How did so many pterosaurs come to die at sea? Not all of them did. Some are found with the neck bones locked into positions often found in birds when they dry out after death. These may already have been dead when they

were washed into the lagoon; their bones are jumbled, and often incomplete. Where the whole skeleton remains intact, we can only assume that its owner was caught by a change of weather, and that it was blown or beaten out of the sky.

Pterosaurs had a fine covering of hair over their bodies, as is shown by some Solnhofen specimens, but even better by new material from Kazakhstan. Fur is a reliable indicator, and most paleontologists accept that the pterosaurs were warm-blooded. This would provide the high metabolic rate that was necessary for an active flying lifestyle. There is no doubt that pterosaurs were efficient flapping fliers, although some of the later Cretaceous forms were so huge that gliding seems more likely. It has also been argued recently that they could have run about on the ground like small dinosaurs, with their wings folded, rather than fumbling along as a grounded bat has to do. The old belief that the pterosaurs could not fly very well is not tenable – why have such highly evolved wings if you can't fly? – but the bipedal running theory is contested.

The undisputed star of Solnhofen, now known from six skeletons and a single loose feather, is *Archaeopteryx*. The fossils were found between 1860 and 1988, and no doubt many more still lie waiting to be discovered. The pigeon-sized skeleton has broadly the shape of a small theropod dinosaur's, with theropod features like a long bony tail, claws on its fingers, and teeth in its jaws, and some of the specimens were labeled as such, until the feathers were noticed. According to most paleontologists, the possession of feathers makes *Archaeopteryx* a bird.

The origin of feathers is a puzzle that *Archaeopteryx* does not solve, since its own are fully modern. Feathers, like hairs and reptilian scales, are made from the tough protein keratin, so they probably derive from a theropod dinosaur that happened to develop "ragged" scales and then retained them, probably as a form of insulation. After that, longer feathers on the forearm may have helped the wearer to leap after prey, or to escape. Most evidence suggests that *Archaeopteryx* was a tree-climber, using the sets of claws on both its "hand-wing" and its toes, and that flight arose as a development from gliding, which would have enabled the

FLIGHT STYLE

Birds are so well adapted for flight that it might seem impossible that they could have evolved from flightless reptiles. Cladistic analysis of modern and fossil reptiles, however, gives a plausible picture. The closest relatives of birds are the dinosaurs, and in particular the theropod (flesh-eating) dinosaurs. Among theropods, some small agile bipedal dinosaurs, the dromaeosaurids, known from North America and Mongolia in particular, come very close indeed. They share with birds the three-toed foot, with a fourth reversed toe at the back, long slender hind legs, and long arms equipped with three long fingers. A detailed comparison of the skeleton of *Dromaeosaurus* (a dinosaur) with *Archaeopteryx* (a bird) shows how close the two groups are.

Archaeopteryx is, of course, more dinosaur-like than any living bird, having just evolved from the dinosaurs. *Archaeopteryx* retains the long slender tail, heavy skull, teeth, and fingers of its immediate ancestors. These have all been lost in modern birds.

It was once debated whether such a reptile-like bird as *Archaeopteryx* could have managed to fly. For example, *Archaeopteryx* does not have a deep breastbone, which, in modern birds, is the site of attachment of the powerful flight muscles (see main illustration below). Without a breastbone, surely it lacked

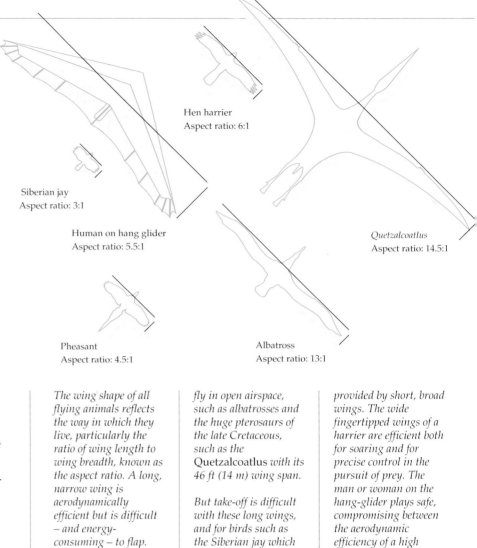

Siberian jay
Aspect ratio: 3:1

Human on hang glider
Aspect ratio: 5.5:1

Pheasant
Aspect ratio: 4.5:1

Hen harrier
Aspect ratio: 6:1

Quetzalcoatlus
Aspect ratio: 14.5:1

Albatross
Aspect ratio: 13:1

The wing shape of all flying animals reflects the way in which they live, particularly the ratio of wing length to wing breadth, known as the aspect ratio. A long, narrow wing is aerodynamically efficient but is difficult – and energy-consuming – to flap. Long wings are therefore most suitable to animals that live and

fly in open airspace, such as albatrosses and the huge pterosaurs of the late Cretaceous, such as the Quetzalcoatlus *with its 46 ft (14 m) wing span.*

But take-off is difficult with these long wings, and for birds such as the Siberian jay which live in wooded country, rapid, powered lift and maneuverability are

provided by short, broad wings. The wide fingertipped wings of a harrier are efficient both for soaring and for precise control in the pursuit of prey. The man or woman on the hang-glider plays safe, compromising between the aerodynamic efficiency of a high aspect ratio and the maneuverability of the shorter wing.

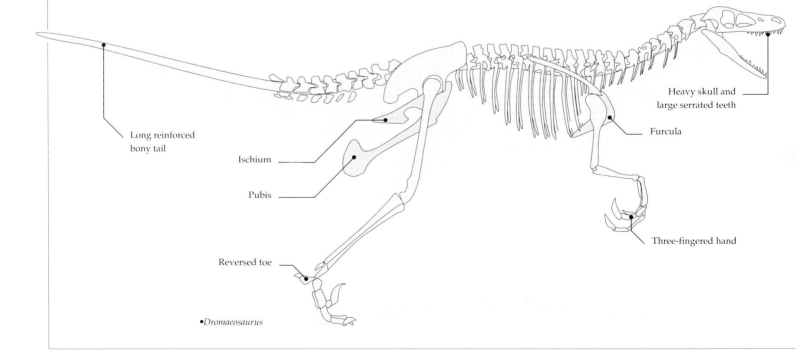

Long reinforced bony tail

Ischium

Pubis

Reversed toe

•*Dromaeosaurus*

Heavy skull and large serrated teeth

Furcula

Three-fingered hand

the muscles needed to beat its wings? The whole question now seems rather ludicrous; after all, why should *Archaeopteryx* have had feathers and wings if it could not have flown? Indeed, the feathers are those of a flier. Further, modern bats do not have a deep breastbone, and yet they fly perfectly well.

The other debate, which is not resolved, concerns whether birds evolved flight from the trees down (i.e. by leaping from branch to branch, and flapping as a means to leap yet further) or from the ground up (i.e. by running and hopping after insect prey, and flapping to achieve greater leaps). The former idea seems most likely; *Archaeopteryx* has strong wing claws which could have been used in climbing.

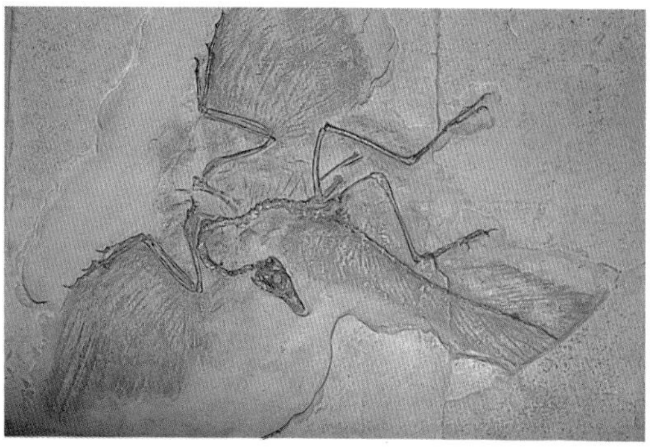

This famous fossil of Archaeopteryx was found in the 150-My-old Solnhofen limestone in Bavaria, southern Germany. It is plainly an animal that could fly. The wing feathers of modern flightless birds are symmetrical about the central quill, but feathers that are used to create lift have narrow leading edges and broad trailing edges,. That is the sort of feather which this and other equivalent fossils have.

The Solnhofen limestone preserves exquisite fossils such as this, and is one of the classic Lagerstätten deposits. Indeed, its properties are so exceptional that creationists have claimed that the Archaeopteryx fossils are forgeries, produced by pressing modern feathers and bones into a paste of plaster, which then set to look like a fossil. This assertion is clearly falsified by microscopic examination of the feathers and by the presence of hundreds of such soft-bodied Solnhofen fossils: jellyfishes, worms, fish guts, and so on.

•*Relative size*

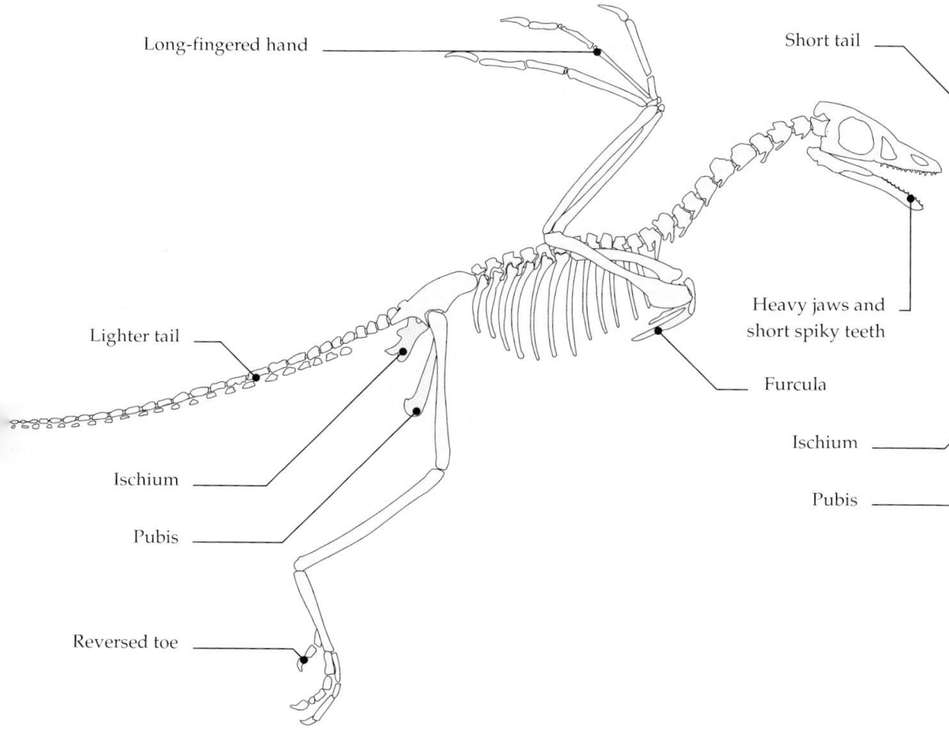

Scale | 20 in (50 cm)

Long-fingered hand

Short tail

Reduced and fused fingers

Light skull and no teeth

Reinforced shoulders

Lighter tail

Heavy jaws and short spiky teeth

Furcula

Ischium

Furcula or wishbone

Pubis

Ischium

Large breast bone

Reversed toe

Pubis

Reversed toe

Strengthened pelvis to withstand impact of landing

•*Archaeopteryx*

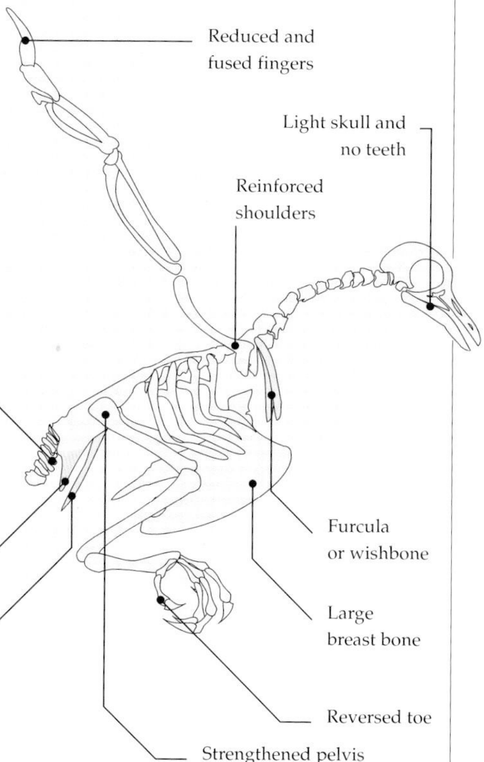

•*Columba* (the modern pigeon)

proto-bird to move rapidly about the upper story of forests, without having to touch the ground. The wing is essentially much the same as a modern bird's, so there is no evidence to support the early notion that it was a poor flier.

Recent close studies of archosaur physiology show that birds are most closely related to small theropods like *Compsognathus* and the early Cretaceous *Deinonychus*, but the timing of bird origins is not at all clear. Both groups have a suite of features in common, as well as a number of obvious differences that must have developed after the point where they diverted from common stock, and some cladograms date that common point to sometime in the middle Jurassic, at the latest. However, a supposed bird has been reported from the late Triassic of Texas. It has yet to be solidly confirmed as a bird, and the cladograms make the identification look unlikely, because it would hugely distort the otherwise fairly coherent family tree of the dinosaurs, and of the rest of the birds. Huge gaps would appear where the common ancestor had left no trace of descendant bird or theropod fossils for tens of millions of years. On top of that, this same common ancestor, arising so early, would have to have been extraordinarily advanced in design over its dinosaur contemporaries.

Following the frog newcomers in the Triassic, the other two main modern groups of amphibians had appeared in the Jurassic. The newts and salamanders (order Urodela) are now known from the middle Jurassic of England, following discoveries reported in 1990. The Mesozoic now had all but a full cast of modern tetrapod groups, following the frogs, turtles, crocodilians, and mammals

recruited in the Triassic. Faunas and floras of both land and sea had grown immensely rich and diverse.

Yet continental shifts in the Cretaceous were to bring still more favorable climatic conditions and a greater variety of habitats. Although no change can be universally favorable to life, the end of the Jurassic announced flourishing new times.

Mesozoic countdown

The last episode of the Mesozoic era was the Cretaceous period, which lasted from 144 to 65 My ago – longer than the whole of the succeeding and as yet unfinished era, the Cenozoic. The Cretaceous is the longest single period of the entire Phanerozoic, the eon of "visible life." It kept the mammals alive for 80 My, created a new flora, and insect groups to coexist with it, established the Atlantic Ocean, decisively broke up Gondwana, and pushed most of the continents into positions not too perplexing to a modern eye.

Efforts to grasp the pattern of Cretaceous events tend to be complicated, if not contaminated, by the inseparable privilege and handicap of hindsight. In its light, the period can be seen as a time of decline for the typically Mesozoic life forms – ammonites, belemnites, gymnosperms, ichthyosaurs, plesiosaurs, pterosaurs, and dinosaurs. This view is misleading. All of these groups radiated and diversified during some or all of their long afternoon, and toward the end of the late Cretaceous they showed a variety of patterns of extinction.

Continental movements and climates

Huge thicknesses of readily accessible sediments have a lot to tell us about the history of Cretaceous times. The most famous are the deep chalk beds laid down during the later Cretaceous in many parts of the world, including Kansas and the Gulf coast of the United States, and also stretching along the south coast of England and into northern Germany and Denmark.

Chalk (Latin *creta*) is a pure limestone made up almost wholly of the tiny calcareous platelets of microscopic planktonic plants called haptophytes. These have a free-swimming active stage as golden algae, and a resting stage in which their single cells construct intricate suits of micro-armor out of a varying number of calcium carbonate scales – coccoliths – each about $1/_{5,000}$ in (0.005 mm) in diameter. They floated in the surface waters of warm seas before sinking as billions of geometrical corpses to the bottom, not more than 10,000–13,000 ft (3,000–4,000 m) below. (Not more than that, because at greater depths the dissolved carbon dioxide makes a weak acidic solution strong enough to dissolve their calcium carbonate skeletons.)

During the Cretaceous, sea levels were rising continuously, higher than ever before or since. As the continents moved further apart, shelf and platform seas formed in between and blanketed areas that had once been land. Former desert areas became flood plains, and by the end of the period as much as two-fifths of the world's continental area lay under shallow water.

Temperatures were also changing, rising to a high about 100 My ago. After the mid-Cretaceous came a slow decline that accelerated until in the last few My of the period mean annual temperatures in the North American midwest fell from 68°F to 50°F (20°C to 10°C).

Rifting in the early Jurassic had split Europe, Africa, and the two Americas, but they had stayed fairly close together. Now they were drifting apart. India and Madagascar were moving away from the east coast of Africa. Antarctica and Australia, still joined together, began to drift eastward and to lose contact with South America. The new seaways created included the early North and South Atlantic, the Caribbean Sea, and the Indian Ocean. A major sea, the Interior Seaway, reached from northern polar waters southward through northern Canada down to Mexico and the Yucatán peninsula. Another seaway slanted through Africa across the mid-Saharan region, and the Tethys Sea extended from its former limit in southern Europe to cover all of the British Isles, central Europe, southern Scandinavia, and European Russia.

The effect was to divide the Earth into twelve or more major isolated landmasses, and now there is evidence for the development of distinct floras and faunas, endemic populations of land plants and animals left isolated on individual island continents of the late Cretaceous and evolving to create much of the diversity of land life today. In the bygone supercontinent of Pangaea, by contrast, a herd of big plant-eating dinosaurs consumed so much vegetation that it had to keep moving like a scavenging horde, with carnivore raiders harrying its fringes. It might cut a long swath across the supercontinent, leaving forests and fern prairies to recover in its wake.

Extensive Cretaceous coalfields were laid down at what were then latitudes of around 50 degrees and upward, both north and south. All along the Tethys Seaway, microplankton multiplied between about 120 and 75 My ago, to be buried in shallow shelf seas in anoxic sediments where they could not decay. They turned into oil, and rather more than half of the world's known oil reserves lie in Tethyan fields of the Persian Gulf and North Africa, the Gulf of Mexico, and Venezuela.

Dinosaurs of the English Wealden

The richest evidence for early Cretaceous land life comes from the Wealden of southeast England and the Isle of Wight – not only dinosaurs, but crocodiles, turtles, mammals, insects, and plants, as well as fishes. The dominant dinosaurs were the ornithopods, a group that had arisen in the late Jurassic. There were few sauropods here, and the age of the herbivore giants had clearly passed.

The Wealden was then a region of ferny plains cut by tree- and cycad-lined watercourses draining into marshes and lagoons. The staple large ornithopod was *Iguanodon*, a biped 33 ft (10 m) long with a horselike skull, long jaws, and the eyes set well back. It seems to have prospered because of its feeding abilities. The jaws are lined with ranks of replacing teeth that kept up a constant grinding rasp. There are no teeth at the front of the jaws, only a bony pad, but this device – shared with modern sheep – is a precision tool for nipping off leaves and other vegetation. The real advance was the ability to chew. The first of the ornithopods,

OPPOSITE *The Wealden of southern England in the early Cretaceous heyday of the ornithopod plant-eaters shows* Iguanodon, *the dominant dinosaur in much of Europe. It had a powerful array of grinding teeth and chewed by flapping its cheeks in and out as it moved its jaws. The spiked thumb may have been for defense or for pre-mating contests between males.*

Dinosaur mating can be imagined in simple, unarmored bipeds such as these; the only problem is to get the female's tail out of the way. Mating in Stegosaurus *is impossible to reconstruct, since they had sharp spines all the way down their backs.*

such as *Xiaosaurus*, may have had some crude equipment, but *Iguanodon* boasted a highly effective system, which has been worked out by David Norman. The side of the snout formed a separate unit from the rest of the skull, and could pivot along a single line running from front to back of the snout. The teeth were angled in such a way that when the jaws closed on some plant food, they pushed the lateral snout plates upward and outward, and this gave a sideways tearing action across the teeth. *Iguanodon* was able to reduce its food to a digestible mush, unlike the sauropods, which merely swallowed plants whole and relied on a cast-iron digestion.

Iguanodon generally walked on two legs, and this allowed it to reach vegetation high in trees, but it was also capable of walking on all fours. This shows in the hand; most of the claws are actually small hoofs. The thumb-claw is a large pointed spike, so different in design that when it was first found last century Richard Owen reconstructed it as a nose horn. Perhaps it was a weapon, used against predators.

The success of the ornithopods is measured by their great abundance throughout the Cretaceous, and by their diversity in individual faunas. The second most common Wealden dinosaur was *Hypsilophodon*, a smaller animal, 10–16 ft (3–5 m) long, with slender sprinter's legs and a stiffened tail for balance on the run. A whole new group had to be recognized in 1983 with the discovery of *Baryonyx*, a big theropod still known only from a single skeleton whose striking feature is its immense sickle-like claw.

The sickle-clawed meat-eaters

A number of theropods seem to have converged independently on the design of curved claws for disemboweling prey. *Baryonyx* had a claw measuring 1 ft (30 cm) round the curve – and this in an animal 30 ft (9 m) long. Oddly, it is still not clear whether the claw was attached to the hand or the foot.

It turns out that *Baryonyx* has roughly the body of a mainstream flesh-eater like *Allosaurus*, but the claw of a monster and the skull of a crocodile. The skull is the oddest feature, since the jaws are long and narrow, the exact opposite of the design in other theropods of similar size, which have deep jaws for crushing bones and tearing flesh. Perhaps the namers of *Baryonyx*, Angela Milner and Alan Charig of the Natural History Museum, were right when they suggested that it was a fish-eater; after all, fish scales were found in the rib-cage area, and modern bears use their massively clawed hands to scoop fishes out of the water. Did *Baryonyx* lurk in the Wealden undergrowth and flick fishes out of the rivers, or did it stand stock-still in the water and spear big fishes like a heron with its bill?

The slightly comic image presented by this 2-ton (tonne) angling thug is stressed by a comparison with other clawed dinosaurs. The dromeosaurids and troodontids of the Cretaceous of North America and Mongolia were generally smaller animals with smaller claws, but they used these to tear away at relatively large prey animals. The best-known of the dromeosaurids is *Deinonychus*, which was excavated from the Cloverly Formation of Montana by John Ostrom of Yale University in the 1960s. *Deinonychus* was about 10 ft (3 m) long and 40 in (1 m) tall, and it weighed between 130 and 165 lb (60 and 75 kg). The skeleton is completely known, except for the skull.

A central point in Ostrom's study of *Deinonychus* is evidence that such bipeds were swift in pursuit and attack. He noted that the tail was bound by long bony rods that kept it from bending. Out went the older reconstructions of such bipeds bounding like kangaroos, with the backbone close to vertical and the tail dragging on the ground. Three lines of evidence show that *Deinonychus* held its tail and backbone virtually horizontal:

1 This posture gives perfect balance, unlike the tail-heavy kangaroo mode.

2 The vertebrae in the neck have a permanent S-shaped curve which would leave *Deinonychus* staring forever skyward in kangaroo position.

3 There appear to have been strong ligaments holding the back stiff and horizontal, as in birds.

The small carnivore Deinonychus *from the early Cretaceous of Montana made up for its small size with a slashing toe-claw. These dinosaurs probably behaved like pack-hunting dogs; they cooperated in singling out a large herbivore – in this case the ornithopod* Tenontosaurus *– from its fellows and then slashed and harried the animal until it fell exhausted.*

Deinonychus had muscular shoulders and strong grasping hands with three long-clawed fingers, but the feet are the key feature. There are four toes, all of them clawed, but the second equipped with an outside scythelike claw on special joints that enabled it to be held well clear of the ground while running, so as not to grow blunt. *Deinonychus* may have hunted in packs. When they attacked a large ornithopod they could leap through the air and slash down with the claw or claws, perhaps while clinging on to the victim with their sharp hand-claws. The foot-claws could swing through an angle of 180 degrees and produce a gash more than 40 in (1 m) long.

Theropod dinosaurs like *Deinonychus* have turned out to offer important insights into an elusive problem, the origin of the birds. In the 1970s most paleontologists were vague about bird origins and preferred to evade the question. Ostrom noted the close similarities to types like *Deinonychus*, and resurrected an older notion that the birds are really just two-legged meat-eating dinosaurs with wings. Recent cladistic analyses have strongly backed this view. In fact *Deinonychus* shows so many similarities to *Archaeopteryx* that parts of the skeleton are almost identical except for their size.

The rise of the flowering plants

Few changes have affected the landscape and ecology of the Earth more than the arrival of the angiosperms, or flowering plants. During the early Cretaceous gymnosperms continued to dominate the floras of the land, but the cycads and ginkgoes were in severe decline, the bennettitaleans heading for extinction, and only the conifers were really prospering. The angiosperms – flowering plants, hardwood trees, and very much later the grasses – arose sometime in the mid-Cretaceous, about 100 My ago. They spread and diversified steadily, so that today there are at least 250,000 species of angiosperms, compared to only 550 of conifers.

Unlike the gymnosperms, which bear naked seeds, the angiosperms enclose their seeds inside an ovary. This protects them from fungal infection, drying up, and unwelcome attentions from insects. The associated flower is the most obvious characteristic of angiosperms, with its extraordinarily inventive range of color, smell, and shape, sometimes inconspicuous, sometimes a flare of scent or brilliance. It is crucial to the mode of reproduction, which shows several advantages over the gymnosperms. Petals are the most familiar feature of flowers, a clear signal to insect pollinators, but many angiosperms (most notably grasses) have less typical flowers, and there is no single design. The essential reproductive parts are located in the middle, and consist of the carpel, a bottle-shaped structure with the ovary and contained ovule (the seed precursor), at the base and the stigma (pollen receiver) at the top. Around this female part, which is the megasporophyll, lie the anthers, the male microsporophylls, a series of thin stalks capped by pollen-bearing structures at the top.

The angiosperm ovule is efficiently protected from damage, deep inside the ovary. This is the last stage in the series of plant developments that started with free-swimming spores in the early primitive plants. The pollen grain is also the final version of the male gametophyte that once grew as a separate plant. It no longer releases sperms to make their way to the ovule over moist surfaces. As with the conifers, it now grows a pollen tube that delivers the sperm directly to the ovule, but in the case of the angiosperms a special surface is designed to receive the pollen grain and make room for the tube that reaches to the ovule.

Another advanced feature of the angiosperms is the mechanism that is often called double fertilization. There are two nuclei at the head of the pollen tube, not a single male nucleus as in the sperm of lower plants. The first nucleus fuses with the ovule nucleus in the embryo sac, which is the ultimate version of the primitive female gametophyte that once also grew as a separate plant. At the same time the second nucleus fuses with another nucleus in the ovary, and divides rapidly to become a food supply for the developing embryo once it is shed as a seed. This is a very economic design because it means that the plant invests very little energy, beyond making the ovary, until fertilization is guaranteed. (By contrast, the female cone of a conifer represents a considerable investment, fertilized or not.) The flower now withers away, and the seed matures, holding both embryo and food supply, in all sorts of forms – for instance, the edible parts of a pea or nut, or inside or outside a wide range of fruits, "false fruits," berries, capsules, and all kinds of devices that enable the seed to hitch a ride from passing animals or glide or float through the air.

The oldest angiosperm fossils date from the middle of the early Cretaceous, about 130 to 120 My ago. They are grains of pollen called *Clavatipollenites*, from the Wealden of southern England. Pollen and leaves become more common around 120 to 100 My ago, in eastern North America, Russia, and Israel. Some remarkably preserved early flowers and fruits have also been recovered from these areas, and appear to be related to modern magnolias and sycamores.

The earliest habits seem to have been areas where streams or floods kept the ground disturbed and fast-growing weedy herbs and small shrubs could gain a foothold, especially if they could set seed quickly. One problem for the conifers was that following fertilization they needed to overwinter before releasing their seeds in the following year. Pines take two years. Perhaps in the wake of a migrant dinosaur herd, weedy opportunism was also encouraged in the soils that they trampled and fertilized.

In looking at the possible co-evolution of dinosaurs and plants, a classic chicken-and-egg dilemma appears. According to Bob Bakker, Jurassic plants developed defenses against high-feeding dinosaurs such as the sauropods and stegosaurs (which he sees as standing on the tripod of their hind legs and tails, cropping at trees). The conifers developed spines, poisons, or unpleasant tastes to deter high-level destruction, but did not need ground-level protection for their seedlings. The extinction of the high feeders and rise of all kinds of efficient low-feeding species now put a premium on the ability to grow and set seed in rapid time. Angiosperms could do that; conifers could not; and so the new dinosaurs had "invented flowers"! Other experts view the process the other way round, with the spread of the angiosperms encouraging the specialist low feeders at the expense of high feeders whose methods took in less nutritious wholesale foods, twigs and all, and devoted far more energy to processing them than was required by more efficient eaters such as the ankylosaurs, ceratopsians, and others.

Do the up-and-coming angiosperms outgrow and outnumber the slower-developing, slower-reproducing seedlings of the conifers and cycads? Or do the fast-evolving ornithischian plant-eaters create the conditions for these speedy gaudy newcomers to radiate? An answer to that would require accurate statistics for the early history of the angiosperms and a census of dinosaur species and their populations for 100 My and more. Even then we would need to know more about soft tissues – seeds and young shoots, tongues and digestive tracts – than the fossil record may have to tell. Circumstantial evidence is always likely to support the more persuasive advocate, but making it stick and getting it right may be very different stories.

During the late Cretaceous, the angiosperms diversified further, until by the end of the period about 50 modern families (out of a total 500) had made their appearance, among them beech, birch, fig, holly, magnolia, oak, palm, sycamore, walnut, and willow. The commonest fossils are leaves, and these have enabled paleobotanists to track the way that angiosperm designs developed, and to draw conclusions about the climatic conditions that produced them. At first most leaves were small, with simple smooth-edged outlines and irregular veins. By the start of the late Cretaceous there were many more leaves with toothed or wavy edges (technically, "serrate" or "lobate"), and the vein patterns were much more regular.

Plants before the angiosperms: a model of a late Jurassic fern that lived in dry soils and formed part of the diet of many dinosaurs.

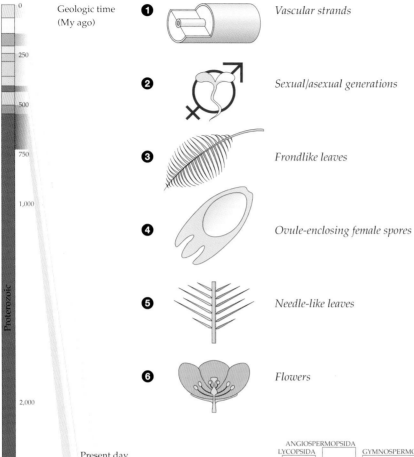

① *Vascular strands*

② *Sexual/asexual generations*

③ *Frondlike leaves*

④ *Ovule-enclosing female spores*

⑤ *Needle-like leaves*

⑥ *Flowers*

PLANT PHYLOGENY

The simplest living organisms, such as viruses, blue-green algae, and bacteria, are shown on this chart, even though they are generally now classed in distinctive kingdoms, equally separate from plants and animals. This is also true for the Fungi.

Plant evolution is best known from terrestrial records. The first plants moved on to land during the Silurian period, with simple small forms like *Cooksonia* that had vascular conducting strands to carry water and nutrients through the stems. A variety of club mosses, horsetails, and primitive "ferns" radiated in the Devonian, and some became large and almost treelike.

The Carboniferous was dominated by giant club mosses, horsetails, cycads, ginkgoes, and primitive conifers. These groups continued through the Permian, Triassic, and Jurassic, but declined thereafter. In the early Cretaceous the flowering plants, the angiosperms, emerged to dominate much of the Earth.

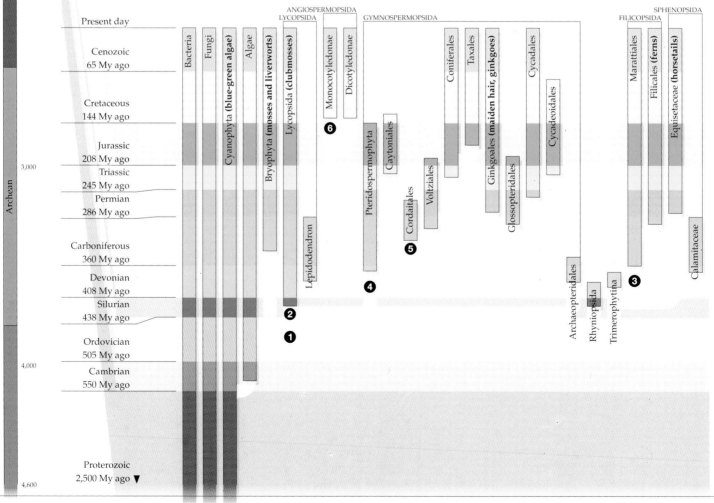

DARWIN'S FLOWERS

Charles Darwin (1809–82) published the book usually called *The Origin of Species* in 1859. (Its full title is *On the Origin of Species by Means of Natural Selection, or the Preservation of Favoured Races in the Struggle for Life*.) For some time he had been troubled by illness, possibly picked up during his voyage on board HMS *Beagle* in 1831–6, and by fits of deep anxiety. Thomas Carlyle called his book "A Gospel of Dirt." The Duke of Argyll pronounced: "I venture to think that no system of Philosophy that has ever been taught on earth lies under such a weight of antecedent improbability." Despite these attacks on him, Darwin continued his scientific work unabashed.

The precise observation of plants offered an area in which theory could be tested against the workings of the natural world. Darwin became fascinated by orchids and their fertilization by insects, as well as focusing on the ways plants climbed, the appearance of different sorts of flowers on the same plant, and the plants that ate insects. For days he would lie in his garden watching the

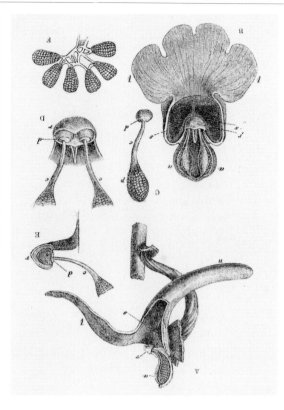

LEFT *One of Darwin's own illustrations from his work on* "The Various Contrivances by which Orchids Are Fertilised by Insects," *1862. These are the parts of* Orchis mascula. *For the first time, Darwin explained the mechanism by which the pollinium is detached from one flower by a visiting bee, which he imitated with a sharp pencil, and is then carried by the bee to fertilize another flower.*

BELOW LEFT *A caricature of Charles Darwin from the London periodical* Vanity Fair *in 1871.*

carnivorous *Drosera*, the sun-dew, catching and eating flies. In 1860 he wrote to his friend, the great botanist Joseph Hooker: "I have been working like a madman on *Drosera*. Here is a fact for you which is as certain as you stand where you are, though you won't believe it, that a bit of hair $1/78,000$ of one grain in weight placed on a gland will cause one of the gland-bearing hairs of *Drosera* to move inwards." This was a marvelously fine-tuned adaptation to provide nitrogen to plants growing on poor soils. The tiniest fleck of matter would trigger the plant's eating mechanism.

Throughout the years in which his theories of natural selection had been germinating, Darwin had pursued his private studies. He would have perceived no gap between his detailed botanical studies and the more general theories. The one relied on the other.

In 1872 Darwin was turned down for membership of the Zoological Section of the French Institute. One of its members alleged that "the science of those of his books which have made his chief title to fame . . . is not science but a mass of assertions . . . often evidently fallacious." Eventually in 1878 Darwin was elected a Corresponding Member of the Institute's Botanical Section.

Average leaf size had also increased. During the same time, angiosperm pollen levels rose from less than 1 percent of the total pollen content to 40 percent – proof of a riotous growth of flowering plants in a span of 20 My.

Leaf shape has been used to reconstruct late Cretaceous climates. Smaller leaves indicate lower temperatures and less rainfall. Features such as large leaf size, serrate edges, thicker leaves, and drip tips (pointed ends to downward-hanging leaves, for fast drainage of water) are associated with humid tropical forests. Wood is another indicator. Tropical

woods with their regular climate are less likely to show growth rings than temperate woods that undergo marked growth seasons in summer and a frozen standstill in winter.

Both wood and leaves document high mean temperatures throughout the late Cretaceous in North America, with a drop followed by a sharp rise about 5 My before the end of the period. Temperatures fell again within the first 5 My of the succeeding Tertiary, before rising again. Such changes produced big geographical shifts among plants. Before the temperatures fell, tropical

CO-EVOLUTION OF PLANTS AND INSECTS

Insects had been around on land for 300 My, and many of them fed on plants, seeking nourishment from the sap and from protein-rich spores or fruits. The origin of flowering plants (angiosperms) during the early Cretaceous seems to have triggered a second great radiation of the insects; new groups, such as butterflies, moths, ants, and bees, arose and flourished. These insects gorged themselves on nectar from the flowers, and they developed in the case of the ants and bees highly complex colonial structures (nests, hives).

The interaction was not all one-way, however. The flowering plants relied on these insect groups for pollination. Indeed, colorful flowers, with their complex arrays of petals and breeding structures in the middle, adapted that form precisely to attract the pollinating insects. Bees can see colors (they can even see them in the ultraviolet range), and the beauty of flowers is solely to attract pollinators. The insect lands on the flower, inserts its body deep into the flower in order to reach the nectar, and in so doing picks up some dustlike pollen (the floral equivalent of sperm) on its hairy back. When the insect enters another flower of the same species, the pollen is wiped off, and this triggers the growth of a pollen tube down the stigma to the ovary (the "egg"), where fertilization takes place. The evolution of flowers and pollinators ran in parallel.

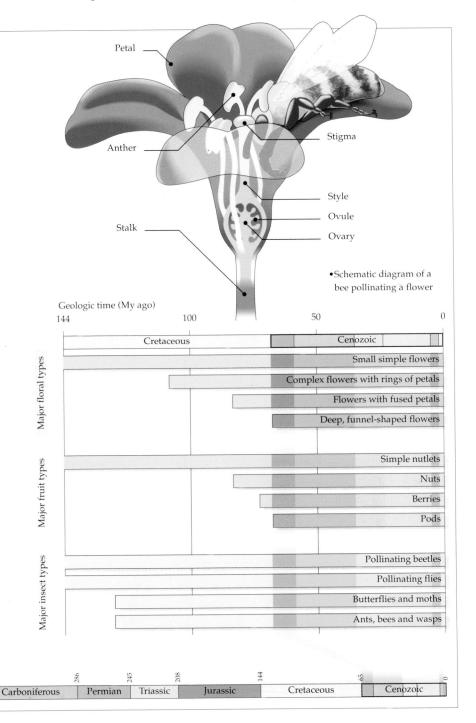

Schematic diagram of a bee pollinating a flower

forests extended north to what is now the United States, while polar broad-leaved deciduous forests covered Canada. The deciduous forests moved south over the Rocky Mountains and the Great Plains region of the United States.

A new career for insects: pollination

Beetles probably pollinated some of the plants thought to be close to the source of the flowering plants, such as the bennettitaleans whose flower-like reproductive organs may have attracted pollinators by odors or pale colors, and therefore "trained" some insects to respond to these lures. Other pollinating insects in the Jurassic and early Cretaceous may have been Diptera (flies such as crane flies and fungus gnats), Hymenoptera (wasps and bees), and even small moths.

Some of the best early Cretaceous insects come from the Wealden of southern England, where siltstone beds may be packed with their carcasses. Ed Jarzembowski of the Booth Museum of Natural History, Brighton, found a concentration of 385 insect specimens in a 20 in (50 cm) square. The quality of preservation is outstanding down to the finest details of wing colors and veining, and eye lenses. There were dragonflies and damselflies, cockroaches, crickets, bugs, lacewings, beetles, scorpion-flies, true flies, caddisfly-like insects, wasps, a termite, and a snake-fly.

Today, the principal pollinators of flowering plants are Hymenoptera. The first to appear in the fossil record were the saw-flies, Xyelidae, which must have been buzzing from the Triassic onward. Some fossil specimens have masses of pollen grains preserved in their guts, sure evidence of their diet. The sphecid wasps that arose in the early Cretaceous have specialized hairs and leg joints that show they collected pollen. Other wasps, the Vespoidea, appear to have arisen in the late Cretaceous; but until recently true bees were only known from the Eocene onward, less than 60 My ago. Although it had been assumed that true bees appeared about the same time as the angiosperms, the first Cretaceous bee was reported only in 1988, found in amber from New Jersey. The preservation shows all details except for the wings, which are badly crumpled. It is a species of *Trigona*, a bee found today all the way from the Amazon Basin to Panama.

Early magnolias were probably pollinated by a range of insects, and had not specialized on one form. More selective plant–insect relationships probably grew up during the late Cretaceous, with the origin of the vespoid wasps, which today pollinate small radially symmetrical flowers. Such closer relationships reward both partners by guaranteeing high-quality food to the visitor and a specialist pollinator to the host. Evidence from the late Cretaceous plant record shows that angiosperms such as roses then had specialized features that catered for pollinators that fed on nectar as well as pollen. Flowers became more and more adapted for one kind of insect pollinator, including bees, during the latest Cretaceous and the Tertiary.

The appearance of termites in the early Cretaceous and of bees and ants in the late Cretaceous signals a key advance in insect behavior. These are the first known social insects – those that live in closely integrated colonies, and which have a division of labor with specialized forms for the various tasks. Typical highly social bees, ants, and termites have a single fertile "queen," a number of fertile males, and huge numbers of infertile "workers" that tend the nest, gather food, provide defense or attack, and care for the young. Such social insects are highly successful to this day, and their origins may be related to the radiation of the angiosperms. Certainly the long relationship between these plant and insect groups has given rise to a range of forms and behaviors so various, and so intricately interlocked, that it illustrates the sheer inventiveness of evolution, given two different gene banks and 100 My.

Revolution in Cretaceous seas

New groups of predators cruised the oceans during the Cretaceous, and their devastating impact has been characterized as the Mesozoic marine revolution. The newcomers included teleost fishes, which continued to do well in the Cretaceous after their Jurassic beginnings, as well as many new groups of neoselachian sharks.

The late Cretaceous Niobrara Sea in the North American midwest teemed with life. Fishes small and large, as well as ammonites, provided the base for a substantial reptilian food chain. Pterosaurs, such as the giant Pteranodon, dived into the water to seize fishes, while turtles and plesiosaurs fed on fishes and shelled ammonites. The pliosaurs, with their heavy jaws, were top predators that attacked other reptiles.

Some lethal new predators also roamed the seabed. They included neogastropods, a new and rapidly spreading group of coiled mollusks which were nearly all armed with new abilities, whether to bore through shells and suck out the flesh inside, to poison prey animals, to shoot out deadly "darts," or to swallow small fishes whole. Other hunters were the crustaceans, represented by lobster-like creatures in the Jurassic, and now scuttling around as true crabs and lobsters. These crabs could break open shells with their claws. Many of the teleost fishes also nibbled and crushed seabed shellfish as a new source of food.

The effects of these new arrivals on seabed life were dramatic. The brachiopods and crinoids (sea lilies) fell into a steep decline. Bivalves increasingly buried themselves deep in sediment to avoid the crabs and mollusks. Others evolved massive shells, or spines, in order to deter attackers. One bivalve group, the inoceramids, grew shells 6 ft (1.8 m) in diameter. For a while, corals were even sidelined as the main reef-builders by another unusual bivalve group, the rudists, which fixed themselves to a hard surface on the sea bottom and grew up to heights of more than 40 in (1 m). Tight clusters of these large conical shells formed massive reefs in tropical seas worldwide. In fact, their size was probably their best protection; it may well have been because they were so large that they escaped being eaten.

Above the seabed swam ammonites and belemnites, fishes of various sorts, and a new crop of aquatic reptiles. The ichthyosaurs had virtually faded out by the end of the early Cretaceous, and their place was taken by the new sharks and by some large teleosts $6^{1}/_{2}$–13 ft (2–4 m) long, such as *Xiphactinus*. In the waters of the Cretaceous Interior Seaway swam the giant fishes, sharks such as the cretoxyrhinids, 20 ft (6 m) long and weighing $1^{1}/_{2}$ tons (1.5 tonnes).

There were large marine turtles such as *Archelon*, over 10 ft (3 m) long, with paddles spread wider than its length. There were pliosaurs and the long-necked plesiosaur group, the elasmosaurs, which could reach lengths of 40 ft (12 m) and more, with upward of seventy vertebrae in their necks. But the most powerful marine reptiles were a new group, the mosasaurs, which ranged from 10 to 33 ft (3 to 10 m) long. Surprisingly these are lizards – the largest ever – related to the monitors of today, whose best-known species is the Komodo dragon of Indonesia, a modern giant at 10 ft (3 m) long. The mosasaurs – *Clidastes, Mosasaurus, Platecarpus, Tylosaurus* – had long jaws and strong teeth, a deep tail, and paddle-like limbs. They may have fed on fishes, but a number of fossil ammonite specimens display rows of tooth marks that exactly match the dental layout of certain mosasaurs.

Birds and pterosaurs of the Cretaceous

The final predators in the Niobrara Chalk, which traces what was once the Cretaceous Interior Seaway all the way from Kansas to Manitoba, were flying animals, and in particular the diving bird *Hesperornis*, which stood about 40 in (1 m) tall and probably looked rather like the much smaller and quite unrelated modern cormorants. *Hesperornis* was flightless – a feature that has arisen independently in many different bird groups since their origin – and its range of features places it, in evolutionary terms, neatly between the first known bird, *Archaeopteryx*, and modern birds. Some advances over *Archaeopteryx* are the reduced tail, a fused breastbone that could anchor the downward pull of wing muscles, and the fusion of the lower leg bones to form a single element. Primitive features include teeth, the lack of a keel on the breastbone, and solid instead of hollow wing bones.

Hesperornis was a diver that swam by kicking its feet, possibly steered with its wing-stumps, and fed on fishes, as we know from the contents of its coprolites.

Fossil birds are otherwise rare in the Cretaceous, except for isolated finds in the early Cretaceous from Spain, and a bird

described in 1992 from China. Not until the Tertiary do modern birds make a definite entrance. At present we have a total of about 155 families. Yet their fossil record is patchy, and 47 families have no known fossil representative.

Cretaceous pterosaurs developed a great variety of forms, but a number of trends seem clear. Compared with their Jurassic ancestors, they are generally larger, and all through the period they grow larger still. At the same time they progressively lose teeth. With their greater size, they also needed to lose weight, and this was achieved by economies that removed bone material anywhere that the lines of force did not run: in the vertebrae, and in some of the wing bones, which were hollowed to a thumbnail thinness crisscrossed inside by bony struts. All these features are moves toward greater specialization, which can often be an evolutionary Catch-22: fail to evolve, and you fall behind the competition and become less and less adapted to a changing environment; stay on the crest of the evolutionary wave, and when conditions change radically your specialization leaves you with no equipment for a different lifestyle. Where less specialized animals can still use their more generalized, old-fashioned designs, or even modify them in a new direction, your royal road turns into a blind alley.

By late Cretaceous times, two families of giant pterosaurs were dominant: the pteranodontids in North America, and the azhdarchids seemingly worldwide. (The family takes its name from *Azhdarcho*, a big pterosaur found in Uzbekistan in 1984 and called after a dragon in Uzbek mythology.) *Pteranodon*, found in the Niobrara Formation,

The oldest flightless bird, Hesperornis, *from the late Cretaceous of the mid-American seaway. This was a human-sized bird that had virtually lost its wings, but could presumably dive and swim using its partially webbed feet as paddles. It dived for fishes from the surface of the water and pursued them, just as many seabirds do today.* Hesperornis *retained teeth, but had many features of modern birds that* Archaeopteryx *lacked.*

is the best known of the giants, and reached a wingspan of 16–26 ft (5–8 m). Its massive toothless head consisted of a pointed triangular beak at the front, balanced by a rearward structure sometimes of a similar shape, sometimes rising as a sculpted aerodynamic crest. The body was tiny, the animal all wings and head.

Pteranodon was surpassed as the largest flying animal of all time by the discovery, inevitably in Texas, of the azhdarchid *Quetzalcoatlus* in 1975. The wingspan of this monster has been variously estimated as from 36 to 50 ft (11 to 15 m), which is four times the span of the largest modern birds, the condors. In terms of its size and flight characteristics, this biomechanical marvel had more in common with a light airplane than with any living flying creature. Unfortunately, the skeleton of *Quetzalcoatlus* is incompletely known. It had long toothless jaws, a long and quite inflexible neck, and a human-sized body – though much lighter because of the hollow bones. Because it was found some 250 miles (400 km) away from the coastline of the former Interior Seaway, it is hard to interpret as a fish-eater; but is impossible to verify an alternative suggestion, that *Quetzalcoatlus* might have scavenged dinosaur carcasses like a colossal vulture.

The late Cretaceous dinosaurs of Mongolia

Dinosaurs are known in plenty from the late Cretaceous of North America and Mongolia, as well as in many other parts of the world.

Teams from Poland, the former Soviet Union, and now from Mongolia itself have made extensive excavations. The Nemegt Formation of the Gobi Desert, which contains remains from close to the end of the Cretaceous, is especially well known.

The Formation is dominated by hadrosaurs. These were duckbilled dinosaurs and the chief ornithopod descendants of the early Cretaceous hypsilophodontids. Hadrosaurs were all large, though at 33–50 ft (10–15 m) long not gigantic, with very similar bodies on much the same two-legged lines as the iguanodontids. Yet dozens of species have been described, each one distinctive because of the flamboyant array of headgear sported by these animals. In all cases, the front of the snout is flattened to give the typical "duckbill" shape, and the teeth are arranged in batteries of multiple rows, often up to 500 teeth per jaw segment. Some hadrosaurs possessed no crest at all, while others had ambitious extensions of the premaxillae and nasal bones of the skull, running back to form a range of crests – some like plates, others like snorkels, and others like spikes.

The premaxillae and nasal air passages in hadrosaurs follow the line of the crest, and often become convoluted. The function of the crests now seems clear. They were species-recognition signals, like the special displays of feathers in many birds, or the complex calls of frogs, birds, and some mammals. David Weishampel of Johns Hopkins University suggests that the crest gave cues both in vision and also in sound.

Weishampel modeled the shapes of the air passages in different hadrosaurs and found that if he blew into them he could produce different honking and trumpeting notes. Just like the brass instruments in an orchestra, the various layouts of the air passages produced different qualities of sound. Weishampel found that every species made a different noise, that males and females of the same species had their own signatures, and that the note changed during growth. So the hadrosaurs broadcast their species identity, their sex, their position in pecking orders, their herd identity, and other information both by the shape of the crest and by the noise produced. It is easy then to understand how several hadrosaur species could live side by side in the late Cretaceous lowlands, like antelope in the African savanna; and with hadrosaur lungs pumping air through resonating tubes in crests up to 40 in (1 m) long, the boom and blare of all that piping, hooting, and honking must have raised a carnival of noise.

The late Cretaceous faunas contain another group of bipeds that have ornithopod bodies but whose outstanding feature is their massively thick skulls. These are the pachycephalosaurs, first known in the Wealden but spreading in the late Cretaceous. *Stegoceras* was only 6½ ft (2 m) long, but its skull roof was exceptionally thick; in larger varieties the skull roof would be up to 10 in (25 cm) of solid bone, equivalent to about half the length of the whole skull. Looking for an explanation of this built-in helmet, Peter Galton of the University of Bridgeport, Connecticut, suggested that the pachycephalosaurs had fought for mates in the style of modern mountain sheep and other similar herbivores. They may have developed the same practice of ritualized contests during which the males charged each other with a powerful clash of heads that did no permanent damage. In the pachycephalosaur breeding season, they would have provided powerful percussion behind the honks and howls of hadrosaur bands.

The remaining late Cretaceous plant-eaters were quadrupedal armored animals carrying their own clubs, shields, and lances, which could only have evolved in defense against the most formidable of predators. Their tanklike size and heavy weaponry form some of the classic images of the age of dinosaurs.

The ceratopsians, from the mid-Cretaceous of Mongolia, had large beaked heads that broadened into triangular shapes when seen from above. Later varieties had a varying number of horns and a bony "frill" to protect the shoulder region. The frill may also have served as an attachment area for powerful jaw muscles. The bigger ceratopsians grew up to 30 ft (9 m) long, with horns almost 40 in (1 m) long, and weighed up to 6 tons (tonnes). The best known is *Triceratops*, known like the rest only from western North America, west of the Interior Seaway. It was the largest in the group, three-horned as the same suggests, and with heavily muscled legs that enabled it to charge like a rhino. (A bull rhino weighs 3–4 tons/tonnes.) *Styracosaurus* had a big nose

horn and six thick, pointed spines running backward from the frill. *Torosaurus* is known only from skulls: at 8½ ft (2.6 m) long, the largest skull of any land animal that ever existed.

Ceratopsians no doubt used their horns for deterring predators. There is evidence that they lived in herds, and this would have enabled them to stand in defensive circles, just as musk oxen do today, to protect their young behind a palisade of bristling horns. Individually, a charging *Triceratops* must have offered a serious threat even to *Tyrannosaurus* itself; but carnivore tactics of the present day do not include head-on confrontations with the strongest prey, and there are no grounds for believing that the big flesh-eaters of the Mesozoic were dumb enough to attack at the enemy's strongest point. Ceratopsians' horns may also have been used in head-wrestling between rival males – a form of non-injurious fighting used by many deer, which sorts out the strongest dominant males without putting them badly at risk. When their heads were lowered, the frill would have stood up almost vertically: a huge shield, with its own characteristic insignia of horns and spines, ceratopsian heraldry.

The final group of plant-eaters, the ankylosaurs, arose in the Jurassic but did not really multiply until the late Cretaceous. Closely related to the stegosaurs, they were often large animals, up to 33 ft (10 m) long and weighing up to 6 tons (tonnes). They specialized in armor, with plates of bone set in the skin of the back, neck, tail, and flanks, bony reinforcements over the skull, and even bony shutters over the eyes. Their heads carried short bony horns, and their backs were patterned with a variety of bony spines. Later ankylosaurs also had a fused mass of bone at the end of their long tails. It must have functioned as a tail club, swung at the legs of an attacker while the ankylosaur hugged the ground. To a big predator on two legs, damage to one of them could amount to a lethal threat.

Euoplocephalus and *Ankylosaurus* are the only North American ankylosaurs, but several other species come from eastern Asia. It looks as if the American species are the result of an immigration from Asia.

The last dinosaur group to lumber into this chapter, and certainly the most famous, is the

No other dinosaur grew more horns than Styracosaurus. *The nose horn is typical of most ceratopsians, as are the bony bosses over the eyes and in the region of the cheeks. The bony frill over the neck, unusually, bears numerous additional pointed horns, which acted as a disincentive to the attentions of predators such as* Tyrannosaurus rex.

tyrannosaurs. *Tyrannosaurus rex* of North America and the very similar *Tarbosaurus* of Mongolia were by far the top predators of their day, and of any other day. *Tyrannosaurus* was the biggest meat-eating land animal that this age of monsters ever produced, and much bigger than any that would succeed it. It reached a length of 46 ft (14 m), a height of 20 ft (6 m), and a weight of 5 tons (tonnes). Its heavy jaws were lined with teeth 6 in (15 cm) long, and a seven-year-old child could have stood straight inside its gaping mouth.

Tyrannosaurus had massive three-toed hind limbs, and what look like preposterous arms – scraggy, two-fingered, and so short that they did not reach the mouth. The solid head lacks the weight-saving bone economies of other big meat-eaters, and may have been designed to hit like a gaping hammer, as sharks do. So big a predator could not run fast for long, and

it may have fed on large herbivores that it surprised from ambush, or by careful stalking, or on already weakened or dead animals. It could hold down the body with its heavy clawed foot, and tear out chunks of flesh with viciously serrated teeth.

This was a landscape teeming with dinosaur life, with diverse and widespread faunas and a range of new animals, many of them highly successful. It is hard to believe the view that the dinosaurs were in decline during the late Cretaceous, except perhaps in the short term, as the kind of phase that had happened before, part of the ordinary ups and downs of species. Yet this entire population, a whole level of animal existence, was soon to disappear, and many others with it. The great extinction at the end of the Cretaceous is heavily studied, but it fits none of the inbuilt patterns of life, and it is still unexplained.

The extinction

The mass extinction at the end of the Cretaceous period, 65 My ago, wiped out the dinosaurs along with every other land animal that weighed much more than around 55 lb (25 kg). Yet it did no more damage than four or five other such crossroads in the travels of life, and it was altogether less catastrophic than the end-Permian event. Still, mammals have a proprietary interest in the causes of their own development, and the sudden death

Maiasaura, *the "good mother reptile," was a duckbilled ornithopod. Collected in Montana, it has been studied in more detail than any other of its kind. Nests full of eggs and young tell a story of parental devotion to the babies. Each nest may have held between ten and twenty eggs. After the babies hatched, they were fed on plant food collected by their parents and by older siblings born in earlier years.*

of dinosaurs is bound to look far more dramatic, if only for the evidence it leaves behind, than the deaths of smaller, less familiar creatures. We do not empathize closely with the last farewells of worms, shellfish, trilobites, ammonoids, or even fishes, let alone with plants known only by unlovable Latin names, or micro-organisms that we cannot see. So what is really surprising about the closure of the Cretaceous is that it should have taken so long to capture the human imagination.

We may as well use the standard geological jargon and refer to this episode as the K–T event – from the German *Kreide*, Cretaceous, and from Tertiary, which is the main subdivision of the post-Mesozoic era, the Cenozoic. It was not till about 1950 that this event began to draw general scientific attention. People had long been aware that a major changeover in fauna had taken place, but the facts were taken for granted, part of the punctuation of the geological record. Only a few German scientists had made serious efforts to study mass extinctions, between 1910 and 1950, and their work was ignored in the English-speaking world, partly for long-term cultural reasons, partly for obvious political ones.

During the 1950s, inquisitive paleontologists like Norman Newell and George Simpson started to document the fossil evidence for extinctions, and others sought theories for their causes. Most of the K–T theories in those days focused on the dinosaurs: they were too big, too stupid, too constipated, or too undersexed to survive; they suffered competition for food sources from mammals or insects; they were poisoned by plants, slaughtered by plagues, wiped out by shock climatic change, by extraterrestrial catastrophe, and so on. These theories were mostly beside the point because they took no account of all the other groups that went extinct, failed to explain why others should have survived, or were just untestable armchair speculations, often not based on close study of the circumstances, which in any case were only vaguely known.

Our knowledge of the available data has increased fantastically since than. Numerous highly detailed studies have been carried out on sequences of rock that span the K–T boundary, looking at evidence for

HADROSAURS

The duckbilled dinosaurs (hadrosaurs) were immensely successful dinosaurs during the last 20 My of the Cretaceous. They all had virtually the same bodies, but were distinguished by an astonishing array of crests on their heads. What was the function of these?

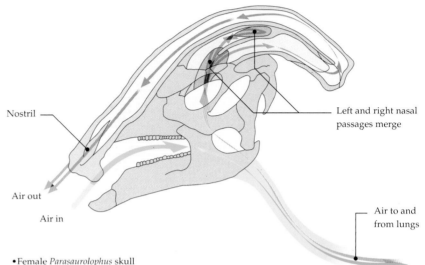

• Male *Parasaurolophus*

Nostril

Left and right nasal passages merge

Air out

Air in

Air to and from lungs

• Female *Parasaurolophus* skull showing internal air passages

The crest is made from the premaxillae and nasals, bones of the top of the snout, and it contains the normal air passages, extended enormously over the top of the head. As it breathed, the hadrosaur drew air in and out through the convoluted tubes. Experiments have shown that these arrangements of tubes

would produce musical resonances, just as orchestral wind instruments do.

Huffing and puffing, the hadrosaurs made different musical notes depending on the shapes of their crests. It seems that males and females had different crest shapes. And each species had clearly different

crest forms. Just as with antelope today (same body, various horns), species-specific visual signaling took place by displays of the crests. Added to this was the cacophony of honks and melodious hootings that accompanied head displays between males and females, or threatening blaring between rival males.

☐ Premaxilla
☐ Nasal bone

• *Kritosaurus*   • *Saurolophus*   • *Lambeosaurus*   • *Corythosaurus*

• Comparative size of modern wildebeeste and *Edmontosaurus*

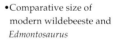

Scale   80 in (2 m)

The pachycephalosaur Stegoceras and its relatives are known for their thick skulls, which were designed to withstand massive impacts. In most other respects, they were standard-shaped, ornithopod-like dinosaurs. Their skulls probably evolved in response to pre-mating contests between males. No doubt, the male with the thickest skull won most mates and ensured yet thicker-skulled offspring in the next generation.

environmental change, geochemical oddities, and distribution, decline, and replacement of fossil groups, both on land and in the sea. The data base is huge now, with detailed information on more than 150 K–T sections worldwide – much more substantial than can be found on the end-Permian event. And we now have the assistance of computers that can be programmed to correlate the data and search for patterns and anomalies, able to conduct studies that would once have taken decades of research time. So what happened?

Paleontologists have documented the groups that went extinct: the dinosaurs, pterosaurs, and some families of birds and marsupial mammals on land; and the mosasaurs, plesiosaurs, some families of teleost fishes, ammonites, belemnites, rudist, trigoniid, and inoceramid bivalves, and well over half of the various plankton groups in the sea. Some of these groups show clear long-term reductions in diversity in the last 10 My of the Cretaceous, whereas others seem to vanish at full diversity right on the boundary, switched off. Yet others that are often supposed to be K–T departures, such as the ichthyosaurs, were already long gone. Survivors include most land plants and land animals – insects, snails, frogs, salamanders, turtles, lizards, snakes, crocodiles, placental mammals – and most marine invertebrates – starfishes, sea urchins, mollusks, arthropods – together with most fishes.

The evidence of the rocks, and of fossils such as angiosperm leaves, show that there was a significant cooling of the atmosphere toward the end of the Cretaceous. Steven Stanley of Johns Hopkins University has argued that extinction patterns match the likely effects of a major cooling shock; in the seas, only tropical faunas were hit, with extinctions of rudists and other Tethyan groups, while high-latitude faunas were untouched. A possible cause of cooling is the plate tectonic changes that were pushing Australia away from Antarctica. Deep cold currents from the Southern Ocean funneled toward the equator's warmer Tethyan seas might have altered the carbonate compensation level, the depth at which calcium carbonate is removed from animals or sediments. Warm waters are supersaturated, and cannot hold any more, but colder waters can dissolve calcium carbonate and so kill any animal that uses it. Colder water, combined with falling sea levels, would affect equatorial temperatures and remove the moderating influence of warm seas. The resulting climatic changes would certainly produce cooler conditions generally, and greater extremes of climate in continental interiors.

The most sensational candidate for the cause of the K–T extinction is some sort of extraterrestrial body. The idea is not new, but it is only since 1979 that evidence has begun to emerge. In that year a team from the University of California at Berkeley was studying a K–T boundary sequence in rocks at Gubbio in Italy, looking for chemical signatures that might help to identify K–T locations elsewhere in the world. What the team found was levels 100 times higher than usual of the rare metal iridium in a clay layer right on the boundary. Iridium is rare on the Earth's surface, but it is often found in meteorites. Further investigation led the geologist Walter Alvarez and his father, the Nobel physicist Luis Alvarez, and their California colleagues, to outline a catastrophic model that is still under examination.

These levels of iridium at one site suggested a total amount, worldwide, of something like 200,000 tons (tonnes). It would take an asteroid at least 6 miles (10 km) in diameter to contain that much iridium; and an asteroid of that size striking the Earth, traveling at a possible 10–13 miles (16–21 km) per second, would create a crater 40 miles (65 km) across, and would cause worldwide havoc. The Alvarezes argued that the impact had blown a cloud of dust into the upper atmosphere large enough to blanket the world for a year or more, and to blot out the Sun. Other scientists had been researching the possible effects on the Earth of a nuclear war, and had found that it would raise enough dust to shroud the whole Earth beneath a "nuclear winter." An asteroid of the size projected would strike with the force of 100

million single-megaton H-bombs – enough to cause global refrigeration.

Since 1979, the iridium "spike" has been found in over 100 K–T boundary sections all over the world, clinching confirmation of the Alvarezes' predictions made on the basis of a single site. Other clues to a violent impact have been identified in many of these sections, including glassy spherules created by tremendous heat, shocked quartz, and the mineral stishovite. Quartz particles showing as many as nine intersecting fracture layers caused by shock are found only at the sites of impact craters and of nuclear tests.

A swarm of arguments have whirled around this theory. Where is the impact site? Possibly on the coast of northern Yucatán, a candidate crater located in 1991, possibly in Iowa, where the Manson crater seems too small for the reported effects – but suppose there was more than one impact? Or perhaps the asteroid landed in the ocean, and the crater somewhere in the floor has since been subducted beneath a tectonic plate, or else covered by sediment. Shocked quartz should indicate a land impact, but there is also evidence for tsunamis, huge tidal waves, which would require an ocean impact. A hit in the shallows of a continental shelf might produce both sets of effects.

Crater counts on the Moon and on the surface of the Earth, together with observations of the number of large asteroids whose orbits could intersect with Earth's, have produced estimates that a 6 mile (10 km) asteroid should strike once in every 50 or possibly every 100 My, with smaller or larger bodies hitting more or less often. Soot particles have been discovered in several

ABOVE *A herd of* Struthiomimus *gallop across the open plains of North America during late Cretaceous times. These ostrich-dinosaurs could probably run as fast as a racehorse at full tilt. They had no teeth and presumably used a sharp-edged, horn-covered beak to feed on flesh or eggs.*

LEFT *The ankylosaur dinosaur* Nodosaurus *was invested with efficient chain mail of small bone plates. These formed a firm, but slightly flexible, shell over its back, from the tail to the roof of the skull. Ankylosaurs even had bony eyelids.* Nodosaurus *had surprisingly strong legs, which were designed for rapid trotting and for galloping at times, rather like a present-day rhinoceros.*

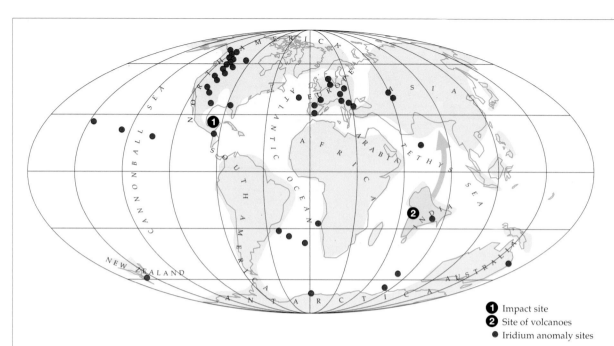

● Impact site
● Site of volcanoes
● Iridium anomaly sites

There are now two main contenders to explain events at the K–T boundary. One is the asteroid impact model, and in 1991 the crater was found at last. The site is called Chicxulub, on the Yucatán Peninsula in Central America (1), and it was detected by studying old oil-exploration boreholes. The other theory is that the iridium spike and the rest were produced by massive outpourings of lava, which were occurring in India; the Deccan Traps (2) belched out over much of India at the time.

THE K–T EVENT

The mass extinction at the end of the Cretaceous period, 65 My ago, has attracted more research than any of the other extinction events, and yet a resolution is still elusive. No one can say for sure just why the dinosaurs died out, and that is despite the involvement of many hundreds of earth scientists of all shapes and sizes in this research field over the past fifteen years in particular.

The recent flurry of research interest was triggered by a remarkable article published in the journal *Science* in 1980 by Luis Alvarez and his collaborators. The authors presented a single piece of rather tenuous evidence, and concluded from it that the Earth had been struck by an asteroid (a large meteorite) 6 miles (10 km) in diameter 65 My ago, and that this had wiped out much of life catastrophically. Their tenuous evidence was an iridium spike (bottom right) recorded from one locality in Italy (Gubbio), and they claimed that the same phenomenon would be found worldwide.

Since 1980, their prediction has proved startlingly accurate. Iridium spikes are known from 150 or more K–T sections worldwide ("K" stands for the German *Kreide* (chalk), thus Cretaceous, and "T" for Tertiary, the following geological period), in rocks deposited on land and under the sea. Was it an asteroid?

❶ Impact theory

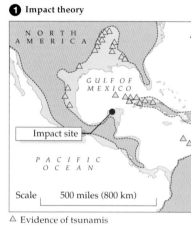

△ Evidence of tsunamis
Coastline at time of impact

❷ Volcanic theory

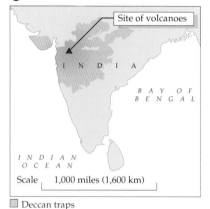

▨ Deccan traps

The iridium spike (bottom) is the key indicator of catastrophe at the K–T boundary. In dozens of localities worldwide, geologists have discovered this unusual geochemical anomaly, usually associated with an ashlike clay later.

Iridium is a rare element in the Earth's crust, present only in minute quantities which have been disseminated from small-scale meteor showers. Iridium is a platinum-group metal, believed to be present in the core of the Earth, and in meteorites, but nowhere else. The vast enrichment of iridium

in the K–T boundary clay, then, is taken to imply the arrival of a vast asteroid. The theory is greatly enhanced by studies of the K–T boundary rocks around the proto-Caribbean. They show evidence of hot glasses in the sediments that were melted on impact and thrown out for great distances round about, and these glasses apparently contain a geochemical signature that matches the Chicxulub bedrock.

The Deccan Traps represent massive volumes of sediment, covering an area equal to half of Europe, spanning the K–T

boundary. According to the vulcanologists, such vast flood basalt volcanoes could send up clouds of dust enriched in iridium, which could encircle the globe, and deal killing blows to dinosaurs and plankton alike. There is much research yet to be done.

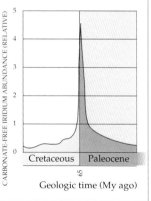

places just above the iridium layer. On the (unproven) assumption that they fell less than a year after the impact, catastrophe scenarios see them as evidence of continental wildfires possibly caused by the re-entry into the atmosphere of debris blasted into space by the original impact and falling again as meteoric sources of radiant energy all across the globe.

The dustcloud turned day into night, month after month, suppressing photosynthesis in plants and so removing the base of food chains both at sea and on the land. Naturally the most lethal damage would be suffered by the largest animals with the largest food requirements, and by the largest of their predators. Temperatures fell below freezing point in the heart of continents. Atmospheric nitrogen and oxygen reacted in the heat of impact to form nitric oxides, falling in rain as nitric acid, and killing the faunas of the ocean's upper levels.

Yet these are speculations. If asteroids have struck so often, why are there no other sharp iridium spikes connected with other major extinctions? If their impact is so devastating, why have other known impacts caused so much less damage? The Manicouagan crater, about 336 miles (540 km) northeast of Québec, Canada, has a diameter of $43^1/2$ miles (70 km), about right for a 6 mile (10 km) asteroid; and it appeared about 220 My ago, near the end of the Triassic. But the Triassic extinctions came in two waves, not one, and there is no iridium record there. The even larger crater at Popigai in Siberia, 60 miles (100 km) in diameter, records an impact about 40 My ago, but with no iridium layer, and no extinctions either.

Bold theories built on controversial statistics need to be treated with caution. Physicists, but also some geologists and paleontologists, perceive the sheer power and statistical probability of an asteroid impact and are certain that the Earth must have been struck more than once by objects at least as massive as the Manicouagan asteroid. They pursue the likely effects of so much energy so

suddenly released, and the picture they assemble is of a catastrophe so terrible that it might easily have been the cause of a mass extinction and the trigger of a new age.

Paleontologists, but also some skeptical physicists and geologists, argue about the extinction statistics, knowing that the fossil record cannot be narrowed down to focus on a single event, and seeing the signs of decline appearing in many groups long before the K–T boundary event. They know that there are powerful forces locked inside the Earth, but not always contained there. They are reluctant to have a single dramatic episode take over as the driving force in a sequence where its role may have been simply to hasten or to further complicate a shift that was already under way. It may be that there is a potential in the Earth's ecology that can cause it to tip over into a radically different state, as when a cell turns cancerous or when tropical weather gives birth to a hurricane. If science is going to deal with that possibility, then it cannot afford merely to file all the data under "extraterrestrial causes" and lose track of complex patterns that may be deeply embedded in the relationship between life and the only home it appears to possess in our local solar system.

The balance of opinion concerning K–T seems to be about equal between the impact-catastrophe model and the gradual-global-cooling model. There is evidence for both, and both may be significant. A third set of causes lies inside the plants and animals of the Mesozoic. What features in their biology caused some to die and others to survive? These and other lines of investigation converge on the K–T event and then weave into a monumental tangle. Scientists also converge, knowing that the riddle from 65 My ago is bound to contain answers to important questions, including a lot that no one has yet thought of asking. Until the K–T tangle begins to unravel, the job of science is to make more observations, conduct more experiments, and keep on arguing.

Mammals were thoroughly at home in the dinosaur world as long as they stayed small and kept out of sight, which mostly meant a habit of nighttime living. Here in the latest Cretaceous of Mongolia a family of Deltatheridium, rat-sized therian mammals, forage freely around a drowsing group of Protoceratops, not long before the fall of the dinosaurs. The Deltatheridium stock were neither true marsupials nor placentals, but an independent line related to marsupials that became extinct at the end of the Cretaceous.

Chapter Five

VICTORS BY DEFAULT
THE MAMMALIAN SUCCESSION

Christine Janis

The dinosaur dynasties were gone. Only the disappearance of humankind might match the impact of that event on other species, and on the surface of the planet. The dinosaurs had shaped the Mesozoic in all but its smallest dimensions, monopolizing most of the larger body sizes and most of the living space beyond the horizons of insects and bacteria. They had been the road makers and the landscape gardeners. The daytime had been theirs. Yet the dinosaurs had not been driven out. No animal was treading on their tails, hastening their extinction.

It used to suit some humans to believe that mammals represented a simple progression from inferior to superior forms, and that they must somehow have "outcompeted" the supposed brute force and ignorance of their dinosaur rivals. One theory even suggested that bold mammal predators crept out at night while the bigger creatures slept, and gobbled up their eggs. But there is no evidence whatever to suggest that mammal populations had been causing damage to the dinosaurs before the K–T event.

What evidence we do have suggests that the mammals had been the midnight ramblers of the Mesozoic, and that their long-delayed rise was the effect, not the cause, of the dinosaurs' fall.

Mammals and their evolution, including human evolution, are the final subjects dealt with in this story. Most books save them till last, and there are good reasons for doing so; the tellers of the story cannot abandon their own perspective. But the story is not over, and in any case it has hundreds of thousands of other plot lines, woven across the framework of a changing world. To focus on our own recent history does not require us to believe in some ladder of perfection that only humans could have climbed. The story does not converge upon ourselves.

In the early Paleocene, dense forests extended to higher latitudes. This scene is from the Early Paleocene of Wyoming. The vegetation included sequoia trees, with a dense undergrowth of shrubs such as tea and laurel, with the addition of ferns and horsetails. On the ground is Chriacus, *a racoon-like omnivore. Facing* Chriacus *on the tree is* Ptilodus, *a surviving member of the multituberculates, primitive mammals often termed the "rodents of the Mesozoic." Higher up in the tree is* Peradectes *(the name means "persisting biter"), an early opossum-like marsupial. Marsupials became extinct in North America by the Oligocene, and did not reappear until true opossums invaded from South America in the Pleistocene.*

For example, mammals were far from being the last vertebrate class to evolve. Their appearance in the fossil record comes about 60 My before the birds; and seeing that the synapsid reptiles, the mammal ancestors, were the first to branch off from the amniote radiation, it would be quite logical to deal with the whole lineage *before* going on to turtles, lizards, snakes, crocodiles, and dinosaurs. Nor can the mammals claim to be the most numerous animals, in terms either of species or of individuals. There are many more kinds of fish, reptiles, and birds than there are of mammals, and the total number of all vertebrate species is dwarfed by the numbers of other types of animal, such as insects or even mollusks.

One reason for placing humans on a peak of evolution is their intelligence, although in evolutionary terms there is no clinching reason for arguing that a greater number of brain cells is an adaptation more magnificent in the long run than, say, a greater number of legs. But intelligence can be defined as a practical, measurable ability, because it gives humans the tools and the vision to change their environment, much as the social insects such as ants and termites can transform their smaller worlds. Who is to decide whose "group intelligence" is of an inferior variety? Perhaps not a group with a tendency to define its own specialties (such as a foot designed for upright bipedal walking) as superior and to label other specialties (such as a bird's wing) as mere interesting deviations.

All the same, this book is written by humans for other humans to read, and humans have become the most powerful occupants of this planet in the latest tiny fraction of galactic time. We appear to be the only animals with an awareness that there is a future after the end of the "final" chapter, and that we may play a part in it.

With our growing knowledge of the fossil record it has become obvious that no main highway stretches from the distant past into the present. No life form has reached the present day by an effortless advance, and neither has it survived by eternally outfighting or outcompeting creatures less "fit" than itself. Paleontologists such as Stephen Jay Gould at Harvard University and David Raup at Chicago have shown how often the sheer force of events has dictated the pathways of life, beyond the power of any life form to control. No animal has a manifest destiny to stay alive when its food runs out, its habitat is destroyed, or a local or worldwide disaster occurs. "Bad luck" is as common as "bad genes"; the best adaptation to one set of circumstances can be the worst for the unpredictable next; and the history of life has been punctuated by mass extinctions that wipe out species utterly at random. The remaining competitors find themselves playing a different game with different rules in changed conditions.

This is not to say that species are to be compared with pinballs fired into a pin-table Earth and left to roll, till they lose their momentum or hit an occasional jackpot. Natural selection, acting over millions of years, works to improve the odds on survival by the constant invention or reinvention of forms and strategies. And there is "good luck," as well as bad. In the case of the mammals, the good luck may have begun with their early failure to match their strength against the dinosaurs in the late Triassic. Part of their rivals' undoing was their size and weight. The mammals and their ancestors were forced to live on a small scale which may have allowed them to retain more options. A small animal could be flexible; it could swim, climb, dig, run, or jump as its conditions required. A larger animal had to specialize – and the greater the specialization, the harder to alter its body plan to adapt to a changing environment.

By the time of the late Cretaceous, many mammals had given up laying eggs and were able to deliver their small young alive. Another vital innovation of the mammals was the variety and efficiency of their teeth, with designs for cutting, shearing, piercing, gnawing, grinding, and otherwise grasping and processing food. Small animals use more energy in proportion to their size than large ones do, and there were greater benefits for species that could develop more economic ways of using food. The so-called "insectivores" probably ate any small animal that would offer a meal – grubs, worms, and centipedes included. Some mammals became omnivores, able to eat animals and plants.

As potential prey for the small fast dinosaur carnivores, the mammals were unable to compete in size and speed (no hippos, no gazelles, in the Mesozoic), but 150 My of selective pressure encouraged greater agility, and probably better smell and hearing. In order to run a so much more efficient nervous system, receiving and sending more information, there had to be a more efficient brain. No matter what problems a mammal species might face, if the solution lay in design the improvement must not push its body size into the dinosaur dimension. Pressure of circumstance compelled the mammals to specialize in small-scale efficiency, and locked them into the role of artful dodgers, where most of them remain to this day.

Getting under their feet

The small mammals of the Mesozoic account for two-thirds of the total evolutionary history of the mammals. If we trace their development from the first mammal-like reptiles of 300 My ago, then the Age of Mammals – traditionally dated from the fall of the last dinosaurs – covers about one-fifth of their history.

The fossil record suggests that the cynodont ancestors of the first true mammals were a group declining in both numbers and diversity throughout the Triassic after an early dominance, with the later species growing progressively smaller and more specialized. It is as if they were a last-ditch experiment in miniaturization, barely squeaking through the late Triassic extinctions to continue the lineage. Fragmentary remains of possible mammals have recently been reported from about 225 to 220 My ago.

The first group of true mammals were the morganucodontids of the latest Triassic and Jurassic. Megazostrodon ("large animal-tooth") was a mouse-sized species from southern Africa. Like similar small insect-eaters of today, it was probably quite solitary, and a nighttime specialist. Like living monotremes, it probably laid eggs, gave birth to immature young, and had neither nipples nor external ears.

The earliest true mammals that we know in reasonable detail were the morganucodontids of the latest Triassic, around 210 My ago. Their most notable feature was that they were considerably smaller than even the latest ancestral cynodonts, shrew-sized (about 1 oz, 28 g) rather than rabbit-sized (about 2 lb, almost 1 kg). This small step for mammal-kind took them into a different ecological niche, probably as nocturnal insectivores rather than more generalized small carnivores.

Morganucodontids possessed comparatively larger brains, more completely encased in bone, than their reptilian predecessors. This was partly an effect of their smaller body size; large animals have proportionally smaller brains than small ones. The comparative difference may be related to the fact that it probably takes no more brain cells to move a large leg than a small one; an animal fifty times larger than another does not need a fifty-times-larger brain.

Small mammals perceive the world as a more diverse and problematic place than the large ones have to face; a tree root is a log, and a pebble a boulder, for a shrew-sized mammal, while the bark of a tree is a ladder to the canopy. Many of the skeletal adaptations of the morganucodontids, so critical to later evolution at a larger scale, may have offered a response to the need for a more maneuverable and scampering type of locomotion in a style that is not available to a larger, heavier animal. Small mammals, by virtue of their size alone, can also climb fairly easily. The fossils show a streamlined shoulder and pelvic girdle and a more flexible type of backbone and ankle joint. The backbone of early mammals allowed them to move with a bounding gait, flexing the spine up and down, in contrast to the sideways-curving reptilian walk.

The smaller the animal, the larger its surface area in proportion to its mass, and the faster it loses heat. It burns more energy, and needs more food to replace it. Morganucodontids evolved larger jaw muscles and precisely interlocking cheek teeth (molars and premolars) that would shear food more rapidly, so that the gut could work faster and make room for more food. A higher rate of food processing goes with a higher metabolic rate, regardless of size, and these early mammals may have run a faster metabolism than cynodonts. It is impossible to take full advantage of an interlocking teeth design if its precision is disrupted by the continuous tooth replacement practiced by cynodonts and present-day reptiles. So early mammals have only two sets of replacing teeth, like ourselves.

This altered design and single replacement of teeth indirectly tell us something else about the biology of early mammals. They must have evolved milk glands and lactation before they made these changes. Once milk is available, the young can be born with few or no teeth, and teeth only need to be added as the jaw grows closer to its final adult size. Without lactation, a newborn morganucodontid would need a full set of teeth in order to eat, and the dental production line might have kept on rolling as before. We can be fairly sure that these early mammals produced milk, but they most likely did not have nipples, as these are absent in present-day monotremes (the egg-laying platypus and the echidnas, or spiny anteaters), the most primitive mammals still surviving.

Lactation must have been vital to early mammalian evolution. With the capacity to produce food for the young directly from their own bodies, the females no longer faced the hard choice between staying in the nest, keeping the new brood warm, and going out to find them food. A female morganucodontid could have laid down layers of fat before giving birth and used them to feed both herself and her offspring without having to leave the nest too often. Birds solve the same problem of warm-blooded newborns needing protection from heat loss by involving the father in parental care – perhaps the strategy of building up fat would handicap an animal that relied on flight.

The earliest mammals and their immediate cynodont ancestors had a double jaw joint, combining both the old reptilian jaw and the future mammalian one, as described in Chapter Three. The articular and quadrate bones of the reptilian jaw probably functioned as a type of ear, with the eardrum housed in the lower jaw. Later mammals developed this cruder layout into the more compact true middle ear region, and some evidence suggests that the final steps took place independently in the chief lineages soon to be described: multituberculates, monotremes, and therians. (Multituberculates take their name from the numerous rounded cusps on their teeth, which contrast with the simpler pointy design of therian teeth.)

Therian mammals (marsupials and placentals) went a step further by having a longer cochlea, which is the part of the inner ear involved in detecting and analyzing sound. This increased length is achieved by coiling the cochlea to fit the space available in the skull, and it seems to confer improved powers of sound detection and pitch discrimination. More primitive mammals have a straight cochlea, and we know that the living monotremes have only a partially coiled cochlea and lack an external ear.

This raises an interesting question: is it really correct to reconstruct early mammals such as morganucodontids with outer ears? Without the evidence of soft tissues, when we equip these early forms with ears we may be merely expressing a bias about what a mammal is supposed to look like.

With or without their ears, it has to be said that the most interesting feature of the Mesozoic mammals is their status as forerunners of a great radiation in the wake of the dinosaurs' extinction. Till then, a modern eye might class them as vermin, mostly very small, and few of them bigger than a rat. Most groups were insect- or flesh-eaters, though the multituberculates formed a separate radiation that added plants to their diet.

There seem to have been two chief periods of diversification among Mesozoic mammals. An early Jurassic radiation produced various lineages that did not outlast the early Cretaceous. (Also in the early Cretaceous came the rise of the flowering plants, associated with the arrival of new insect species.) In the late early Cretaceous, alongside a similar turnover in the dinosaur fauna, came more advanced multituberculates and several lineages of true therians – marsupials, placentals, and others that possessed the key therian features but were not closely related to either of the living groups and did not survive the end of the Cretaceous. ("Therian" comes from the Greek *thérion*, a wild beast, and will recur in many suffixes in this chapter: *Sivatherium*, for example, combines Sanskrit and Greek to

USES FOR THE MAMMAL LIMB

The basic mammalian limb belongs to generalized ("scansorial") mammals that can do a bit of everything – cling, climb, jump, run, hold. More specialized mammals focus on some particular purpose at the expense of all-round skills.

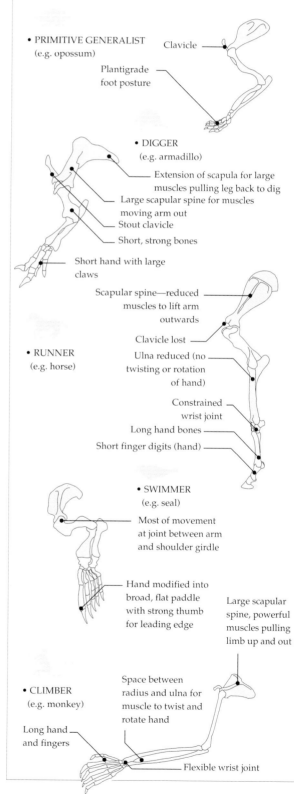

• PRIMITIVE GENERALIST
(e.g. opossum)
Clavicle

Plantigrade foot posture

• DIGGER
(e.g. armadillo)

Extension of scapula for large muscles pulling leg back to dig

Large scapular spine for muscles moving arm out

Stout clavicle

Short, strong bones

Short hand with large claws

Scapular spine—reduced muscles to lift arm outwards

Clavicle lost

Ulna reduced (no twisting or rotation of hand)

Constrained wrist joint

Long hand bones

Short finger digits (hand)

• RUNNER
(e.g. horse)

• SWIMMER
(e.g. seal)

Most of movement at joint between arm and shoulder girdle

Hand modified into broad, flat paddle with strong thumb for leading edge

Large scapular spine, powerful muscles pulling limb up and out

• CLIMBER
(e.g. monkey)

Space between radius and ulna for muscle to twist and rotate hand

Long hand and fingers

Flexible wrist joint

PRIMITIVE scansorial small mammals have loosely arranged wrist and ankle joints and paired bones in the forearm (ulna and radius) and lower leg (tibia and fibula), to allow movement in many directions. The shortish hands and feet have flexible, widely spread digits. The whole hand or foot is on the ground ("plantigrade") for standing or walking, the human method.

DIGGERS have strong thick claws, especially on the front feet, and short stocky limbs with distinct projections to anchor powerful muscles. They keep the plantigrade stance.

RUNNERS need longer, less flexible limbs that channel energy into fore-and-aft movement. The forearm and lower leg lose or reduce the double, swiveling bone design. The base of the hand and foot is lengthened, but the digits are more tightly compressed and/or reduced in number. They move on tiptoes ("digitigrade"), like a dog, or on the tip of one digit encased by a hoof ("unguligrade"), like a horse.

SWIMMERS' front limbs spread into flattened paddles, with the bones of hands or feet extended to make a flipper. Movement is focused on the joint at the shoulders.

CLIMBERS' limbs grow longer, their wrists and ankles more flexible, and the double bones in the lower limbs arc outward from each other. All joint surfaces are rounded and shallow, to allow maximum freedom of movement.

mean "gracious beast.") These later mammals tended to be slightly larger (though still not more than cat- or rabbit-sized). Their more complex tooth arrays provided for crushing and grinding, as well as simple shearing – adaptations to new types of diet, maybe connected to the changes in flora and insects.

There are three main subdivisions of more advanced mammals, counted as subclasses in the class Mammalia: the allotherians (the extinct multituberculates), the prototherians (monotremes), and the therians (marsupials and placentals). Multituberculates were around from the late Triassic until the later Eocene, about 35 My, which gives them the impressive evolutionary span of about 160 My, compared with the 120 My of more modern mammal development since the early Cretaceous. Their fall was probably linked to the rise of therians adapted to a similar lifestyle – at first the early primates and ungulates, and then the diversifying rodents that they resembled. Although some multituberculate forms were ground-living and rather wombat-like, mostly they were small (mouse- to squirrel-sized) tree-dwelling mammals; they would eat almost anything and were probably active at night. Their skeletons modified more primitive mammalian designs mainly by adapting for life in the trees, with ankle joints that enabled the feet to be rotated backward, and in some forms a grasping tail. Their main specializations were in the skull and teeth; they developed a fairly broad flat skull, with flat multicusped teeth and rodent-like gnawing incisors.

The pelvis of multituberculates looks too narrow to make room either for eggs or for large newborn young to emerge. They probably gave birth to very immature young, in the style of living marsupials. As monotremes lack nipples it does not seem likely that the multituberculates would have possessed them – here again, they would have had to evolve this fairly specialized equipment along a separate path of development.

Living monotremes are known today only from Australia and New Guinea, which sit on the same tectonic plate and were joined by land until fairly recently. The evidence for their existence, once found exclusively in Australia, now extends to South America. In 1992 Rosendo Pascual, of the Museum of La Plata in Argentina, was able to confirm the discovery in Patagonia of the tooth of an ornithorhynchid (platypus) about 62 My old. This find makes it sensible to suggest that monotremes were originally a radiation of Mesozoic and early Tertiary Gondwana (Australia, Antarctica, South America).

The present-day monotremes, the platypus and the echidna, are highly specialized animals. It is possible that the group was once broader and more varied, but the rocks that would record their early history do not exist. Both living forms have rather heavily built limbs with a reptilian sprawl, but the platypus swims and the echidna digs for termites or soil-dwelling invertebrates, so their akimbo stance is more likely to be a special development than an original "primitive" mammalian feature – moles have a similar posture, for instance.

Several new features identify the therian mammals, both marsupials and placentals. They possess the tribosphenic type of molar teeth, which add a crushing function to the original mammalian shearing capability. Where the more primitive animals had retained the reptilian coracoid bone in the shoulder girdle, which was jointed to the breastbone, the therians developed a distinctly different type of shoulder girdle that reduces the coracoid to a knob fused with the lower part of the shoulder blade or scapula. The clavicle ("collarbone") remains as the only connection between shoulder blade and breastbone. Alterations to the scapula are correlated with a changed design of the shoulder muscles; the shoulder girdle is now free to swing as an extra limb segment, giving a greater range of movement, and it can act more effectively to cushion the body against the impact of landing on the front legs during bounding movements.

Both marsupials and placentals give birth to live young, but marsupials have very immature newborns that they usually place in a pouch, while placentals carry their young in their bodies until a later stage of growth, and give birth to relatively more mature newborns. Marsupials tend to have smaller brains than placentals, and a lower metabolic rate. While many of the more primitive placentals operate at night, larger and more specialized forms are creatures of the day. The

TEETH: THE MAMMAL'S VITAL INSTRUMENTS

An animal's teeth are the tools by which its diet is made suitable for its stomach. The evolution of precise food-processing mechanisms in mammalian teeth, giving access to a wide variety of foods, probably played a crucial part in mammals' radiation into a great diversity of types and body sizes.

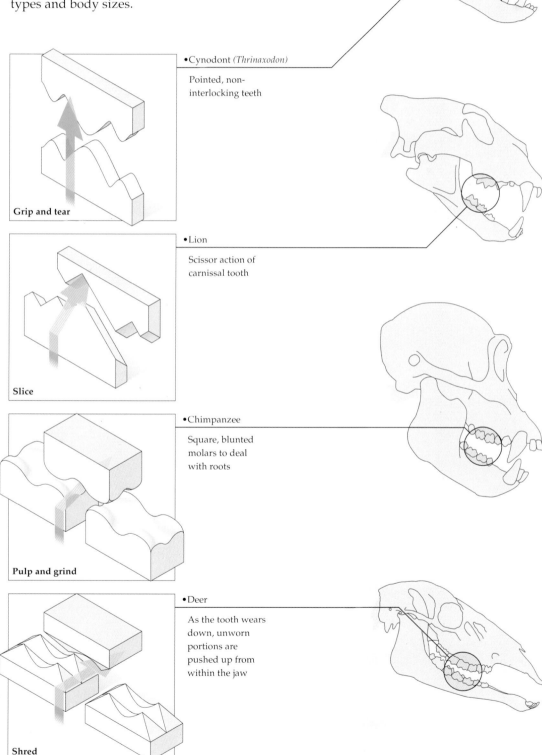

Grip and tear

• Cynodont (*Thrinaxodon*)

Pointed, non-interlocking teeth

Slice

• Lion

Scissor action of carnissal tooth

Pulp and grind

• Chimpanzee

Square, blunted molars to deal with roots

Shred

• Deer

As the tooth wears down, unworn portions are pushed up from within the jaw

CYNODONTS' cheek teeth had three cusps that could rip and tear the food, but did not precisely interlock like teeth of true mammals.

TRUE MAMMALS, emerging about 210 My ago, have slightly narrower lower than upper jaws. When the lower teeth move to contact the uppers, they move not just upward but also slightly inward. Thus, when the jaws close, the teeth shear past each other, slicing up the food.

THERIAN MAMMALS, first seen 120 My ago, added more cusps to their interlocking tooth pattern to add pulping and crushing to the original shearing function. Modern mammals with all-purpose diets, such as the opossum, still have this basic therian triangular-type tooth.

CARNIVORES, such as the lion, have modified a pair of matching cheek teeth into elongated shearing blades. These teeth are known as carnassials. Carnivores usually retain some unmodified cheek teeth for more generalized functions.

SPECIALIZED OMNIVORES, including primates such as ourselves and the chimpanzee, make the basic tooth shape flatter and more square to pulp and grind more fibrous food such as roots.

HERBIVORES, usually descending from omnivores, have run the distinct cusps together to produce high ridges on the surface of the tooth. As the lower ridges run across the upper ones the vegetation is shredded.

only true daytime marsupial is the numbat, or marsupial anteater. Judging from the more primitive forms alive today, early marsupials were probably tree-dwellers with a grasping tail and the ability to use their forepaws to hold food. Early placentals were probably more terrestrial, and their habit of burying their feces suggests that they may originally have been burrow-dwellers.

The chief difference between the marsupial and the placental skeleton is that placentals have lost the epipubic bones, a pair of struts that grow forward from the pubic bone in marsupials. These bones were once viewed as a specialized feature developing to support the marsupial pouch, but the fossil record shows that they are present in most mammals and that it is the placentals that have the rarer design. Placentals probably lost these bones because they would limit the expansion of the abdomen in the later stages of pregnancy.

Scientists used to assume that the marsupial way of birth was inferior to the placental, but John Kirsch of the University of Wisconsin and Jason Lillegraven of the University of Wyoming have recently put forward a more balanced argument. They suggest that both these mammal groups evolved from a pouchless form whose very immature newborn young completed their development fastened on the nipples. The marsupial and placental types of nurturing would then represent "equal and opposite" branching from the same original condition, with the marsupials developing a pouch to protect their immature young, while the placentals opted to keep them inside for a longer time.

A disadvantage of the placental route is that it is harder and perhaps more dangerous to cut short the pregnancy when conditions are harsh, while the marsupial mother can usually choose to eject the tiny young from the pouch when there looks like being too little food available to feed both herself and her offspring. (This strategy would not be viable for those marsupials that breed only once or twice in a lifetime, such as some Australian marsupial "mice.") On the other hand, nutrients are more efficiently delivered across the placenta than through the milk, and this fact appears to have had two chief impacts on therian history. First, placental young grow faster than marsupial young, so placentals can have a faster reproductive turnaround. Second, this nutrient input may be expressed in brain development, so that even the more intelligent marsupials are no match for their placental counterparts.

A last key difference is that the way that the young marsupials are born, having to make a steep climb up the mother's fur from the birth canal to the pouch, requires them to have precociously developed forelimbs. This may block the evolution of marsupials with highly modified forelimbs of the kind found in bats or whales, or of four-legged single-toed horselike runners. Although the marsupial and placental reproduction methods probably offered equal advantages in small Cretaceous mammals, it seems that once the size ceiling was removed placentals have outcompeted marsupials over the long haul where the two groups have been found together. The exception appears to be the case of South America, where placentals and marsupials took different ecological roles, and diversified in tandem on a continent that was cut off from the rest of the world for upward of 50 My.

A patchwork Earth

In the latest Triassic, about 210 My ago, the continents were grouped together in the supercontinent Pangaea, and the earliest mammals could radiate fairly easily to all parts of the world. Fossils from this epoch show very similar mammals in North America, South Africa, and China. During the rest of the Mesozoic the continents started to drift apart, leaving the mammals established on the different continental blocks to populate the "island continents" of the Tertiary, the chief and much the longest division of the following Cenozoic era.

Telling the story of the Mesozoic has a few things in common with tracing the rise and fall of the Roman Empire. A single dominant group occupies a single stage. Broad movements are clear, or appear to be clear, because distance obscures the details but preserves big patterns. On the other hand, the evidence comes from the "ruling" group. The dinosaurs leave more bones because they are larger and more likely to be preserved; they write their documents in bone as the Romans wrote theirs both in their words and in their monuments. The history, culture, and language of lesser groups are harder to unravel; they appear as mere footnotes to the main plot.

Then the empire collapses. New powers rise, the next age begins, but instead of an overall history with at least one common theme we find centrifugal forces pulling and actually moving in different directions. All sorts of subplots develop. As time passes, more information is preserved, but of course more contradictions emerge. When Rome falls, communications begin to disintegrate, and goods, information, and people travel more slowly. The outer regions lose contact with the center. When the dinosaurs fall, there is no longer a center, and the Pangaea supercontinent reverts to separate continental blocks, each inaccessible to the animals of other regions and to the genetic information they carry.

No Cenozoic "dark age" followed the K–T extinctions. Yet the history of life becomes fragmented, and with our greater knowledge of the changing climates and continental movements, provided by an improving geological record, comes a more detailed picture of the power of these global forces to influence the pathways of evolution. The rest of this chapter will follow these pathways – some of them. They weave into a complex pattern.

To begin with, the late Mesozoic distribution of the major groups of mammals appears to be as follows. Monotremes probably originated in Australia, although recent evidence from South America suggests that they may have had an early radiation throughout the original southern supercontinent. Multituberculates appeared in both Asia and North America. Marsupials originated in South (or possibly North) America. Placentals arose in Asia. By the late Cretaceous, marsupials and placentals had crossed paths in North America, and although marsupials did not fade out in the Northern Hemisphere their diversity of species shrank. Marsupials must have dispersed from South America to Australia by way of Antarctica sometime in the late Cretaceous or early Tertiary, before these continents lost contact.

Both placentals and marsupials are familiar residents of the early Paleocene of South

America; but because no placental fossils have turned up in Australia until the late Miocene (when rodents and bats appear to have immigrated from southeast Asia), it has been commonly assumed that it was only marsupials that first colonized Australia. In the absence of evidence to disprove a biased notion of placental superiority, it was almost taken for granted that Australia became the cradle of marsupial evolution only because placentals somehow failed to enter across the original land bridge by way of Antarctica.

A new fossil site from the early Eocene of Australia has exploded this myth. Mike Archer and his colleagues at the University of New South Wales have recently described the Tingamarra fossil locality of 55 My ago, stating that it contains a single placental mammal tooth. The animal – called *Tingamarra*, after the site – has been identified as a primitive type of omnivorous ungulate. ("Ungulate," another frequent mouthful in this chapter, derives from the Latin *ungula*, hoof.) It is probable now that the marsupials did not cross Antarctica alone, and that at least one type of placental arrived with them, only to go extinct for reasons as yet unknown.

In addition to the now extinct multituberculates, it was the therian mammals that emerged as heirs apparent to the dinosaurs. Some subgroups had already disappeared before the late Cretaceous, others crossed the divide or arose in the Cenozoic, and some of these survive to this day. Rather than introducing them in fits and starts as

Some extinct placental mammals do not fit into any of the six surviving major groups.

ABOVE *This wombat-like creature is from the taeniodont group, compact Paleocene and Eocene herbivores equipped for digging up roots and tubers. Some later species had ever-growing teeth that coped with constant wear.*

ABOVE *A member of another such group, the plagiomenids of the North American Paleocene and Eocene. They seem to have been gliding mammals that could spread a sail of fur, but they may not be related to the flying lemurs of today. Their presence inside the Arctic Circle during the Eocene indicates a year-round jungle habitat unusually far north.*

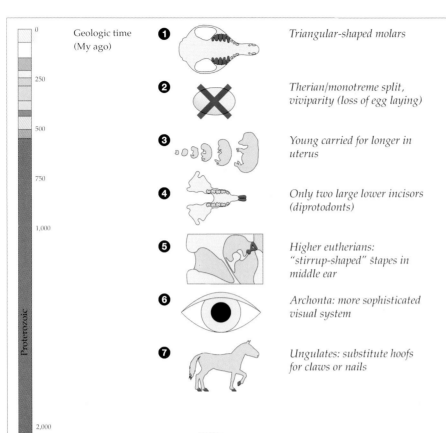

Geologic time
(My ago)

❶ *Triangular-shaped molars*

❷ *Therian/monotreme split, viviparity (loss of egg laying)*

❸ *Young carried for longer in uterus*

❹ *Only two large lower incisors (diprotodonts)*

❺ *Higher eutherians: "stirrup-shaped" stapes in middle ear*

❻ *Archonta: more sophisticated visual system*

❼ *Ungulates: substitute hoofs for claws or nails*

MAMMAL PHYLOGENY

In modern times, most of the world's largest animals are mammals, from the African elephant to the blue whale, the largest animal that has ever lived. Although it may have required the extinction of all the dinosaurs except the birds before mammals could expand into larger sizes, they had been a successful Mesozoic group for 150 My before the dinosaurs departed. The chart shows how most of their key features had been developed well before the end of the Cretaceous – with the exception of hair which they seem to have inherited from the cynodont reptiles. No class of animals has been more deeply affected by the rise of humankind. Many mammals ranging from the saber-tooth cats to giant mammoths may have been hunted to extinction as recently as 12,000 years ago, and mammals' return to the sea has been marred in the last few centuries by the commodity value of seals and whales.

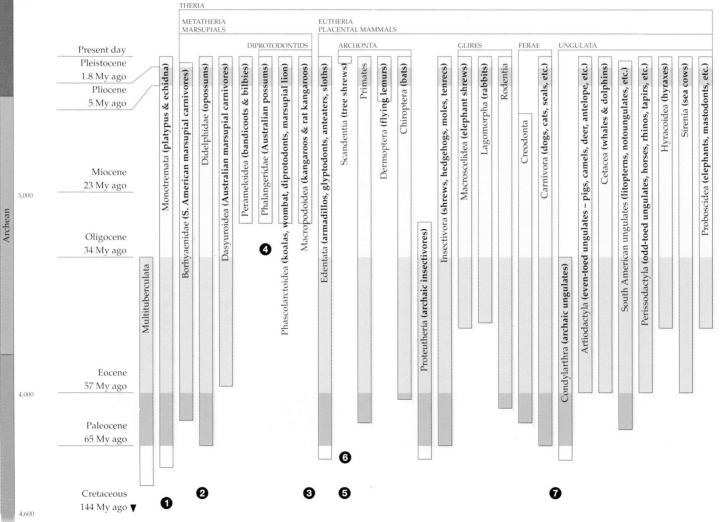

GEORGES CUVIER AND *PALAEOTHERIUM*

After the Revolution in France, science applied the rational lessons of the Age of Enlightenment to examining the natural world. At the National Museum of Natural History, Georges Cuvier (1769–1832) showed the value of combining geological and fossil evidence to devise new theories of an ancient and changing world. At the turn of the century he was receiving specimens collected during voyages of exploration, such as the big fossils of South American giant sloths. Cuvier was a gifted and methodical anatomist. He valued the gathering of evidence far more highly than high-sounding theories, and used his talents for reconstructing fossil animals – to make comparisons with living species.

Cuvier's ambition to establish a rational system of zoological classification led him to divide animals into four categories – Vertebrata, Mollusca, Articulata, and Radiata – based on their body architecture. Cuvier was not an evolutionist. He thought that revolutionary episodes in the past had caused local extinctions that made room for fresh populations of animals to immigrate from other regions.

To consolidate his theories, he excavated the early Tertiary gypsum quarries near Paris, where he found a previously unknown fossil

Georges Cuvier (1769–1832).

mammal that combined the characteristics of the modern tapir, rhino, and pig. He called this specimen *Palaeotherium*, "ancient beast," and used it to back his view that the world of the past had been very different. He went on to identify seven formations in the Paris region, geological beds that contained alternate marine and freshwater fossils. To him they suggested regular episodes when the land had sunk and disastrous floods had caused breaks in the fossil record.

A German engraving made in about 1850 of the beast named Megatherium *by Cuvier, a giant ground sloth.*

they straggle into the story, it is worthwhile to review the key groups here.

Marsupials and placentals had entered the fossil record by the late early Cretaceous, around 120 My ago. Marsupials fall into four major groups. The didelphids include opossums and a few extinct animals that also had double wombs. They are chiefly South American, although didelphids (true opossums and various opossum-like animals) ranged into North America and the Old World during the early Cenozoic and reinvaded North America in the Pleistocene.

The other three groups, found exclusively in Australia and New Guinea, are assumed to derive from common ancestors that invaded Australia in the early Cenozoic. They consist of the dasyuroids (marsupial carnivores and marsupial "mice"), perameloids (bandicoots), and phalangeroids (possums, koalas, wombats, and kangaroos, plus the extinct rhino-like diprotodonts and the marsupial lion). Phalangeroids include most of the herbivorous Australian marsupials; their key feature is their two prominent lower front teeth. There are four separate possum families, all closely related and including specialized gliding forms (a parallel with the placental "flying" squirrels).

Placentals can be divided into six major groups: Edentata (South American armadillos, sloths, and anteaters); Archonta (including primates, tree shrews, bats, and flying lemurs); Glires (including elephant shrews, rodents, and rabbits); Insectivora (true shrews, moles, hedgehogs, and related groups); Ferae (extinct creodonts and true carnivores, including the seals); and Ungulata (a multiform crew whose survivors include the familiar hoofed mammals and the whales, as well as the elephants and the hyraxes).

The sloths and their kin appear to be an early, more primitive offshoot; the links among the other groups are an unsolved maze, though with some mapped sections. Insect-eaters were once considered to be the founder-members of the placentals, but obviously the living members of the order Insectivora are no more truly primitive than most other placentals, even though their appearance and lifestyle may echo early forms. A number of fossil "insectivore" families crop up in the early Tertiary: leptictids, palaeoryctids, pantolestids, apatemyids, stuck with these ponderous academic labels because no human being ever set eyes on them to call them dog or cat.

An assortment of early mammals classed as "condylarths" have been considered to be ancestors to the ungulate order, the hoofed mammals. In fact, though, different groups of condylarths have turned out to represent quite separate lines of development, some possibly ancestral to later ungulate groups, while others are evolutionary dead ends. Not all of them are more primitive than some later groups of ungulates, and they appear to have branched off the evolutionary line *later* than

The largest carnivorous land mammal so far known was Andrewsarchus, *which was a lot bigger than a grizzly bear, with a head over 3 ft (1 m) long, a body length of 16 ft (5 m) or more, and a mass of close to 1 ton (tonne). Contrary to the docile image of its distant modern relations, it was technically an ungulate, a member of the family Mesonychidae, the group that also gave rise to the whales.*

The mesonychids appeared in the early Paleocene, but Andrewsarchus *prevailed in the late Eocene of Mongolia, where it was discovered by the fossil prospector Roy Chapman Andrews. Here it is feeding on the chance discovery of a dead* Gobiotherium *("Gobi beast"), an archaic ungulate. The teeth and skull of* Andrewsarchus *indicate that is was not an expert killer or meat specialist but had habits more like a large bear or wolverine, partial to whatever came along, including carrion.*

RECONSTRUCTING EXTINCT ANIMALS

The teeth of a fossil animal can tell us whether it was a carnivore, omnivore, or herbivore, and the limb bones identify the basic mode of life – runner, climber, etc. From teeth and limb bones we can estimate the animal's weight. Breaks or distortions may record diseases, and perhaps the cause of death, but the clues may give a broader picture, especially if the animal has close or at least comparable living relatives. *Sinclairomeryx* was a dromomerycid ("cud-chewer/runner") of the early North American Miocene (17 to 14 My ago). Its skeleton was roughly as big as a fallow deer's, and it weighed 100–120 lb (45–55 kg). Dromomerycids were related to deer, and had unbranched horns rather than branching antlers. But which of the varied deer lifestyles can we detect for *Sinclairomeryx*?

Some mature skulls have horns, others not. We can assume that the horned ones are male. The horns have no break-point near the base, as deer antlers do, which tells us that they were not shed. The horned skulls also have small paired "bosses" of the bone on the nose, and though we have no fossil canine teeth for them, the bigger socket in these skulls makes room for a bigger upper canine. This much

male/female difference suggests frequent fighting among the males, as seen today in deer or antelope groups whose males either hold territories for attracting females, and defend them against rivals, or stage yearly "rutting" tournaments that decide which males gain access to the females. In this structure the females would live together in small groups with their young, while the males would lead a solitary life, except when fighting other males. We can be confident about the fighting, because all of the fossil males have horns that have been broken and then healed. Because they have healed, we also know that the horns must have been covered with skin; naked antlers have no blood supply, and cannot heal.

The cheek teeth of *Sinclairomeryx* are moderately high, and the muzzle is quite narrow. These features would nowadays indicate a browser like the mule deer, which also takes some grass. Their legs reflect their techniques for escaping predators – do they run or take cover? – and the distances they go to find food. The legs of *Sinclairomeryx* are neither long nor short, and this suggests a halfway habitat, part forest and part savanna, which fits the suggested diet. Now the paleontologist can offer a description, and a habitat, and using these bones the artist can aim to re-create an animal that no one has seen before.

Male

Female

LEFT *A reconstruction of* Synclairomeryx, *based upon the fossil evidence, by Marianne Collins, one of the major contributors of paintings to this book. Her work combines a scientific training with artistic ability. Working in conjunction with scientists, using fossil data, and with much experience in the field, she is able to re-create, as faithfully as is possible, the extinct creatures of the past.*

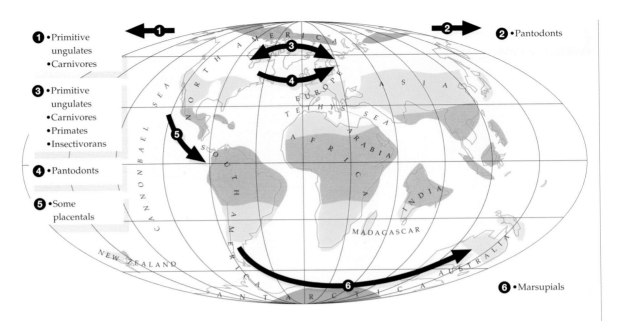

Tropical rainforest

Paratropical rainforest
(with dry season)

Subtropical woodland or
woodland savanna
(broad-leaved evergreen)

Polar broad-leaved
deciduous forest

1 •Primitive
ungulates
•Carnivores

3 •Primitive
ungulates
•Carnivores
•Primates
•Insectivorans

4 •Pantodonts

5 •Some
placentals

2 •Pantodonts

6 •Marsupials

*THE EARLY TO
MIDDLE
PALEOCENE
(63 My ago)
Our knowledge of past
vegetation is patchy,
and the broad patterns
shown in this chapter's
maps represent
informed guesswork.
They should be seen as
crude snapshots of a
more complex reality.
During the Paleocene a
warm, equable, and
much damper climate
enfolded the Earth, with
no great contrast
between the equator and
the poles. The forests of
North America appear
to have become more
dense and swampy than
during the late
Cretaceous. Tropical
and paratropical forest
types extended much
further north and south
than they do today.
Subtropical woodlands
spread to the polar
regions, where they
gave way to a broad-
leaved deciduous forest
cover unknown today,
and adapted to seasons
of perpetual night or
continual day.
A number of animals
managed to follow ice-
free Arctic routes
between North America
and either western
Europe or eastern Asia.
No migrations seem to
have been attempted
between Europe and
Asia, which were
separated by inland
seas. Africa, India, and
Madagascar were
probably cut off. South
and North America
were not joined. Yet
some mammals moved
south to found the
ancestral stocks of a
unique fauna. South
American marsupials
had access to Australia
by way of the Antarctic.*

the even-toed ungulates, the artiodactyls,
which are today's most successful and diverse
ungulate group and include pigs, hippos,
deer, and antelope.

We tend to think of the hoofed animals as
herbivores – another main group are the
perissodactyls, odd-toed ungulates, which
include horses, rhinos, and tapirs. But one
condylarth group, the extinct mesonychids,
were carnivorous, and their skulls have
features in common with the skulls of early
whales. Biochemical studies of living
mammals show that the whales can be
classed among the ungulates! Among living
mammals, they are most closely related to
even-toed ungulates like pigs and cows.

The Paleocene world

The world of the earliest Tertiary, from about
66 to 57 My ago, was a much more equable
place than it is today, with a tropical or
subtropical type of climate reaching to the
polar regions. Evidence suggests that patterns
of rainfall may have changed dramatically after
the extinction of the dinosaurs, with much
higher levels spread more evenly through the
seasons of the year. The forests of the North
American Paleocene seem to have been more
dense and swampy than in the late
Cretaceous.

Our knowledge of past vegetation is very
patchy, and the broad bands shown in the
maps for this chapter represent a few good
sample points extrapolated into a general

view. Nor can any one map describe an epoch
that lasted for millions of years. Climatic
fluctuations are certain to have shifted the
boundaries between the chief vegetation types
as they are still doing today; these are quick
snapshots of a changing scene.

In the early Tertiary there were deciduous
forests at the poles. These contained plants
adapted to a much warmer environment than
is found north of parallel 66 today, but
probably subject to the same regime of light
and dark. The trees grew very large leaves to
capture light and energy during the summer's
constant daylight, then probably shed them in
the winter night. Although tropical kinds of
forest covered the rest of the globe, they must
have differed from today's because practically
all of their inhabitants were small tree-
dwelling mammals. There were none of the
larger carnivores or herbivores seen in today's
tropical forests – no version of the jaguar,
leopard, okapi, or tapir.

It seems that the mammals took several
million years to evolve into even moderately
large body sizes, let alone the upper range of
the niches left vacant by the dinosaurs. One
reason may be that the forests were more
dense in the Paleocene than they had been in
the Cretaceous, with no massive dinosaurs to
trample or browse them, and a dense forest
habitat favors small tree-dwelling animals
over larger ground-dwellers.

Another peculiarity of the Paleocene
mammals is that hardly any had teeth that
would suggest a leaf-eating diet. Most of the
mammals seem to have been either insect-

eaters or generalists, perhaps adding fruits, berries, and shoots.

Why ignore the vast supply of food in the leaves of plants? In the forests of the modern world, leaf-eating mammals are most common in seasonal environments where the temperature and/or rainfall fluctuate considerably through the year. These environments harbor many deciduous plants that shed their leaves during the hardest season. If plants are going to discard their leaves regularly, there is less point in taking major precautions to protect them from consumption by mammals or by insects. As a result, the leaves of plants in seasonal habitats are more edible than those in non-seasonal ones, where the plants will protect their investment by putting all sorts of nasty chemical compounds in the leaves. The shortage of leaf-eating mammals implies a world with little seasonal change.

The first flush of mammal radiations in the Paleocene contained mainly groups that are termed "archaic" because they were not the direct ancestors of any surviving animal group. That does not mean that these animals were somehow inefficient or ill-adapted, and in fact their very rapid spread shows that they were highly successful for their time. "Archaic" conditions require "archaic" designs, and the replacement of these early designs by mammals that we consider more "modern" is likely to reflect the sweeping changes produced by a later, more seasonal world. There is no foresight in evolution; adaptation can only respond to the here and now.

At the start of the Paleocene the mammals consisted of holdovers from the late Cretaceous: multituberculates, didelphid marsupials, primitive types of raccoon-like ungulates, plus a tribe of insectivorous

mammals. With no large true carnivores in circulation, the small carnivore role seems to have been played by some of the condylarth ungulates such as *Oxyclaenus*.

Some new types of mammals saw the start of the Paleocene. Squirrellish early primates (plesiadapiforms) appeared in the Northern Hemisphere. Some of the proteutherian families were making their debuts, and there were "true" insectivores in Europe and North America distantly related to present-day shrews and hedgehogs. The North American paleanodonts were burrowing and anteating animals that looked like the South American edentates but had probably developed the same adaptations by an independent evolutionary pathway.

Some larger, more herbivorous types took to the ground. The taeniodonts ranged from the early Paleocene through the middle Eocene in North America – from around 66 to 40 My ago – and are also found in the early Eocene of Europe and the later Eocene of Asia. The best comparison is with giant heavy-jawed wombats; like wombats they had powerful limbs for digging, prominent incisors, and high-crowned cheek teeth, and they probably had a similar habit of grubbing for roots and tubers. Some later forms, such as *Ectoganus* from the middle Eocene of North America, developed the wombat feature of ever-growing cheek teeth, suggesting a parallel solution to the tendency for dirty, gritty food to wear out ordinary teeth.

With so much ecospace to be filled, an account of the Paleocene cannot help turning into a roll call of mammalian diversity. Many of its characters are unknown today. The condylarths explored a number of different roles in the form of the piglike peryptichids; the squirrel-like hyopsodontids; the

All of today's placental carnivores belong to the single order Carnivora, but larger carnivores of the Paleocene and Eocene were members of a related order, Creodonta ("meat-eating tooth"), now extinct. There were two main types, the hyaenodontids and the oxyaenids ("hyena-tooth" and "sharp hyena"), only some resembling modern hyenas. Evolution has a basic pack of carnivore types that it keeps shuffling and redealing. Here we show four standard types.

BELOW, TOP LEFT *The "ferret" design produces a small generalized carnivore that eats small prey such as rodents and lizards, and may be slim enough to raid burrows. The version illustrated is the hyaenodontid Tritemnodon ("three-part cutting tooth").*

BELOW, BOTTOM LEFT *The "cat" is a carnivore specialized for a diet of almost pure meat, in this case the oxyaenid Patriofelis ("father of cats").*

BELOW, TOP RIGHT *The "dog" is a carnivore that takes some meat as part of a broader menu, and is often more specialized for running than other types. This "dog" is the hyaenodontid Arfia, named after the sound that a dog makes.*

BELOW, BOTTOM RIGHT *The "hyena" or "wolverine" type is large and often heavily built, with strong blunt teeth well suited for a scavenging, bone-crunching role, here Oxyaena.*

Carnivore types not shown include bear- or raccoon-like omnivores and the saber-toothed "cats."

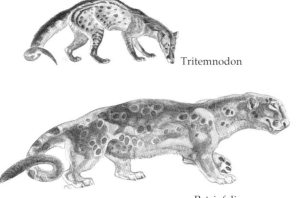

Tritemnodon

Patriofelis

Arfia

Oxyaena

Barylambda belonged to the order Pantodonta, found in the Paleocene to middle Eocene of North America, Asia, and western Europe. Plant-eaters, they were probably part- or full-time water-dwellers.

In the early Paleocene of North America, mammals had begun to prey on each other. Here, a Purgatorius prepares to defend her young against the carnivorous Oxyclaenus. Purgatorius was an omnivore, about the size of a small bushbaby, a primitive member of the plesiadapiform group that were either the earliest primates or their close relations. Oxyclaenus was an arctocyonid ("bear-dog"), related to today's hoofed mammals.

omnivorous raccoon-like or bearlike arctocyonids; the ground-dwelling omnivorous or herbivorous phenacodontids; and the carnivorous mesonychids, resembling wolverines or hyenas. The main carnivores belonged to the order Creodonta, now extinct. The earliest members of the Carnivora were the small genet- or ferret-like miacoids, also extinct. They were probably opportunistic general feeders rather than true specialists; it would take another 25 My for members of the order Carnivora to emerge as the leading predators on land.

Much of our knowledge of the early Paleocene comes from North America, but sparse evidence from the rest of the world reveals some unique regional differences. Asia seems to have introduced the large, clumsy-looking pantodonts and tillodonts, omnivores or herbivores that migrated into North America later in the Paleocene. The pantodonts grew to be probably the biggest mammals of their day. They ranged from dog- to bison-sized, and included semi-arboreal

forms and semi-aquatic hippo-like forms, although most stayed on the ground. One of the most unusual pantodonts was *Barylambda* of the late Paleocene of North America, about 58 My ago, a variation on a ground-sloth design.

North America alone provides a fossil record of the opossum-like marsupials, not yet found in South America until the middle Paleocene, when they appear in such diversity that they must have been there earlier. Rather later in the early Paleocene, and sometime after their first appearance, came the unique radiation in North America of the plesiadapiform primates, branching into several distinct and different families. For example, the carpolestids had specialized teeth for husking fruit, like some present-day Australian possums, and the picrodontids appear to have been gum-eaters, like some bush babies and marmosets among modern primates. Plesiadapiform primates were not found elsewhere in the world at this time.

We know almost nothing about Africa during Paleocene times. But sparse fossil evidence from late in the epoch suggests the possibility that Africa gave birth to the foxlike members of the extinct Creodonta order of carnivores, as well as to some more advanced true primate forms such as the tarsiers found today in Asia.

Our knowledge of South America comes only from the mid- to late Paleocene. Here all the equivalents of insect-eaters and flesh-eaters in the Northern Hemisphere carried their young in pouches: for example, the opossum-like didelphids and the carnivorous borhyaenids. But with its hoofed animals, South America pushed ahead with placentals such as the litopterns and notoungulates, carrying their young in the womb much longer. It is possible that climatic conditions were different enough in South America to give rise to a more seasonal and open type of forest habitat that made room for a variety of ground-dwelling leaf-eaters.

The later Paleocene of the Northern Hemisphere saw a further wave of expansion, particularly of larger ground-dwelling animals. The main carnivores were mesonychid "condylarths," found throughout the northern world. North America also saw the evolution of various catlike members of the Creodonta, the oxyaenids. Among the

larger plant-eaters, more kinds of pantodonts appeared, especially in Asia, and it was also in Asia and North America that the early uintatheres found a foothold. (The later uintatheres of the Eocene were spectacular rhino-like animals with an assortment of bony horns on their heads, and saber-like upper canine teeth.)

Yet these wonderfully diverse large animals might have had a lumbering look to a modern eye whose mammal images are formed by dogs and cats, antelope and horses. These were not the fleet, long-legged, graceful forms that bound across the grasslands of today, and it would be many millions of years before those grasslands evolved, along with the mammal forms adapted to their open conditions.

Oxyaenids were long-bodied and short-legged, like an outsize weasel, and would have been forest-dwelling ambush hunters that could pounce on their prey, but not give chase. Mesonychids had rather longer legs, but their teeth suggest a hyena's lifestyle, scavenging rather than active hunting. Taeniodonts grubbed in the forest floor. The condylarths were small and stocky creatures, and the more ponderous pantodonts and uintatheres may well have been semi-aquatic riverside dwellers; their teeth do not look capable of dealing with anything but soft vegetation.

There was a limited exchange of mammals between various parts of the world in the Paleocene. Some primitive ungulates, carnivores, and insectivores must have been able to migrate over an ice-free Arctic route between North America and Europe, on the one hand, and North America and Asia, on the other. There seem to have been no migrations between Europe and Asia, as these areas were separated by inland seas. Africa, India, and Madagascar were probably isolated from the rest of the world, and although South America was also cut off, some mammals found their way there from North America to found the unique mammal fauna.

A key event of the later Paleocene was the way that the earliest rodents such as *Paramys* invaded North America from their first known habitats in Asia. Their arrival in North America coincides with a fall in the numbers of the multituberculates and the local squirrel-like primates.

The early and middle Eocene

Toward the end of the Paleocene and until about 50 My ago in the early Eocene, the global climate grew noticeably warmer. The range of the tropical type of vegetation expanded, pushing the paratropical rainforest inside the Arctic Circle to create one of history's more unusual environments: jungles at the poles. The forests of the higher latitudes contained a growing diversity of mammals.

Many of our present-day orders made their first appearance in the early Eocene, among them the true primates and the even- and odd-toed hoofed mammals (artiodactyl and

THE EARLY TO MIDDLE EOCENE (50 My ago) Around 55 My ago the global climate grew noticeably warmer. Tropical vegetation spread inside the Arctic and Antarctic circles, and brought with it a much greater mammalian diversity than the Paleocene could support.

Many of our present-day orders made their first appearance in the early Eocene, including true (lemuriform) primates and even- and odd-toed hoofed mammals (artiodactyls and perissodactyls). Archaic mammals such as plesiadapiform "primates," condylarths, proteutherians, and multituberculates were now on the decline, and would be extinct by the close of the epoch.

Waves of mammalian migration swept to and fro across the northern continents, whose regional populations grew more alike, without losing their individual flavor. A key development was the movement into and out of Africa that may have begun in the very latest Paleocene. Migration appears to have tailed off toward the middle Eocene.

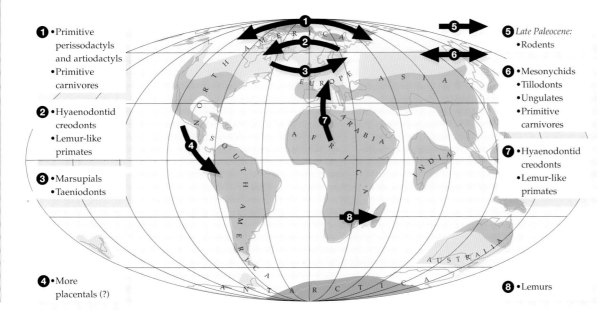

1
• Primitive perissodactyls and artiodactyls
• Primitive carnivores

2
• Hyaenodontid creodonts
• Lemur-like primates

3
• Marsupials
• Taeniodonts

4
• More placentals (?)

5 *Late Paleocene:*
• Rodents

6
• Mesonychids
• Tillodonts
• Ungulates
• Primitive carnivores

7
• Hyaenodontid creodonts
• Lemur-like primates

8
• Lemurs

☐ Tropical rainforest
☐ Paratropical rainforest (with dry season)
☐ Subtropical woodland or woodland savanna (broad-leaved evergreen)
☐ Polar broad-leaved deciduous forest

ABOVE Icaronycteris *is one of the earliest known bats.*

BELOW *Whales and sea cows first appeared about 50 My ago. Basilosaurus, 60 ft (18 m) long, was a "prototype" whale with a tiny head. The heads of modern whales may make up one-third of body length. With an eel-like motion,* Basilosaurus *pursued smallish prey, and retained hind limbs that later whales shed.*

perissodactyl ungulates). Many of the more archaic mammals of the Paleocene, such as condylarths, proteutherians, and multituberculates, were on the decline and would be extinct by the end of the epoch. Yet while a lot of new types appeared, they did not immediately replace the older mammals. The early Eocene world made room for them both, with no dramatic extinctions but an increasing tilt toward newcomers that expanded and radiated while the old-timers dwindled and mainly disappeared by the end of the middle Eocene. This period also covers the rise of the present-day rodent families.

The Eocene, "dawn of new times," takes its name from the introduction of many of the living orders and even some living families of mammals between about 57 and 34 My ago. Three of the most specialized living mammalian orders are the bats (order Chiroptera), first found in the early Eocene, and the whales and sea cows (orders Cetacea and Sirenia), mammals returning to the sea in the middle Eocene. Other newcomers were more modern types of primates reliably

classified among the order Primates, and dividing into the lemur-like adapids (such as *Notharctus*) and the tarsier-like omomyids (such as *Teilhardina*, named after the French paleontologist and philosopher Teilhard de

Chardin). There is some doubt about the true primate nature of the plesiadapiforms (like *Purgatorius*), though they were surely close relations of some living types of primates.

Early members of the orders Artiodactyla and Perissodactyla are the tiny mouse-deer-like *Diacodexis* and the earliest horse,

Hyracotherium. Both of these orders are known across the Northern Hemisphere, though few artiodactyls have been found in Asia. Primates turn up only in North America and Europe.

The Eocene was a particularly successful time for the perissodactyls. They diversified into a variety of families, some of them survivors (horses, rhinos, tapirs), some now extinct (brontotheres, chalicotheres, palaeotheres). Brontotheres thrived in the early and middle Eocene. They were large rhino-like beasts found in Asia and North America, with later members sporting elaborate bony horns on the nose, whereas true rhinos have horns made of keratin, which is compressed and hardened skin. Other successful ungulates of the earlier Eocene were the uintatheres of Asia and North America, which had now evolved into rhino-sized horned forms. The diversity of the early Eocene artiodactyls was limited to very small and mostly omnivorous forms, the dichobunids, such as *Diacodexis.*

Among the carnivores, hyaenodontid creodonts, about the size of foxes, spread throughout the Northern Hemisphere in the early Eocene. The wolverine-like oxyaenid creodonts spread to Europe from North America, then went extinct there by the end of the early Eocene and also vanished from North America by the end of the middle Eocene. The true carnivores, the miacoids, abundant throughout the Northern Hemisphere, were still small animals, like our weasels and ferrets.

Africa and South America appear to have been completely isolated from the Northern Hemisphere for much of the early Eocene, although there may have been some exchange between Africa and Europe at the start of the epoch, as we shall see. Some groups probably originated in Africa: the extinct rhino-like arsinotheres, some more advanced types of primates, early hyraxes, proboscideans, and sea cows.

The first proboscideans ("trunk-feeders"), such as *Moeritherium,* were hippo-like water-dwellers weighing about 1/4 ton (0.25 tonne) and only just beginning to extend their incisors into tusks. The original long-nosed style was to develop these short tusks in both their upper and their lower jaws, and many of their descendants retained and lengthened

this full set. The deinotheres, distant cousins to mastodons and elephants, lost the upper tusks and kept a pair of lower tusks that curved down and back. The familiar type of tusk seen today, curving forward from the upper jaw only, was a late invention in late Tertiary elephants and mastodons.

It was these same comparative newcomers that evolved the familiar domed forehead and short face that humans identify with elephantine wisdom. Earlier proboscideans had long faces, more like a horse. Their most extraordinary invention is the trunk that enables them to handle food with such delicate precision. It is an ingenious solution to the handling problem presented by modifying the front legs for running and load-bearing with hoofs rather than claws.

The fossil record from South America is much better, and shows how the Paleocene animals continued to flourish there. As in the Northern Hemisphere, the numbers of tree-living insect-eaters declined (here they were all marsupials), and the ground-dwellers multiplied, with a variety of armadillos and glyptodonts – extinct, armored, giant armadillo-like creatures, using defensive strategies rather like those of the ankylosaurian dinosaurs, sometimes including the use of the tail as a club, complete with spikes. Hoofed animals diversified, and in particular the notoungulates expanded to comprise six different families, though in those days all of them were small- to medium-sized and stocky, like a pig or a hyrax in design. Finally, the middle Eocene introduced the early ground sloths.

How could a changing Eocene climate extend the Earth's vocabulary of mammal types? Climatic warming may have brought a more phased distribution of annual rainfall, and this would result in a looser spacing of forest trees, with roots more widely spread so as to absorb enough water in the dry season. As the forest floor opened up, more light would fuel the lower levels, encouraging undergrowth vegetation that would feed a larger population of ground-dwelling plant-eaters, free to develop new types and to promote themselves to sizes that no tree-dweller could safely maintain. New types of primates in the trees and of ungulates on the ground were eating more leaves, or so their

ABOVE *Steller's sea cow, also called* Hydrodamalis *("water-maiden"), was a giant dugong about 26 ft (8 m) long. It lived in the Arctic in the region of the Bering Strait and was hunted to extinction by humans within twenty-seven years of its discovery in 1741. Sea cows first appeared in the Eocene, where they fed on the sea grass that grew in the warm shallows of the Tethys and other seas. They reached their peak diversity in the Miocene. Present-day sea cows are tropical animals that eat sea grasses or floating vegetation in rivers and estuaries – sluggish, apparently highly vulnerable creatures that have retreated to a perilous safety.*

OPPOSITE, MIDPAGE *The puppy-sized* Hyracotherium, *also called* Eohippus *("dawn-horse"), the earliest known horse, belongs to the paratropical forests of North America and Eurasia in the early Eocene. The teeth suggest a diet of soft leaves and fruit, rather than grass, and instead of running in herds on open plains it probably lived alone or in pairs, slipping through forest shadows on small clustered hoofs, four on each front foot, three on each back foot, supported on a fleshy foot pad.*

A scene from an early Eocene forest, from around 55 My ago, which supported a diversity of mammals.

Diacodexis *drinks from the swamp, an even-toed ungulate and omnivorous browser. Beside it on the ground is* Palaeosinopa, *a primitive insectivore resembling a modern otter shrew. Approaching them with bared teeth is* Palaeonictis *(literally "ancient weasel"), a scavenging, bone-crushing carnivore belonging to the extinct order Creodonta.*

On the trunk of a tree is Cantius, *a type of primitive omomyid primate and fruit-eater, and in the fork of the tree* Apatemys, *an early insectivorous mammal.*

In the distance are two Coryphodon, *members of the archaic ungulate-like order Pantodonta.*

This reconstruction is based upon the fossil evidence found at the Abbey Wood site in England. The wide range of vegetation indicates a dense, paratropical forest, and the presence of mangrove swamps. Tropical trees with large, juicy fruits and abundant insect life were prominent, as were lianas and ferns. The site's fossils also indicate open areas of ground vegetation. This variety of food provided a habitat for many kinds of mammals, and most of the plant species found grow today in tropical Asia.

teeth suggest, than their Paleocene ancestors, and this implies a seasonal environment where leaves were less well protected against consumption.

One parallel is with the rainforest of present-day Central America, whose rainfall is rather more seasonal than in Old World rainforests. All of today's tropical forests contain a diversity of leaf-eating monkeys, but Central America houses a much greater variety of ground-living herbivores, including not only tropical deer and tapirs but also large rodents such as pacas, agoutis, and capybaras, which seem to have expanded into a role played elsewhere by hoofed mammals.

The early Eocene brought a great wave of migrations between the continents of the Northern Hemisphere, except where the Turgai Strait divided Europe from Asia. There seems to have been an open frontier across the Arctic between Europe and eastern North America and between Asia and western North America, though each region retained its own distinctive brand of mammals. For example, where North America had all sorts of small tapir-like browsers, in Europe that

role fell to palaeotheres, more closely related to horses. The rhino-like brontotheres were found in both North America and Asia but were unknown in Europe, which appointed its own variety of largish rhino-like tapirs.

In the early and middle Eocene, Asia had a great diversity of tapir types, many of them unique to that continent, including tiny dwarf species and long-legged forms. The chalicotheres, strange odd-toed "ungulates" that later substituted claws for hoofs, are first found in the early Eocene of Europe, but turn up in the middle Eocene of Asia and North America.

One notable Eocene migration was the spread of opossum-like marsupials from North America into the Old World, where they persisted as rare animals until the early Miocene. There also appears to have been migration between Africa and the Eurasian continent at the start of the Eocene or in the very latest Paleocene. Lemuriform primates, the ancestors of the later primates, probably grew up in Africa and gained access to the rest of the world. The lemurs may have rafted from Africa to Madagascar at this time.

Middle to late Eocene

The late early Eocene represents the peak in global temperatures for the Tertiary. During the middle and late Eocene a cooler and drier trend was particularly marked in higher latitudes, where winter frosts may have been felt in the later epoch. At the end of the Eocene, about 34 My ago, the downhill trend grew suddenly steeper. Perhaps in the space of a million years, northern latitudes experienced a plunge in annual temperatures and a great increase in their seasonal range.

During the late Eocene the tropical types of vegetation were squeezed toward the equatorial latitudes, while higher latitudes developed a new type of vegetation, temperate woodland of mixed coniferous and deciduous trees, similar to the present forest cover seen in Canada and across northern Europe. The climatic changes led to a host of extinctions, especially in the higher latitudes; many once common mammals became rare or vanished entirely.

These extinctions were probably not a direct effect of colder weather, but happened when the increased seasonality and cold winter disrupted the growth patterns of vegetation. Food items such as fruit and berries were no longer available all year round outside the remaining tropical forests. The scales now tipped in a different direction. As the northern forests became more seasonal, it was the old and more fruit-dependent herbivores and tree-dwellers that died out or were restricted to the tropics. At the same time larger herbivores appeared, with teeth more prominently ridged ("lophed"), specialized to deal with the coarser vegetation more typical of seasonal woodland.

What caused the change? A likely explanation stems from the splitting of the southern continents; at this time Australia separated from Antarctica. In the Northern Hemisphere the passage between Greenland and Norway opened up and allowed circulation between the Arctic and the North Atlantic oceans. The clash between the cold and warmer waters of the poles and tropics would have altered the patterns of global deep water circulation, cooling the continental masses of the higher latitudes. The build-up of the icecap on Antarctica may have begun during the middle or late Eocene.

The onset of such great changes in climate and vegetation had to be reflected in the mammalian faunas of the latter half of the Eocene. This revolution gave its own stern definition of the so-called "archaic" mammals: it made most of them extinct. The start of the late Eocene saw the worldwide loss of the following groups: primitive (plesiadapiform) primates; miacoid true carnivores and nearly all the condylarth carnivores, the mesonychids and arctocyonids; most multituberculates; herbivorous condylarths (though the squirrel-like hyopsodontids reached the end of the Eocene); and the clumsy large archaic herbivores – uintatheres, tillodonts, taeniodonts, and pantodonts, though some survived in Asia into the Oligocene. There were still lemuriform

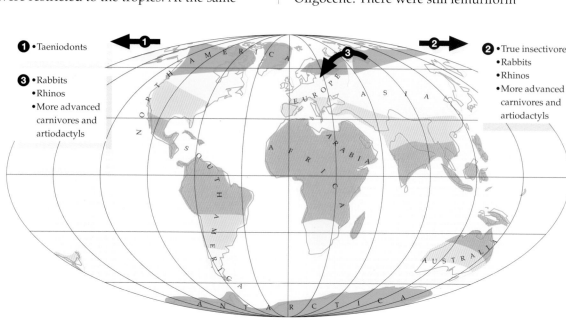

1 •Taeniodonts

3 •Rabbits
•Rhinos
•More advanced carnivores and artiodactyls

2 •True insectivores
•Rabbits
•Rhinos
•More advanced carnivores and artiodactyls

■ Tropical rainforest
□ Paratropical rainforest (with dry season)
□ Subtropical woodland or woodland savanna (broad-leaved evergreen)
□ Temperate woodland (mixed coniferous and deciduous)
■ Polar broad-leaved deciduous forest

primates in Europe and North America, but they had dwindled in numbers and did not outlast the Eocene in the Northern Hemisphere.

A tide of new types rose, forerunners of the mammalian lineages that took hold in the later Tertiary. Rabbits appeared in the late Eocene of Asia and spread fast to North America (but did not reach Europe until the Oligocene). The rodents were diversifying, and the earliest members of many living families now appeared, including animals related to present-day gophers, beavers, dormice, and hamsters.

With the departure of many archaic carnivores came the arrival of new families of the order Carnivora (which contains all the living placental carnivores). Amphicyonids (extinct bear-dogs) and early rather ferret-like members of the bear family were seen across the Northern Hemisphere, while Europe yielded genet-like carnivores (viverrids) and weasel-like mustelids, North America early dogs, and Asia early nimravids (the extinct "false" saber-tooths). Most of these early true carnivores were still small-bodied in the late Eocene. The hyaenodontid creodonts still played a larger role all over the world, reversing the fate of the other archaic carnivores in the late Eocene by producing large wolf-sized forms such as *Hyaenodon* itself.

A dramatic turnover among the herbivores was headed by the even-toed ungulates, whose teeth showed more and more evidence of adaptation to the fibrous diet provided when greater seasonality favored the supply of coarser, older vegetation over an abundance of new young growth. In the late Eocene, we can see the emergence of more piglike specialized omnivores and more deerlike herbivores.

Although true pigs had not yet evolved, peccaries appeared in North America. Soon after piglike animals first appeared in Asia, the now extinct pig-related anthracotheres spread into Europe, North America, and Africa, and the equally porcine entelodonts to North America. The more herbivorous types of even-toed ungulates included species related to camels, but the earliest ruminants (animals that chew the cud) were the traguloids whose only modern survivors are the mouse deer or chevrotains (family

Tragulidae). A few early true ruminants – ancestors to the later horned families – turn up in the late Eocene of Eurasia.

The odd-toed ungulates, perissodactyls, that had been the leading browsers on leaves of the early and middle Eocene still remained prominent among the fauna, despite the rise of the new even-toed types. There were many kinds of primitive horses and horse-related paleotheres in Europe, though the number of horses had declined in Asia and North America. (They bounced back at the end of the Eocene in North America with the evolution of members of the larger and more herbivorous horse subfamily Anchitheriinae, specialized browsers such as *Mesohippus*.)

The small tapirs so successful in the early Eocene were also vanishing, but the rhino-like lophiodontid tapirs were still doing well in Europe, as were the small tapirs of Asia, looking almost like gazelles. Around the middle of the Eocene a new odd-toed group of browsers appeared, one that would give rise to a creature fabled in our own day for power and belligerence: the rhinoceros. But its earliest ancestor lacked horns, and was about the size of a small horse.

Finally a great success story for the late Eocene was the spread in both Asia and North America of the brontotheres. Middle Eocene brontotheres had ranged from the size of a tapir to that of the small present-day Javan rhino, and they had no horns. Later Eocene brontotheres were as big as or bigger than today's white rhino, with a medley of big nasal horns made of bone, instead of the compressed skin (keratin) that forms the horns of living rhinos.

The very end of the Eocene brought another wave of extinctions that finished off both Paleocene relics such as the condylarths and newer types from the early Eocene such as the lemur-like primates in the higher latitudes. Humanity was reprieved when the primates survived in the more tropical regions where forests could still provide a year-long fruit supply.

A blow to several European mammals was the drying up of the Turgai Strait that had previously formed a barrier between the mammals of Europe and Asia. The disappearance of this inland sea was linked to a worldwide fall in sea level as water was locked into the polar ice. Animals once shut

BIRDS: VARIATIONS ON A VERSATILE PLAN

The most familiar and successful birds are small fliers of the kind that thrive even in the heart of the cities, so fragile that few have reached the fossil record. The Andean condor of today has the most advanced design for long, high flight: a 10 ft (3 m) wingspan, with a big "keeled" breastbone to anchor powerful flying muscles. The birds have lost what used to be another spectacular type, the flightless running carnivore, with a massive hatchet of a beak, powerful legs, and deadly claws, and at least 7 ft (2 m) tall at the shoulder.

Diatryma belongs to the Eocene of North America and Europe. *Phorusrhacus* comes from a family that were top predators in South America from the Oligocene to the Pliocene (38 to 2 My ago). Too tough to tackle when full grown, they may have been unable to protect their fledgling young, hatched on the ground, from later breeds of small fast mammal carnivores.

Birds have developed two main designs for swimming. They can "fly" through the water using wings transformed into flippers, mounted on solid breastbones, as penguins do. Or they can convert their hind legs to broad-webbed paddles. Many amphibious birds do this today, but *Hesperornis* was a full-time water-dweller of the Cretaceous whose propulsion came from powerful legs.

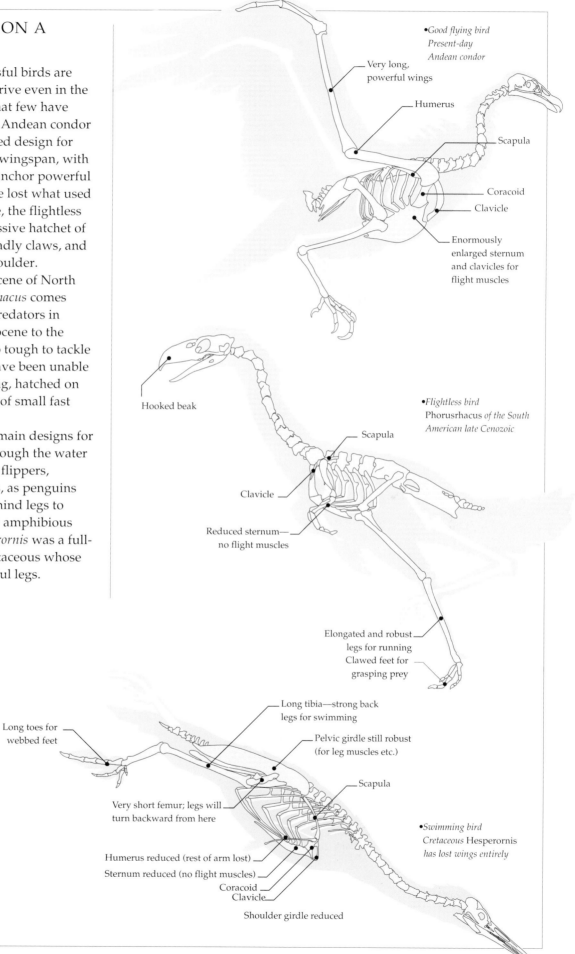

•*Good flying bird Present-day Andean condor*

Very long, powerful wings

Humerus

Scapula

Coracoid

Clavicle

Enormously enlarged sternum and clavicles for flight muscles

Hooked beak

•*Flightless bird* Phorusrhacus *of the South American late Cenozoic*

Scapula

Clavicle

Reduced sternum— no flight muscles

Elongated and robust legs for running
Clawed feet for grasping prey

Long toes for webbed feet

Long tibia—strong back legs for swimming

Pelvic girdle still robust (for leg muscles etc.)

Scapula

Very short femur; legs will turn backward from here

Humerus reduced (rest of arm lost)

Sternum reduced (no flight muscles)

Coracoid
Clavicle

Shoulder girdle reduced

•*Swimming bird* Cretaceous Hesperornis *has lost wings entirely*

THE LATEST
EOCENE TO THE
EARLY OLIGOCENE
(35 My ago)
*A steep decline in
annual temperatures
may have happened in
as little as 1 My at the
end of the Eocene. It
introduced a cooler,
more seasonal world
climate, unfriendly to
the warmth-loving,
older types of mammals.
By the early Oligocene
the polar broad-leaved
deciduous forests had
vanished. The northern
latitudes, all the way
across Asia, North
America, and Europe,
were dominated by
mixed coniferous and
deciduous temperate
woodland, bordered to
the north by a new zone
of broad-leaved
deciduous forest.*

*In the cool world of the
Oligocene there was less
tropical and
paratropical forest even
than today, but not the
same range of climates
between the tropics and
the poles, though a
sparse tundra had
sprung up around the
new icecap in
Antarctica. The Earth
was less dry than today.
Some of the woodland
may have been patchy,
but there were no broad
stretches of grassland,
or of desert or semi-
desert. We have little
evidence of migration
between continents after
the earliest Oligocene,
except for the monkeys
and caviomorph rodents
that reached South
America.*

☐ Tropical rainforest
☐ Paratropical rainforest
   (with dry season)
☐ Subtropical woodland or
   woodland savanna
   (broad-leaved evergreen)
☐ Temperate woodland
   (broad-leaved deciduous)
☐ Temperate woodland
   (mixed coniferous and
   deciduous)
☐ Tundra
☐ Ice sheet

out of Europe now flooded in from Asia, with dire effects on the natives. For example, the European tapirs became extinct, replaced by the Asian true rhinos, and many of the unique small herbivores of Europe were replaced by the ruminants of Asia.

This interval at the Eocene/Oligocene boundary of Europe has taken the name of "La Grande Coupure," the Great Divide, which signals the outbreak of extinctions that resulted from the impact of climatic changes and the Asian invasion. Perissodactyls were hard hit. Horses disappeared outside North America. Tapir-related animals declined everywhere; and only those that gave rise to the modern tapirs survived past the Oligocene. Brontotheres went extinct in North America and lasted in Asia only till the middle Oligocene. Only the true rhinos came through in good order all over the world.

The quiet Oligocene

As it entered the Oligocene the world was growing rapidly cooler and more seasonal. Waves of extinction overtook the mammals that had been better suited to the more tropical world of the earlier Eocene, more than 15 My before. By the early Oligocene of 32 My ago the polar broad-leaved deciduous forests were utterly gone. Antarctica was icecapped, with a sparse tundra type of vegetation spreading around its margins. As yet there was no ice at the other pole, and the vast expanse of the northern higher latitudes

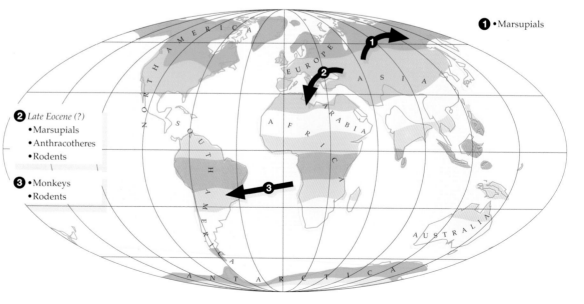

❶ • Marsupials

❷ *Late Eocene (?)*
   • Marsupials
   • Anthracotheres
   • Rodents

❸ • Monkeys
   • Rodents

The Fayum fauna from the early Oligocene of Egypt (around 35 My ago) represents the first major fossil locality in Africa. The fossils reveal animals that were unique to Africa when it was isolated from the rest of the world. Illustrated are two members of the extinct primate family Propliopithecidae: Aegyptopithecus *and* Propliopithecus. *Near the shore is the herbivore,* Arsinöetherium, *which belonged to an order of mammals, related to elephants.*

BELOW Hesperocyon *("western dog") was an early dog from the late Eocene and Oligocene of North America. It was about the size of a small fox, and like present-day foxes and other small members of the dog family it was probably omnivorous rather than strictly carnivorous, and probably lived alone or in small family groups rather than running in packs. Wolflike dog types did not show up until a few million years ago, although before them other types of predators such as the now extinct bear-dogs and dog-bears may have done similar work as large carnivores.*

was dominated by mixed coniferous and deciduous trees. (The northern tundra is very recent.) A zone between this and the subtropical woodland contained a new type of vegetation, a broad-leaved deciduous woodland similar to the forests of present-day western Europe and the eastern United States.

This world was cooler than in the early Tertiary, and the range of its tropical forests was actually more restricted than today's. At the same time there were fewer differences between pole and tropics; at least in the Northern Hemisphere there were vast stretches of essentially temperate woodland, with no evidence of the highly cold-adapted plants we see on today's tundra or taiga. Likewise, the Oligocene world was not as dry

as now. Some woodland may have been patched with shrubby growth and open spaces, but there were no far grassland horizons.

The picture that emerges is of a generally equable environment. Except at the beginning, no great events took place. There is little evidence for animal migrations between continents or for any sweeping changes either among the mammals or even among the fishes. Apparently this was an age of evolutionary stability, a lull between the origins and extinctions of the earlier Tertiary and the new appearances and migrations of more modern types of mammals in the later Tertiary. Few paleontologists have been drawn to the Oligocene – and perhaps for that

RIGHT *Cainotherium ("new beast"), a rabbit-like hoofed mammal distantly related to camels, was active in Europe in the Oligocene and early Miocene. Cainotheres may have gone extinct because of competition from newly emerging advanced ruminants. Alternatively, the cause may have been changes in climate and vegetation.*

OPPOSITE, ABOVE *The gazelle-like camel* Stenomylus, *known from the late Oligocene. The early rise and spread of camels took place in North America, and they are not found anywhere else till the early Pliocene, about 5 My ago.*

RIGHT Protoceras *("first horn") was an early member of the North American protoceratid family, a new group of the Oligocene that looked rather like deer but were related to camels. Only the males carried horns, which grew on both the back of the head and the nose. Protoceras was probably a forest browser, about the size of a roe deer.*

reason it is often harder to imagine what it was like than if a band of enthusiasts were playing its tune.

Some mammals came through from Eocene times. *Cainotherium* was a rabbit-like hoofed mammal distantly related to camels. It hung on for a while in its original role as a small herbivore.

Metamynodon was a North American member of the rhinoceros family Amynodontidae (now extinct), which flourished in North America and Asia from the mid-Eocene to the early Miocene, and seems to have been among the earliest mammals to embark on a hippo-like semi-aquatic career.

The prize of the old brigade were members of the extinct rhino family Hyracodontidae, with roots in the small running rhinos of the American and European early Tertiary, but

ABOVE
Metamynodon *belonged to the extinct amynodont family of rhinos that flourished in the later Eocene and Oligocene of North America and Asia.*

living in a region of southeast and central Asia that includes China and Baluchistan. The largest member, and the largest known land mammal, was the giant "giraffe rhinoceros" *Indricotherium*, hornless like its relations, 18 ft (5.5 m) at the shoulder, and weighing about 15 tons (tonnes) according to Mikael Fortelius, a rhino specialist at the University of Helsinki. Its teeth show that it was a browser, probably feeding in the high tree levels.

Many present-day families of mammals were around in this epoch, but not in the roles we might expect. Bears were common throughout the Northern Hemisphere, in the form of creatures such as *Cephalogale*, which had the look of a small stocky fox. Members of the dog family (Canidae) were diversifying in North America as small foxy beasts such as *Hesperocyon*. There were abundant horses in North America, such as the pony-sized

woodland browsers like *Miohippus* and *Anchitherium*, but there were no large plains-dwelling grazers.

Camels were fairly common in North America, but the Oligocene camels were not the hefty models of the later Tertiary. Instead they were represented by delicate designs such as the gazelle-like *Stenomylus*, whose great similarity to living gazelles, including high-crowned cheek teeth and very long legs, has led to speculation that there were arid open areas somewhere in the Oligocene, as well as the ubiquitous subtropical woodland. Another camel, the llama-sized *Oxydactylus*, had long legs and a long neck and seems to have had the habits of a giraffe.

The seeds were being sown for the coming

boom in new forms, but it would have taken a clairvoyant to identify the eventual "winners."

In contrast, many of the abundant and successful mammals in the Oligocene belonged to orders that are still with us today, but to families now extinct. The bear-dogs (amphicyonids) were predators and scavengers. The more specialized, exclusively meat-eating ambush predators were at this time cat-mimicking "false" saber-toothed nimravids. (True cats of the family Felidae did not arrive until the early Miocene.)

BELOW Indricotherium *was a "giraffe-rhinoceros," the largest land mammal ever found – weighing about 15 tons (tonnes). It could feed from the tops of small trees. Indricotheres inhabited southeast Asia from late Eocene to early Miocene times. Their small companions are the related sheep-sized rhino,* Hyracodon, *known from the same period.*

The most abundant grazers and browsers of the Oligocene were mainly sheep-sized piglike forms, though their teeth indicate a more vegetarian diet than is eaten by modern pigs, which will eat anything. In the Old World the part was played by the anthracotheres, also found in North America but far outnumbered there by the camel-related oreodonts, long-tailed browsers walking on their fingers, their feet more like a dog's than a pig's. A similar role was played by a large cast of hyracoids in Africa and notoungulates in South America. These are the stock players in the Oligocene cast, short-legged and stumpy-bodied, like a present-day hyrax or capybara. They lack the size and grace of our own time's deer and antelope, and while they obviously played a key role the plain sameness of their design offers no vivid clues to their possible function or behavior.

The Oligocene is also hard to picture as a general ecosystem. We can imagine the Paleocene and Eocene as a modified tropical rainforest, or the Miocene and Pliocene as a worldwide Serengeti, though these images admittedly omit, distort, and oversimplify. Somehow the Oligocene resists the restorer's imagination. The predominant vegetation was no doubt a woodland of some sort, but nothing like the temperate woodlands that nowadays house fleet and graceful deer rather than lumbering oreodonts.

Some paleontologists have suggested that one small area of the modern world contains a remnant of something like the Oligocene type of vegetation. The unique "fynbos" vegetation lies at the very tip of the South African cape. It is a lush dense chaparral about 3 ft (1 m) high, a rough cross between the Sonora desert of

Arizona and a close-packed English shrubbery. Sitting in the midst of the fynbos it is not hard to imagine the undergrowth disgorging a herd of snuffling, quarrelsome oreodonts, acting the way hyraxes do today when they vie for sunbathing space on rocky outcrops.

Yet life does not take quiet vacations for 10 million years, and we need to know more about the backstage shifts that were setting the scene for the next great burst of change.

Early to middle Miocene

The early Miocene, starting about 23 My ago, saw the beginnings of a shift toward a warmer and considerably drier climate. It was around then that Drake's Passage opened up between Antarctica and South America, making way for a circumpolar current of cold water. Tectonic pressures were building to produce major uplifts of today's great mountain ranges – the Cordilleras of western North America, the Andes of South America, and in Asia the Himalayas. All these events were bound to disrupt established patterns of circulation in both atmosphere and ocean, bringing changes in the global climate and rainfall, and therefore in the vegetation pattern.

A further influence was the shrinkage of shallow inland seas over the continents; for instance, the Tethys Sea was closed off by a land bridge between Africa and Eurasia that dammed the Mediterranean. With more continental landmass exposed, there was less sea to buffer the global climate from extremes of heat and cold.

These great shifts in the continental plates brought changes in the ocean currents so that nutrients welled upward from deep waters in different places to feed the phytoplankton that form the base of the ocean food chain. Upwelling now occurred in more temperate latitudes, where it is also found today, and the resulting enrichment of the marine food supply lured some mammals into the water. The first seals were flourishing by the early Miocene or latest Oligocene, which also produced the ancestors of the modern whales. One long-term impact of these developments was their effect on the evolution of diving birds. Seals and sea lions seem to have replaced, in the Northern Hemisphere's middle latitudes, various large carnivorous diving birds, among them the larger than human-sized giant penguins.

Warming in the middle latitudes brought some expansion of tropical and subtropical forest, but the drying trend encouraged a new vegetation, chaparral or thorn scrub, which resembled the tough plants seen today in California and in the harsh *maquis* of southern Europe. Chaparral grew mainly on the western side of continents, as it still does, owing to the constant circulation of the atmosphere from west to east, caused by the Earth's rotation which tends to make more rain on the eastern sides of continents. Teeth designs among some mammals found in the early Miocene of Argentina suggest that by now there were some grassland meadows and even savanna areas emerging in South America.

There were now great new radiations of animals that linked with the migrations between continents made possible when falling sea levels left land bridges high and dry. The ocean population was already transformed by the appearance of seals and modern types of whale. Bats appeared in Australia, flying across from southern Asia, and it was probably now too that Madagascar received its rats, and genet and mongoose types of carnivores. Most vitally, with its first connection to the Eurasian mainland the once unique nature of the African fauna was irrevocably changed. Many of the original

African forms could enter Eurasia, and in particular the elephant-like creatures and the spreading hominoids – the apes.

Africa drew in as immigrants from Eurasia various rabbits, cats (including saber-tooths), and modern rhinos. The rodents that are found in the early Oligocene Fayum fauna – located in the Fayum Depression, just south of the city of Cairo – provided the founder members of unique African groups such as the present-day spring hares and mole rats. Later in the early Miocene more northern groups trooped in: bear-dogs and genet-like carnivores, mouse deer, pigs, clawed chalicotheres, primitive types of giraffe and antelope, and true insectivores that diversified into singular African animals such as golden moles and tenrecs, the latter now found only in Madagascar. Among the primates, hominoids such as *Dryopithecus* were descended from Oligocene ancestors, but these times also introduced some other Old World primates, the cercopithecid monkeys (now including macaques and baboons) and the bush babies.

Proboscideans, with their long trunks, had sorted themselves into three distinct groups: the specialist high-browsing deinotheres that retained only the lower set of tusks, the browsing mastodonts, and the omnivorous gomphotheres. All three types of proboscideans migrated over to Eurasia in the early Miocene, soon to be followed by hyraxes, aardvarks, and hominoids. The middle Miocene anthracotheres were the likely ancestors of the hippopotamuses.

THE EARLY TO MIDDLE MIOCENE (16 to 20 My ago) The climate grew warmer again. Prominent mountain ranges were produced – the Cordilleras, the Andes, and the Himalayas. These new ranges combined with changes in ocean currents to alter the circulation of air and patterns of rainfall. Inland seas were shrinking, and the Mediterranean was enclosed between Africa and Eurasia.

A new vegetation type, chaparral, grew mainly on the drier western sides of the continents. Grassy glades may have developed in subtropical woodlands, and there were grasslands in South America, but no expanses of savanna.

Land routes opened up, encouraging migration. Africa received ruminants, pigs, and true carnivores from Eurasia, and sent aardvarks, proboscideans, and hyraxes in the opposite direction. Proboscideans reached North America as well as more modern ruminants and carnivores. Eurasia was colonized by the North American browsing horse Anchitherium.

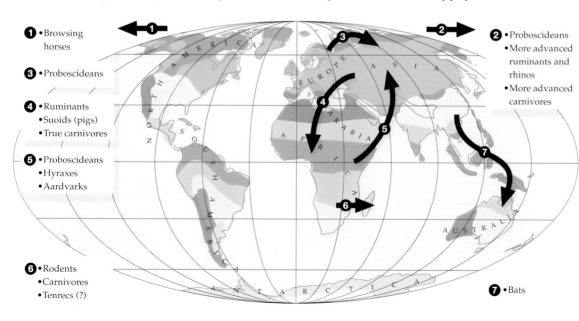

❶ •Browsing horses

❸ •Proboscideans

❹ •Ruminants
•Suoids (pigs)
•True carnivores

❺ •Proboscideans
•Hyraxes
•Aardvarks

❻ •Rodents
•Carnivores
•Tenrecs (?)

❷ •Proboscideans
•More advanced ruminants and rhinos
•More advanced carnivores

❼ •Bats

◼ Tropical rainforest
◻ Paratropical rainforest (with dry season)
◻ Subtropical woodland or woodland savanna (broad-leaved evergreen)
◼ Temperate woodland (broad-leaved deciduous)
◼ Temperate woodland (mixed coniferous and deciduous)
◻ Tropical grasslands or savanna
◼ Mediterranean-type woodland, thorn scrub or chaparral
◻ Tundra
◻ Ice sheet

The faunas of Europe and Asia were broadly similar in the earlier Miocene. The earlier small hornless ruminants had given rise to several distinct and large-sized lineages, each with their own unique versions of horns: for *Giraffokeryx* (an early giraffe), bony cores (ossicones) covered with skin; for *Dicerocerus* (an early deer), branching deciduous antlers of naked bone; for *Eotragus*, true antelope horns consisting of a bony core covered by a keratin sheath. Although larger than their ancestors, none of the ruminants were bigger than a fallow (or white-tailed) deer – and they appear to have been woodland browsers. Some pigs such as *Listriodon* had tapir-like ridged teeth for slicing leaves, rather than the bunodont (bumpy and rounded) teeth that most pigs have. But instead of a delicate tapir-like snout to pick and choose their vegetation, they had an extremely broad mouth, probably to shovel it down wholesale.

With so much prey available, some new types of carnivores developed. Among the most common of the large carnivores were the hemicyonids or dog-bears. These were true members of the bear family, but looked and probably had a lifestyle more like dogs than like modern bears. They were mistaken for dogs when originally discovered, and their name means "half-dogs." The small ferret-like lineage of bears once common in the Oligocene was now extinct, and was succeeded by the present-day families cast in the role of small and opportunist carnivores in the ecosystem: viverrids (genets and civets), mustelids (weasels, badgers, otters, and others), and procyonids (raccoons, now virtually confined to the Americas).

Hyenas also made their debut in the early Miocene, not as large bone-crushers but as small coyote-like scavengers. Cats were spreading, and included some early saber-tooths, the machairodontids. Yet the more archaic carnivores of the Oligocene had not yet bowed out: bear-dogs (amphicyonids) were still common.

The modern types of insectivores – hedgehogs, shrews, and moles – flourished and diversified across the Northern Hemisphere, though hedgehogs became extinct in North America at the end of the Miocene. More modern rabbits and rodents were also becoming common throughout the Northern Hemisphere; the murid rodents (rats and mice) that now head the rodent world made their first appearance in Asia in the middle Miocene. These tough little scavengers were to have a spectacular future, especially in the company of humans. Coincidentally, the same period also saw the appearance of larger hominoids such as *Sivapithecus* (alias *Ramapithecus*), which seems to be a distant ancestor of the orangutan.

The North American scene was enriched at this time by the immigration of hemicyonid bears and true cats. The amphicyonids remained as large bone-crushing scavengers such as *Daphoenodon*, but the hyaenodontids had faded out, and the saber-toothed nimravids did not survive past the earliest Miocene. In the middle Miocene the elephant-related gomphotheres and mastodonts migrated into North America from Asia, although their cousins, the browsing deinotheres, never made it across the Bering Strait.

One of the most important arrivals in the early Miocene was a variety of more advanced types of ruminants: the small hornless deer (related to today's Asian musk deer); the deer-related dromomerycids; and the antilocaprids, which were the only group to survive to this day, as the pronghorn "antelope" (actually more related closely to deer) of the western North American plains.

As these newcomers spread, so several of the original herbivores declined or died out. Primitive traguloid ruminants such as *Leptomeryx* grew scarce, and the piglike entelodonts and anthracotheres were gone by the end of the early Miocene. Oreodonts, the rank and file of the Oligocene, also declined in numbers, but a few larger forms evolved with longer legs and more highly crowned teeth. Peccaries and tapirs were not abundant, but they persisted and they still survive.

The real success story was among the horses. The first equines, ancestors of modern horses, sprang up in the early Miocene. There was an immense radiation of various species of *Merychippus*, horses with longer legs and more highly crowned cheek teeth, abler to face the more open habitat and coarser type of vegetation of the early Miocene.

From the early to the middle Miocene, the whole Northern Hemisphere saw a vast extension of the number of animals adapted

EVOLUTION OF THE HORSE

Horses began as small, four-toed woodland scamperers and ended as big, one-toed grassland gallopers. The sequence recorded here is deceptive. It seems to show a trend toward greater size and a single hoof. Different conditions might have preserved a three-toed, forest-dwelling horse, like the specialized *Hypohippus* of the Miocene.

Hyracotherium, also known as Eohippus, *belongs to the early Eocene and the paratropical forests of the Northern Hemisphere. It was roughly the size of an African mouse deer today, with a four-hoofed forefoot and three-toed hind foot. It supported its hoofed toes with a foot pad, and moved with an agile, scampering gait. Low-crowned teeth with partial ridges enabled it to eat berries, buds, and young leaves.*

Mesohippus belonged to the new subfamily, Anchitheriinae, that emerged in the late Eocene of North America, when cooler climates had removed Hyracotherium. *Mesohippus was about the size of a gazelle, and well adapted to the woodlands of the Oligocene and early Miocene, with longer legs and no fourth toe on the forefoot. Low-crowned teeth with a pattern of ridges (lophes) suggest a diet of more leaves, now that more seasonal woodland offered no year-long diet of buds and berries.*

The genus Merychippus, *first of the surviving family Equinae, the modern horses, appeared in the early Miocene. A grassland contemporary of various woodland anchitheres, which also had their own evolutionary changes and increased in size, it adapted to cooler drier climates and the spread of grasslands, with its longer legs, without foot pads, and more highly crowned teeth, which could cope with grass. Equines were faster movers in more open country. They reduced the side toes and strengthened the central toe, the main load-bearer.*

Equus *is today's survivor of several lineages that spread from the early* Merychippus *stock, all with more highly crowned and well-lophed teeth. Some groups kept their three toes, some grew smaller, and two lines simplified to a single toe – the graceful pliohippines, now extinct, and the more rugged* Equus *ancestors. Other groups had equally efficient teeth for grazing.*

Tooth—crown of upper molar

Front foot

Foot pad

•Hyracotherium   •Mesohippus   •Merychippus   •Equus

The descendants of Hyracotherium *evolved mainly in North America, an island continent during the Tertiary. With nowhere to emigrate as climates changed, they had to adapt or die. Woodlanders such as* Hypohippus *and* Archaeohippus *became extinct. The successful* Hipparion, *a three-toed savanna specialist, migrated to the Old World in the late Miocene and spread across southern Eurasia and Africa. With the exception of*

Equus, *all the different lineages of horses were extinct in North America by the early to middle Pliocene.* Equus *and* Hipparion *overlapped elsewhere, and we do not know whether the survival of* Equus *derives from some competitive edge, or sheer happenstance. What the horse has gained in numbers and breadth of distribution, it has lost in diversity.*

▢ Hyracotherine
▢ Anchitherinae
▢ Merychippines
▢ Hipparioni
▢ Equini

	Habitat in North America	South America	North America		Eurasia	Africa
Present day						
Recent 0.01 Mya	Prairie				•Wild asses	•Zebras
Pleistocene 1.8 Mya		•Hippidion				
Pliocene 5 Mya	Savanna		•Equus	•Hipparion group		•Hipparion
Miocene 26 Mya		•Pliohippus group	•Dinohippus •Merychippus	•Hypohippus •Archeohippus	•Anchitherium	
Oligocene 37 Mya	Woodland		•Parahippus •Miohippus •Mesohippus			
Eocene 53 Mya	Forest		•Epihippus •Orohippus •Hyracotherium			

□ Tropical rainforest
□ Paratropical rainforest
(with dry season)
□ Subtropical woodland or
woodland savanna
(broad-leaved evergreen)
■ Temperate woodland
(broad-leaved deciduous)
■ Temperate woodland
(mixed coniferous and
deciduous)
□ Tropical grasslands or
savanna
■ Temperate grassland
(prairie, steppe, pampas)
■ Mediterranean-type
woodland, thorn scrub or
chaparral
□ Tundra
□ Ice sheet

to living on the more open savanna at the expense of their woodland contemporaries. In Eurasia, antelope diversified, as deer and giraffids declined. In North America, the more specialized horses and the deer outstripped the older oreodonts. Yet still there is no evidence of true open grassland. These herbivores seem to have lived on a mixed diet of grass and browse, but they are not yet true grazers like modern horses and antelope.

Late Miocene

After the middle Miocene, apparently the warmest part of the whole period from 35 My ago to the present day, the world became a cooler, drier place. Why did this happen? Two contributing causes may have been, first, the connection of the Arctic Ocean to the Atlantic Ocean, with an increase in the colder deep waters of the North Atlantic, allowing the Arctic to become a heat sink; and second, the

closure of equatorial ocean circulation in the Indonesian region, affecting circulation patterns in the Pacific and Antarctic oceans.

A key influence was the continued building of new mountain ranges, not only because an increase in land elevation shifts the continents upward toward the cooler upper air, but because mountain barriers force the atmosphere to flow higher, to where it loses heat. A rain-shadow effect on the eastern slope of mountains furthered the spread of grasslands. Another more subtle long-term effect of mountain uplift is that as it exposes more rock it increases the amount of chemical weathering. As the greater area of rock interacts with the atmosphere it removes more carbon dioxide and returns it eventually to the ocean. The loss of carbon dioxide affects the ability of the atmosphere to retain heat, and a kind of reverse greenhouse effect leads to global cooling. The result of all these changes was the appearance in North and South America of savanna-like grasslands: one of the key events of the past half-billion years.

In the Cretaceous, the rise of the angiosperm plants had transformed the world's vegetation and the animals that fed on it. Animal feeding habits and eating equipment influenced the evolution of the angiosperms. One strategy in areas of abundant water and sunlight is to race upward and carry on a high-level struggle in the canopy, leaving a bare floor far below. Angiosperms that need insects to pollinate them have evolved flowers and smells. By contrast, the grasses have pursued a totally

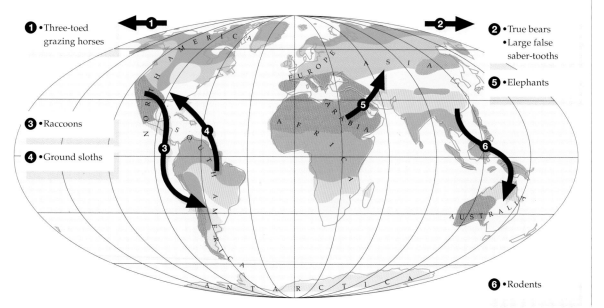

❶ •Three-toed
grazing horses

❸ •Raccoons

❹ •Ground sloths

❷ •True bears
•Large false
saber-tooths

❺ •Elephants

❻ •Rodents

different line, designed for unfriendly exposed conditions. They cling to the ground, huddle together in such numbers that they can rely for fertilization on wind-borne pollen, and thrive by spreading outward rather than upward. Their lowly position makes them easy to eat; but instead of making their growth at the top of the shoot, grasses grow from the base and along the length of the blade. The more they are cropped, the more densely and healthily they grow. Animals able to eat grasses are guaranteed a constant self-replacing supply of a plant that is able to stretch almost unbroken for hundreds of square miles.

One drawback of grass as a food is that its leaves take up silica, an abrasive mineral that wears down teeth much faster than most other kinds of foliage. The evolutionary response of the grazing mammals has been to develop "hypsodont" teeth, whose crowns are taller than they are long or wide, with complex folding patterns of enamel and strong reinforcements of tooth cement. So one effect of these silica particles (phytoliths) has

been to create feeders so adapted to eating grass that they are practically dependent on it: a captive market. (Changes in the gut flora are highly important too; these hoofed armies of the savanna march on their stomachs.) Grass now covers about one-third of the world's surface, and is the staple food of the animal population of the world.

The mammalian faunas of the late Miocene pursued the epoch's earlier trends, reinforcing the predominance of forms adapted for more open habitats. In particular there was a sharp rise in the numbers and types of rodents that eat grass seeds, such as voles and deer mice.

In the Old World, more and more antelope types came on to the scene, though as yet there was no form so specialized for grazing as an expert like the present-day wildebeest. In contrast, the North American faunas produced a diversity of horses that do appear to be adapted for a diet made up almost exclusively of grass, as are modern horses and zebras. The three-toed horse *Hipparion* found its way across the Bering Strait to become common in the Old World. True elephants

ABOVE *Coming from the African late Miocene,* Prodeinotherium *belonged to a type of proboscidean which branched off before the common ancestor of elephants and mastodons. Their long legs and tapir-like teeth suggest they browsed high branches of trees.* Hipparion *had recently invaded from North America. The true, one-toed horse of the genus* Equus *arrived later, in the late Pliocene.*

OPPOSITE, ABOVE Epigaulus, *a primitive rodent of the North American Miocene, and distant relation of the mountain beaver. They burrowed using large claws. Their paired horns could have been a fighting device, or may have aided in burrowing.*

MAMMALS OF THE SAVANNA

In the late Miocene, around 10 to 12 My ago, a community of plants and animals in North America foreshadowed the scene found in present-day East Africa. The former savanna of North Africa was nothing like today's treeless prairie. Pollen records show the presence not only of grasses but also of conifers and deciduous trees (walnut, oak, elm, birch, willow, juniper), plus shrubs and herbaceous plants. A range of larger herbivores and carnivores inhabited a landscape of grassland dotted with trees and shrubs.

Although the form and structure of the North American savanna mammals are echoed in designs among the mammals of East Africa, these later animals are not all closely related to the Miocene species. They illustrate the way that similar environments may result in the evolution of similarly adapted but possibly unrelated species at a distance of thousands of miles and millions of years.

MEGAHERBIVORES
Grass provides much of the diet of a modern elephant. These Miocene forerunners were probably more specialized for browsing or fruit-eating. The mastodon Zyglophodon *had two pairs of tusks. The gomphothere* Ambelodon *may have used its shovel-shaped lower jaw to strip bark. The black rhino is the ecological counterpart of its distant relation* Aphelops. Teleoceras *was a semi-aquatic rhino, resembling a mix of hippo and white rhino.*

GRAZERS
The African buffalo is an important large grazer. Apart from the zebra, which is common but not diverse, other medium-sized grazers come from among the several tribes of antelopes. In the North American savannas, most of these niches were filled by the Miocene's inventive range of horses, one-toed or three-toed, ranging in size from a modern zebra to a Shetland pony. There appear to be no North American piglike grazers such as the warthog.

Elephant   Black rhinoceros   White rhinoceros   Hippopotamus

Zygolophodon   Amebelodon   Aphelops   Teleoceras

Buffalo   Zebra   Roan   Wildebeest   Hartebeest   Kob   Reedbuck   Warthog

Cormohipparion   Neohipparion   Pliohippus   Protohippus   Nannipus   Calippus

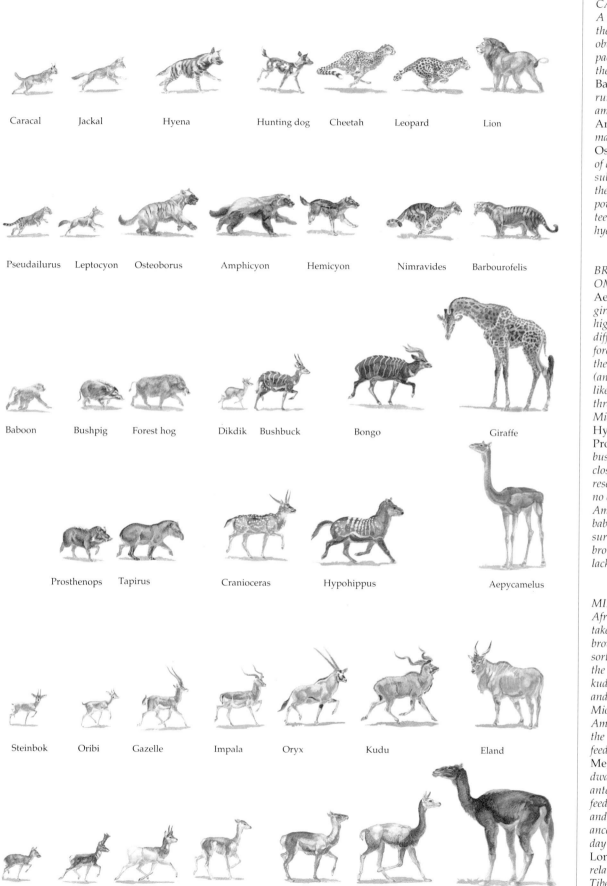

Caracal Jackal Hyena Hunting dog Cheetah Leopard Lion

Pseudailurus Leptocyon Osteoborus Amphicyon Hemicyon Nimravides Barbourofelis

Baboon Bushpig Forest hog Dikdik Bushbuck Bongo Giraffe

Prosthenops Tapirus Cranioceras Hypohippus Aepycamelus

Steinbok Oribi Gazelle Impala Oryx Kudu Eland

Longirostromeryx Cosoryx Plioceras Hemiauchenia Protolabis Procamelus Megatylopus

CARNIVORES

A major difference is the Miocene lack of obvious fast-running pack-hunters. Among the bigger killers, Barbourofelis was no runner, and must have ambushed its prey. Amphicyon was a massive "bear-dog," Osteoborus a member of an extinct dog subfamily. Both have the solid skulls and powerful bone-crushing teeth found in modern hyenas.

BROWSERS AND OMNIVORES

Aepycamelus and the giraffe are long-necked high browsers from different stocks. Large forest ruminants like the African bongo (antelope) have roles like those of some big three-toed horses of the Miocene such as Hypohippus, while Prosthenops and the bushpig seem to have a close ecological resemblance. There is no equivalent in North America to the savanna baboon as no primates survived there. Small browsers are also lacking.

MIXED FEEDERS

African species able to take both grass and browse comprise all sorts of antelopes from the large eland and kudu to the little oribi and steinbok. In Miocene North America, camelids were the larger mixed feeders, and the massive Megatylopus would dwarf the largest living antelope. Smaller-sized feeders like Plioceras and Cosoryx are ancestors of the present-day pronghorn. Longirostromeryx is related to the living Tibetan musk deer.

and hippos also made their first appearance in Africa in the late Miocene.

Meanwhile, for animals more adapted to woodland habitats and a browsing diet, the downward trend went on. North America illustrates what happened. There the early late Miocene had equally encouraged both browsing and grazing savanna-type animals; but by the end of the epoch the fauna had been drained of its smaller browsers, and the remaining hoofed mammals were mainly larger and adapted to a diet of coarser food. They could weather out the harsher climate. The late Miocene of North America saw off the ruminant browsers (blastomerycids and dromomerycids), the remaining oreodonts, the camel-related horned protoceratids, and all of the rhinos. Peccaries and tapirs somehow survived. Of the camels, only small llama-like forms such as *Hemauchenia* and the very large forms such as *Megatylopus* survived.

Similar changes affected the Old World, but not so severely. This was probably because there the animals could migrate to more equable tropical regions in Eurasia and Africa, whereas in the New World North America was not yet connected to South America, and browsing animals trying to follow the receding tropical forests down toward the equator were heading for a dead end.

There was a decline among Old World browsers such as deer and primitive giraffes, but these groups did not become extinct. Instead, new types of giraffids evolved; very large long-necked species like the present-day giraffe appeared in Africa, and the mooselike short-necked sivatheres in Asia. Rhinos did not vanish in the Old World but diversified into more modern types, which included the ancestors of the present-day black, white, and woolly rhinos.

Changes also overtook the larger carnivores in the late Miocene. The bear-dogs and dog-bears (amphicyonids and hemicyonids) so successful in the earlier Miocene dwindled and faded out. On the other hand, the saber-toothed true cats drew upon fresh reserves. North America had two lion-sized catlike carnivores, the true cat, *Nimravides*, and the "false" saber-toothed nimravid, *Barbourofelis*, which both seem to stem from Old World immigrants. True bears

(ursines) put in an appearance when large lumbering animals such as *Agriotherium*, willing to eat flesh or vegetation, left their native Eurasia and headed for North America. Dogs diversified in North America, spawning a number of animals resembling coyotes that migrated to the Old World in the latest Miocene.

One of the most interesting carnivore stories of the Miocene is the origin and diversification of the hyenas, recently researched by Lars Werdelin of the Swedish Museum of Natural History in Stockholm and Nikos Solounias of the Long Island School of Osteopathic Medicine in New York. Hyenas are not very diverse today. Three species are specialized bone-crushers, one of them (the spotted hyena) hunting in packs; and one additional species, the aardwolf, is a specialist termite-eater with reduced jaws and teeth. The first spread of Miocene hyenas was with coyote-like creatures such as *Protictitherium*, but these became rare in the Pliocene when true dogs emigrated to the Old World from North America.

True hyenas diversified into two other distinct types in the middle to late Miocene, again rehearsing roles that other carnivores now play. One type was the heavy-bodied wolflike hyenas, such as *Thalassictis*, that went on till the end of the Miocene. The other was the more slender, very long-legged hyenas, such as *Chasmaporthetes*, that survived into the late Pleistocene and migrated to America in the Pliocene. These animals were more like cheetahs in appearance than the hyenas of today's savanna. Neither type possessed the domed head and large bulbous cheek teeth typical of present-day hyenas, whose bone-crushing career began about 10 million years ago.

There is nothing like a hyena in North America today, but during the Miocene it appears that bear-dogs may have been the first bone-crushers. Many of the mammal roles have persisted throughout the Cenozoic. It is the actors that have constantly changed, and many of today's mammal families started out in very different careers.

Evolution operated inside closed arenas on the island continents of Australia and South America. Our knowledge of the Australian Miocene, once very skimpy, has now been swelled by the work of Mike Archer and his

colleagues at the University of New South Wales, who have excavated and extensively studied the Riversleigh site in western Queensland. Here they have unearthed a luminous window on Australia's past, with fossils that range in age from the latest Oligocene to the Pleistocene, although most of the finds come from the earlier Miocene.

Riversleigh has yielded not only the more primitive ancestors of many surviving marsupials, such as possums, kangaroos, and bandicoots, but also an earlier record of extinct forms previously best known from the very large animals of the Pleistocene. The rhino-like diprotodonts are related to the wombats; the marsupial lions, related to koalas, are an example of reversal to meat-eating among the ranks of the otherwise vegetarian phalangeroid marsupials. There are also a few surprises – some small mammals otherwise totally unknown and not obviously related to any known marsupial

(appropriately awarded names such as *Thingodonta* and *Weirdodonta*), and a giant python immortalized as *Montypythonoides*.

By the early Miocene, South America had received some new arrivals which appear to have found their way over from Africa in the Oligocene. These were the unique South American monkeys and marmosets, the caviomorph rodents (such as the guinea pig and the capybara), and also phorusrhacid birds, formidable flightless carnivores ranging up to 9 or 10 ft (3 m) tall. Yet these invaders do not seem to have disrupted the diversity of the old-time natives. The local ungulates (graceful litopterns and stockier notoungulates) diversified with longer legs and more highly crowned teeth, in parallel with the changes seen in their hoofed Northern Hemisphere relatives in response to the development of the Miocene savannas. The native marsupials (opossums and borhyaenids) and the toothless placentals (sloths, anteaters, armadillos, and glyptodonts) also continued to diversity.

By the end of the Miocene, Australia and South America were becoming a little less isolated. Rodents reached Australia at this time, probably by hopping along the Malaysian island chain. Although the great interchange of faunas between North and South America did not happen until the Pliocene, at the end of the Miocene the continents were close enough together for early messengers to float on logs or even swim across the waters in between. Raccoons appeared in South America, and some of the

Neohelos ("new bump," from the distinctive cusp on a premolar tooth) was a Miocene member of the extinct diprotodontid family of marsupials, often labeled "giant wombats" but really a distinct though related family. Unlike wombats, which eat grass, diprotodontids were specialist browsers. Their teeth were rather like a tapir's, and some species had skulls that may have made room for a tapir-like proboscis.

Neohelos was cow-sized, but some of its Pleistocene relations were the size of a small rhino. It is found in the Riversleigh deposits of Queensland, Australia, and is shown here with a pair of pygmy possums, marsupials found as fossils in the same Miocene deposits, but living to this day in the northwestern rainforest of Australia.

The mammal fauna found in the Santa Cruz Formation in Argentina lived around 25 My ago and reflects the separation of South America from North America. Such mammals include Cladosictis, *a ferret-like member of the carnivorous marsupial family Borhyaenidae ("gluttonous hyenas").*

Also illustrated here are two ungulates: Diadophorus, *a horselike, gazelle-sized member of the order Litopterna, and* Nesodon, *a wombat-like, cow-sized member of the Notoungulata. A unique feature of the South American fauna of the later Tertiary was the phorusrhaciform birds: giant, human-sized, flightless carnivores.*

smaller ground sloths turned up in the more southerly parts of North America.

The Plio-Pleistocene

By the start of the Pliocene, about 5 My ago, the trend toward a cooler, drier global climate had imposed conditions quite similar to today's, and it was then that the major new types of vegetation we are familiar with began to appear. By the start of the Pleistocene, about 1.8 My ago, the world had entered a cooler period of alternating glacial and interglacial phases. Even now, in the present interglacial phase, the climate is colder and drier than in Pliocene times.

Those times began in the Northern Hemisphere with the first appearance of arctic vegetation: tundra inside the Arctic Circle north of the highest forest reaches whose frontier is called the tree line, and taiga – coniferous evergreen forest – in a band below this. (Tundra and taiga also appear today at high mountain levels, even in tropical regions.) Tundra is a world of permanently frozen soil, permafrost, with a very short growing season for plants that are mainly mosses, lichens, and sedges. Taiga forest consists of vast areas of nothing but coniferous evergreens and provides a more meager living for mammals than the mixed coniferous and deciduous temperate forests that previously occupied this region.

In the lower latitudes the drier climate brought the spread of previously unknown desert types of vegetation that occupied much of the areas earlier covered by chaparral and thorn scrub. Grasslands also spread, but instead of the more wooded savanna once typical of higher latitudes in Miocene times, these were more arid treeless plains, now familiar as prairie, steppe, or pampas. Although the modern world has plenty of tropical grassland, such as the savannas of Australia, only Africa still has a mosaic of grassland and woodland able to support a wide variety of mammals. The tropical forest zone has been squeezed much tighter into equatorial regions since the Miocene. In many areas, what used to be tropical forest has been replaced by subtropical woodland or grassland, and this in turn has been replaced in the middle latitudes by more temperate types of woodland.

None of these changes worked in favor of variety, among either the faunas or the floras of the world. Each step from tropical to subtropical woodland and then from savanna to prairie brings a considerable reduction in the number of species and an even greater reduction in the number of possible useful interactions among them. Many of the mammals common in the earlier part of the Miocene became extinct at the end of the epoch, or during the Pliocene. North America was especially hard hit, because as yet it had no access to an equatorial zone which could harbor more tropically adapted species.

The start of the Pliocene saw a trend toward the evolution of very large-bodied

THE PLIO-PLEISTOCENE (5.5 to 0.1 My ago) Cooling and drying produced an Earth much like today's. Tundra grew over permafrost inside the Arctic Circle. Taiga grew in a band to the south, and further south still, drier conditions encouraged the spread of desert and semi-desert over chaparral. A temperate grassland replaced the more wooded savanna at higher latitudes.

Tropical grasslands are plentiful today, but only in Africa does a savanna support a wide variety of mammals. The tropical forest zone is today shrinking fast. None of these changes favor biological diversity. Grassland supports far fewer species than savanna, and temperate woodland lacks the teeming populations of subtropical forest. The start of the Pliocene saw large-bodied herbivores, and with these came large specialized predators. Among the smaller herbivores, grazers fared better than browsers. Finally, in the late Pleistocene, humans emerged from Africa.

□ Tropical rainforest
□ Paratropical rainforest (with dry season)
□ Subtropical woodland or woodland savanna (broad-leaved evergreen)
□ Temperate woodland (broad-leaved deciduous)
□ Temperate woodland (mixed coniferous & deciduous)
□ Tropical grasslands or savanna
■ Temperate grassland (prairie, steppe, pampas)
■ Mediterranean-type woodland, thorn scrub or chaparral
□ Semi-arid scrub
□ Arid desert
□ Taiga (boreal coniferous forest)
□ Tundra
□ Ice sheet

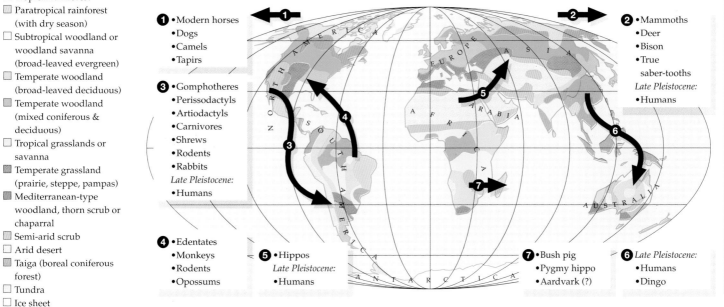

1 •Modern horses •Dogs •Camels •Tapirs

2 •Mammoths •Deer •Bison •True saber-tooths *Late Pleistocene:* •Humans

3 •Gomphotheres •Perissodactyls •Artiodactyls •Carnivores •Shrews •Rodents •Rabbits *Late Pleistocene:* •Humans

4 •Edentates •Monkeys •Rodents •Opossums

5 •Hippos *Late Pleistocene:* •Humans

7 •Bush pig •Pygmy hippo •Aardvark (?)

6 *Late Pleistocene:* •Humans •Dingo

Smilodon ("knife-
tooth") was a lion-sized
member of the
machairodontine true
cats known from the
Plio-Pleistocene of the
Old World and North
America. The classic
"saber-toothed tiger," it
is well known because
of the many skeletons
found in the Rancho La
Brea tar-pit treasury of
Los Angeles. Saber-
tooths had shorter legs
than other types of cat.
They would have been
short-range pouncers
rather than pursuers,
and may have used
their enormous canines
to slash open the belly of
their prey. Living cats
tend to give a bite to the
back of the neck that
plunges their shorter,
rounder canines down
to the bone. This
approach would have
damaged the narrow,
flattened, canine blades
of Smilodon, designed
to slice, not crunch.

herbivores that were better equipped to weather out periods of seasonal food shortages and to make do on coarse diets all the year round. Along with these larger packages of herbivore meat came more specialized predators, such as the saber-toothed true cats. Among the smaller herbivores, the grazers and mixed feeders fared better than the expert browsers. Out of a variety of horses with different feeding habits, only the grazers survived in North America. The Plio-Pleistocene of Africa brought a flourishing variety of grazing antelope on the new East African savannas which spread in the rain shadow cast by the newly uplifted East African ridge system.

The great geologist Preston Cloud wrote that tectonic plates, together with the continents they carry, "glide over the upper surface of the yielding asthenosphere . . . at rates comparable to that at which an anvil might settle through solid asphalt." Not all geological events are as leisurely as that patient 8 ft (2.4 m) or so per century, or as instant as meteor strikes or volcanic eruptions. Between 6.5 and 5 My ago, the

Mediterranean Sea may have dried into an enormous desert basin, in places as much as 3 miles (5,000 m) below sea level, as many as fifty times, to be filled again by a gigantic waterfall pouring out of the Atlantic through the Strait of Gibraltar. Throughout the Pleistocene, vast sheets of glacial ice covered and then partly uncovered the Earth's higher latitudes, particularly in the Northern Hemisphere. As many as twenty of these oscillations may have taken place.

At present the world is passing through an interglacial period that started at the end of the Pleistocene, about 12,000 years ago, and temperatures are between 7°F and 11°F (4°C and 6°C) higher in upper latitudes than during the height of the last glacial period. This particular interglacial is cooler than many have been; others have allowed hippos and macaque monkeys to thrive in England.

The Northern Hemisphere provides some of the icing machinery. It has large landmasses close to the Arctic and able to channel glaciers southward. The Antarctic can be just as cold, but it is separated from the southern continents by circumpolar ocean around latitude 55–60 degrees, and the separation of Antarctica from other southern continents restricted southern glaciation.

What caused the final sudden cooling of the world in the Pleistocene that led to the start of the Ice Ages about 2 My ago? One significant event seems to have been the North–South American connection established about 2.5 My ago. It disrupted patterns of oceanic and atmospheric

circulation, and in particular it sent northward some warmer ocean currents that would previously have flowed further south. It takes plenty of atmospheric water to build new glaciers, and this can be provided by the rainfall associated with warm currents along continental margins. This event seems to be linked with the creation of the Arctic icecap and the formation of cold currents around the North Pole.

The rhythm of these cold periods is related to a combination of three planetary cycles that affect the amount of solar radiation that the Earth receives. There are regular slow changes in the degree of tilt of the Earth's axis, its inclination toward the Sun, and the Earth's orbit around the Sun. Each cycle has its own tempo, with its own maximum and minimum peaks and troughs. Acting together they are known as Milankovitch cycles, all of them

interfering with each other, but combining every few tens of thousands of years to produce thermal maximum and minimum effects. These cycles have probably influenced Earth's climate throughout its history, but it took the formation of the Arctic icecap to enable them to tip the global system into a series of Ice Ages, not yet concluded.

It is tempting to think of the glacial stages of the Pleistocene as a globally chilly time that killed animals by the intense cold and the hardships of seeking food under the ice. However, it would be a mistake to imagine Arctic conditions throughout the northern latitudes during these episodes. The winters were probably not much colder than they are now; what made the difference was the cooler summers, which could not melt the winter snows. In any case, the ice sheets alone were a lesser problem. The resulting patterns of

By the Pleistocene, the tropical rain forests had retreated from the interior of Australia, leaving a land of open scrub with scattered gum trees. The fauna included giant marsupials, such as the rhino-sized, wombat-related Diprotodon *("two front teeth") and the giant monitor lizard* Megalania *("giant butcher"), feasting on a dead* Diprotodon. *The giant flightless bird* Dinornis *("running bird") was herbivorous and reached a height of 10 ft (3m).*

Elasmotherium ("plate-beast," so called because of its flat cheek teeth) was a large rhino of the Plio-Pleistocene, 3 to 0.5 My ago, about the size of a present-day white rhinoceros. Related forms are known from the late Miocene. Unlike the other large extinct rhino of the Eurasian Pleistocene, the wooly rhino, it was not a hairy Arctic creature but lived on the more temperate steppes.

Elasmotherium carried a single very long horn (up to 6½ ft, 2 m) on its forehead, and had longer legs than most rhinos, which gave it a more graceful horselike aspect. It has been speculated that sightings in early human times could have given rise to the myth of the unicorn. Unlike the teeth of other rhinos, its cheek teeth were a maze of enamel infoldings, and were ever-growing – a specialization nowadays found in some rodents and rabbits, but not in any hoofed mammal. It must have munched extremely coarse grass to require such a battery of dental reinforcements.

vegetation made the greatest impact on life.

In the world at the height of the last glaciation, about 18,000 years ago, the global water tied up and useless in a frozen state led to arid conditions in the tropics. The tropical forests shrank to an area less than they occupied just after the last ice age and just before humans became farmers. It is not hard to see how the larger mammals of Australia and Madagascar with their adaptations to tropical forest living would have been pushed to the edge of extinction, and often past it, by the encroaching deserts, while the forest animals in the rest of the tropics would find themselves squeezed into shrinking refuges where scarcity might force direct competition and extinctions.

When tropical grasslands spread in the glacial periods, the grazing animals could flourish. When tropical forests returned in the warmer times, the grazers would be pushed to higher latitudes. Yet these latitudes would now contain not tropical grassland but a temperate prairie, a poorer habitat. This helps to explain how the late Cenozoic of South America first supported its own herbivores, then lost them, then made room for a diversity of northern immigrants such as horses, camels, and gomphotheres, only to lose them in turn in the present interglacial.

The term "Ice Ages" therefore distracts attention from the chief agent of change. The drying of the tropics may have had a much more profound effect on mammalian evolution and extinctions than the freezing of the ice sheets.

Plio-Pleistocene times are famous for the presence of giant mammals often called the "megafauna" and almost all extinct today. All sorts of large animals with trunks traveled throughout the world, with the exception of Australia. In the middle Miocene, gomphotheres and mastodonts turned up in North America, deinotheres were in Africa and Eurasia, and elephants made their first appearance in Africa. In the Pliocene, elephants (including mammoths, which were closely related to the Indian elephant) reached Eurasia, and by the Pleistocene mammoths had arrived in North America. In the Pliocene the gomphotheres migrated into South America.

Large rhinos such as the long-horned *Elasmotherium* roamed across northern Eurasia. Its legs were longer than most

rhinos', giving its body a more graceful, horselike appearance, which together with the single large horn on its forehead has led to speculation that sightings in historical times could account for the myth of the unicorn.

Giant camels such as *Titanotylopus* were found in North America, while Asia and Africa preserved the large mooselike sivatheres. Of these, the largest was

Sivatherium, whose spreading horns may have evolved independently from the simpler skin-covered "ossicones" of the present-day true giraffe and okapi. An archeological artifact from Sumeria suggests that *Sivatherium* too lasted long enough to affect the human imagination.

The island continents also had their giants. Madagascar had a variety of enormous lemurs that lasted till only 1,000 years ago. South America had giant ground sloths such as *Megatherium*, which weighed 3 tons (tonnes) and stood 10 ft (3 m) tall when it reared up to browse. In Australia, *Diprotodon* was as big as a small rhino. Unlike its wombat relations it was a browser, and may even have led a hippo-like existence. *Diprotodon* is the only really large marsupial that ever existed,

The early hominid Australopithecus africanus *is known from the Sterkfontein cave site. The herds of grazing ungulates included wildebeest and* Makapania. *A couple of cheetah-like hyenas,* Chasmaporthetes, *circle the herbivores.*

about four times the mass of one of the giant kangaroos of the Pleistocene.

Why should marsupials not routinely achieve the size of elephants, or at least horses or bison? The answer may lie in the marsupial mode of birth, which obliges the young to climb from the birth canal into the pouch while the mother squats on her haunches. A really big marsupial might struggle to maintain this posture, and its size would give the tiny newborn a sterner climb that might damage its chances of survival.

Some of these Pleistocene giants are more striking for us because we know their living relations. They are relatively young as fossils, and many are discovered preserved so well in tar pits or peat bogs that they make exceptionally lifelike mounted specimens in museums – think of those noble Irish elks. The creator of all these giants, and the cause of their extinction, may have been the rise and fall of the ice age climate, which gave individual large mammals advantages comparable with those of the very much larger dinosaurs.

For a more massive animal, the relatively smaller surface area loses less heat; it can store more fat than a small animal; its longer legs take it further in search of food for a comparatively small energy cost. But *as a species* larger mammals are more vulnerable to extinction, especially in times of environmental stress. They may need less food for their size (an elephant weighs the same as ten horses, but eats less than ten horses), but they still need more food (much more for one elephant than for one horse), so a period of shortage longer than just one bad winter will cause problems. Secondly, large mammals take longer to carry their young, and need longer intervals between giving birth to each young. In severe conditions the species may be unable to rebuild numbers fast enough to survive.

The Pleistocene climate kept alternating between cold and warm stages, and these must have resulted in severe environmental changes. Surprisingly, mammals are more vulnerable to extinction in times of shifting from a cold stage to a warm one, probably because such changes happen faster than the build-up from a warm stage to a cold one. These changes must have driven at least some of the migrations that distributed the

mammals into the patterns of today.

The largest interchange took place between North and South America after the formation of the Isthmus of Panama, about 2.5 My ago. Once the bridge was opened, traffic could pass in either direction, and what happened next has sometimes been interpreted as a success story for the northern and supposedly "superior" mammals, sweeping southward to decimate inferior southern cousins like the marsupials. This version of events may owe more to biases in the history of human colonization of the globe than to the details of a complex episode in mammal evolution.

Certainly more species moved from north to south than from south to north, but North America is the larger continent, and had more species to contribute to the interchange. Further than that, calculations of the fossil record sometimes fail to allow for the presence in the tropics of Central America of many southern species that did not go all the way north, such as tree sloths, monkeys, and caviomorph rodents – guinea pigs, chinchillas, capybaras. Many of the forms that migrated to North America became extinct by the end of the Pleistocene, such as giant ground sloths and armored glyptodonts. Yet northern invaders such as gomphothere "elephants" and horses also went extinct in South America, and these events have more to do with global end-Pleistocene extinctions than with any immediate effect of the interchange.

What seems to be the key feature in the success of the northerners in South America is that their descendants diversified more strongly after their first immigration. There may be more practical factors working here than an inbuilt superiority. First, when the climate became colder and more variable during the Pleistocene, South America with its equatorial zone may have simply retained a greater diversity of climates, while North America's contracted. The immigrants to North America therefore faced harsher conditions than their opposite numbers.

Second, there appears to be little evidence of direct competition between the endemic South American mammals and the northern invaders. Many species of the savanna radiation that took place there in the Miocene had already become extinct: most of the native hoofed animals, for instance. Much of the diversity among the northern invaders in

Megaloceros ("*big antler*"), commonly called the "Irish elk," but really a giant moose-sized deer, is best known from peat deposits in Ireland, though it appears throughout the Eurasian Pleistocene. It is more closely related to the fallow deer than to the elk or red deer.

The enormous antlers could reach a span of 12 ft (3.7 m) and weigh up to 100 lb (45 kg), about one-seventh of its total body weight. A garbled theory used to view them as an example of gross evolutionary overspecialization, a handicap possibly inflicted by sexual selection run rampant, till it caused its owners' extinction. (In all deer except the reindeer, only the males grow antlers.) Studies have shown that for a large deer the Irish elk's antlers were not out of line with those of species living today, because the antlers tend to grow proportionally larger toward the upper end of the scale of size.

present-day South America is among the deer, cats, and dogs, which may have done nothing more than walk into ecological roles that were already vacant. The interchange appears to have done little harm to the endemic rodents and primates, the opossum-like marsupials, or the smaller anteaters, armadillos, and tree sloths. The United States is still home to a number of successful invaders from South America, such as armadillos, porcupines, and opossums.

There were other interchanges. From the Old World, North America absorbed deer and bison, as well as the mammoths and saber-toothed cats that did not outlast the Pleistocene. Dogs, camels, and true horses (one-toed grazers) went the other way from North America. Madagascar probably acquired its bush pig and pygmy hippo (now extinct) at this time. And finally, in the late Pleistocene humans left Africa and spread to the other continents of the world.

By the end of the Pleistocene the following types of mammals were all extinct: all the great trunked animals other than elephants worldwide, and elephantids (mammoths) in North America and northern Eurasia; saber-toothed cats, cheetah-like hyenas, and hyena-like dogs; sivathere giraffids, chalicotheres, and giant hyraxes. South America lost its marsupial carnivores, native ungulates (notoungulates and litopterns), and large edentates (ground sloths and glyptodonts). North America lost the same large toothless animals, present as immigrants, as well as its horses, camels, and tapirs. In Australia the giant browsing kangaroos, diprotodontids, and marsupial lion disappeared, though anecdotes and Aboriginal tales suggest that some of these may have lasted into more recent times.

The faunas of the world today, and especially the larger mammals, are heavily reduced compared with those of only 12,000 years ago. Today only equatorial Africa contains anything like the numbers and diversity of large mammals that flourished until fairly recently throughout much of the

Around 2.5 My ago, Florida had the greatest diversity of animals in what is now the United States. Recent immigrants from South America seen here are the giant ground sloth Glossotherium *("tongue-beast"), the giant, flightless, carnivorous bird* Titanis, *and the capybara* Neochoerus *("new pig"). Today's largest rodents, capybaras, have vanished from the United States and live only in Central and South America.*

BELOW LEFT *There used to be a vacancy for a tall, brawny, large-clawed herbivore able to rear up on its hind legs and drag branches. Two such were the homalodotheres of the South American Miocene and the chalicothere from the Eocene to the Pleistocene in Africa, Eurasia, and North America. Homalodotheres were notoungulates, hoofed mammals of the south; chalicotheres were perissodactyls, odd-toed hoofed mammals. At a height of under 10 ft (3 m), both creatures would have been dwarfed by the giant ground sloths of South America, which survived until about 10,000 years ago.*

RIGHT *Another kind of convergence produced high-level browsers. Only giraffes now play this role. In the Miocene of North America, alticamelines such as* **Alticamelus** *stood on long legs and craned long necks. In the Pliocene and Pleistocene of Australia, the sthenurine kangaroos specialized for the same diet, evolving arms to grasp vegetation – a unique adaptation outside of the primates.*

world. Perhaps it is its relative closeness to us in time that leads us to think of the Pleistocene as a golden age of great mammals.

Evolutionary trends in later mammals

Three trends are prominent in the evolution of mammals during the past 65 million years. First, as well as diversifying into a variety of different species, they also developed a far greater variety of body forms, including those specialized for flying or swimming. This diversity was further amplified by the fact that different continents often produced their own varieties of specialized adaptive body types. A classic example is the parallel evolution of marsupials in Australia and placentals elsewhere: marsupial "wolf" and flying possums; placental wolf and flying squirrels. Second, mammals achieved much larger sizes than in the Mesozoic, when the giants were about the size of an opossum. Lastly, a number of mammalian types developed larger brains – not only primates and dolphins but also many types of carnivores and ungulates.

The great diversification at the start of the Cenozoic must be related to the death of the dinosaurs. Although there is no evidence for direct competition either with them or with the large marine reptiles, it does seem that their presence blocked the way. Once they were gone, the mammals were no longer tied to the role of small generalists, though it took them some millions of years to come fully into their own. The great variety of different types of mammals on different continents relates

directly to the role of tectonic movements. The continents were moving apart in the later Mesozoic, so different "seed groups" of more primitive mammals were able to evolve in varying degrees of isolation from one another, and to extend the range of mammal specializations.

Some other factors have affected the diversity of animal types. The spectrum of body forms seems to have broadened during the Cenozoic, though it is true that we have also lost some classic types, such as saber-toothed carnivores. (The "type" itself may only be dormant, and the absence of human intervention could enable it to rise again. For example, the Asian clouded leopard, *Neofelis nebulosa*, is a practiced tree-dweller with elongated canines. A few million years of non-interference by an industrialized world, and it is quite conceivable that this line of the cat family might "re-evolve" the saber-tooth form.)

Conditions dictated this trend to diversity. The past 65 My have seen a continual climatic deterioration from a warm, wet world covered by tropical types of forest to a cold, dry world with sharp climatic zoning from poles to equator. With the rise of different types of environments, various specialized types of mammals have evolved body forms able to cope with them. The growing variation from the start of the Cenozoic to the present day has been largely a matter of adding, say, camels and polar bears (to take extreme examples) to an original fauna of tropical forest types such as bush babies and mouse deer.

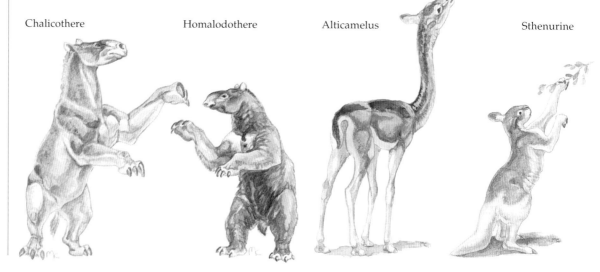

Chalicothere   Homalodothere   Alticamelus   Sthenurine

A final reason for the increased diversity in body form stems from the second general trend in mammal evolution, toward increased body size. The laws of physics dictate that bigger animals experience the world quite differently from small ones. We have seen that a bigger mammal's relatively smaller body area in relation to its mass causes it to lose less heat, and therefore to need less food in proportion to its size, than a smaller mammal does. It also causes the same surface area to carry much more mass; if a mouse and a person fall out of a second-story window, it is the mouse that is likely to get up and walk away, because the impact of its hitting the ground is distributed over a proportionally much greater surface area. (Likewise, children are more likely than adults to survive such falls, not only because their bones are less brittle but because they are smaller.)

The bigger the animal, then, the more cautiously it has to deal with gravity. As with the dinosaurs, its design must give priority simply to supporting its own weight, even when merely standing still, let alone when running.

A consequence of these different experiences is that bigger mammals need to be built in a more specialized way than smaller ones, even if their behavior is not so very different. So as mammals evolved into larger body sizes, the diversities of body forms had to increase, depending on what the body had to do. If we take two quite different activities, such as climbing and digging, we can illustrate this point. A field mouse climbs tall stalks of grass to eat the seeds at the top; a gopher burrows beneath the soil to get at grass roots. These rodents do not look very different from each other. On the other hand, even though they are just as closely related to each other as the field mouse and the gopher (different families within the same order), a tree-dwelling sloth looks very different from a digging armadillo. Much more of their design effort has to go into equipment for dealing safely with gravity. (That is why land-dwelling animals are so much more diverse in form, above the smallest sizes, than water-dwellers, which are all more or less free from gravity.)

At truly large sizes, the differences become even more dramatic. Sea otters and walruses are both sea-dwelling shellfish-eaters that spend some time on land; raccoons and grizzly bears are generalist omnivores; all of these mammals belong to a single order, the Carnivora. Now although the smaller sea otters and raccoons look fairly different, and would be easier to distinguish than a field mouse and a gopher, an untrained eye might still confuse them, especially at a distance. In the range of larger sizes, no one could mistake a walrus for a grizzly bear. Their different functions have led to far more different designs.

A basic question arises: what is a large mammal? We tend to take our own size as the norm, and think of an elephant as large and a rabbit as small, but if we shift our perspective to the average body size of the total range of mammal species (where rodents and bats are much more numerous than antelope or whales), it turns out that the "average" mammal body size is about 1 lb (450 g), which is the size of a small rat. This makes even a rabbit a large mammal, let alone a human being. Seeing that most Mesozoic mammals were small even by this standard, why was there an evolutionary trend to produce so many larger mammals? There is even an evolutionary law, Cope's law, named after one of the legendary nineteenth-century dinosaur collectors, Edward Drinker Cope, which states that in the course of time all animals tend to evolve larger body sizes.

One explanation for this trend may just be pure chance; if mammals started small, then random changes in size would leave "up" as the easier place to go. If we go back to the average late Cretaceous mammal, with a body mass of about 1 lb (450 g), we can follow its diversification in two directions. "Down" leads to the smallest mammal, a $1/6$ oz (5 g) shrew. "Up" to the other end of the mammal scale leads to a 5 ton (5,000,000 g) African elephant. A mass of $1/6$ oz (5 g) is probably a genuine lower limit for a mammal's size (given that vertebrates have body systems too complex to be micro-miniaturized), while 5 tons (tonnes) is certainly nowhere near an upper limit, even for a land mammal, seeing that some extinct ones may have weighed up to 15 tons (tonnes). So from a starting point of 1 lb (450 g) there is simply more opportunity for random evolutionary events to result in a larger mammal than a smaller one – more space for them to work in.

Many of the Pleistocene mammals of prehistoric North America, such as bison and mammoths, are relatively recent immigrants. A scene from the mid-continent plains around one million years ago shows the steppe bison (Bison priscus), larger and longer-horned than the modern bison. While proboscideans were present for much of the later Cenozoic, such true elephants as the mammoths did not appear until the Pleistocene. A recent arrival from South America is the armadillo Holmensina, while native North American mammals are represented by the pine marten and peccary.

This does not mean that random change is the most *likely* reason for greater body sizes; but the notion of available body space is a reminder not to take it for granted that all evolutionary change is strictly adaptive. It may happen because it gives its subject an advantage, but it may also happen because it gives no disadvantage. One major reason for the larger size of Cenozoic mammals compared to Mesozoic ones may again be the removal of a "ceiling" imposed by the dinosaurs. It is also true that many mammal lineages increase in size over the Cenozoic. Here climatic changes may play a part.

Another "law" for mammal body sizes is Bergman's rule, which states that animals tend to be larger at high latitudes than in equatorial regions. This rule makes more sense when applied to closely related species than when looking at mammals in general. Arctic foxes are bigger than the regular foxes in more

temperate regions; and although the elephants that now live only in equatorial lands are larger than any present-day Arctic mammal, they are smaller than the closely related mammoths that inhabited the Arctic tundra in the Pleistocene. The advantages of larger size in harsh conditions have been rehearsed in the story of the Plio-Pleistocene, when mammals faced an increasingly cold and threatening world. Not only did they retain more heat with greater size, but they improved their capacity to last out a rough winter on stored fat.

Although it is possible to identify various broad "laws" and trends, and to list the relative advantages and disadvantages of larger or smaller body sizes, the truth is that we simply do not know in any individual case what selective forces have acted to make changes. Different factors can pull in different directions, and random chance cannot be totally excluded.

Island conditions, for example, can produce some odd effects on size; it appears that the big mammals tend to shrink and the small ones to grow. Elephants stranded during the Pleistocene on Mediterranean islands developed dwarf strains, some no bigger than a St Bernard dog. On the same islands, the rodents grew large. The fossil record shows that these changes could happen very fast, in as little as a few thousand years in some cases, not only in elephants but also in hippos, antelope, and deer.

Herbivores may become small on islands because a confined space will limit the absolute amount of food available. The balance can swing the other way, and human populations are now a few inches taller in some prosperous parts of the world than they were as little as 150 years ago. Small mammals like rodents tend to grow larger only if stranded on islands where there are no predators, which suggests that on the mainland the key advantage of small size was the ability to slip into a variety of cracks and holes to avoid being eaten.

The third main trend in mammals was toward intelligence. Here again, the evolution of larger brain size may relate to the changing climate of the Cenozoic. A more seasonal climate is also a less predictable one, and a mammal needs to remember where there are water holes in the dry season, or where the best winter feeding grounds are. The evolution of large brains might be another product of random adaptiveness if there were some disadvantage to making an animal's brain *smaller*. Here too, any natural trend for variation in brain size over time might be preferentially expressed in an upward direction.

This cannot be a general law, however. Although brain size has increased in some groups, mammals display no *overall* increase in that area. Rather, a few brainy species have evolved alongside others that have changed little or not at all; present-day shrews have brains not much larger than the Mesozoic mammals they resemble.

Not all members of a group will develop in the same way. For example, among the odd-toed ungulates horses have proportionally large brains for mammals of their size, but their relatives, the rhinos and tapirs, do not. (So horses are bright enough to be trained for exacting tasks such as police work, but the world will never see a squad of rhino Mounties – it is the smarter animals that learn how to cooperate with humans.)

Primates are a classic example of increased brain size through time, but this trend among the group as a whole seems to have stabilized by the Oligocene, and not all living primates have large brains for their size, as is shown by the bush babies and lemurs. Since the Oligocene, only hominoids (apes and human lineages) have shown a continued increase in brain size. The rapid increase in the direct ancestor of humans started only a few million years ago, and seems to be an extreme case of evolutionary specialization, for which we lack a real adaptive explanation.

Although we pride ourselves on our intelligence, and some people may believe that smarter means better, still we do not really know what makes greater intelligence an adaptive advantage either in primates or in other mammals. Plenty of mammals manage perfectly well without an enlarged brain. Both small-brained opossums and large-brained raccoons are expert scavengers in suburban habitats in the United States, and both seem equally likely to be run over by cars.

One possibility is that larger brains were *forced* on certain species to protect the young. Even small primates only have one or two young at a time, and usually breed only once a year. If you breed like a rabbit, you can always make more rabbits, but if you breed like a primate you may need to have young smart enough to stay out of trouble and ensure the survival of the species.

Of course, this theory fails to explain why carnivores tend to be large-brained, since they also have large litters. As with so many evolutionary trends, we can document what happened, and dream up all kinds of attractive explanations – especially if they satisfy our deepest biases – but it is rarely possible to trace definite causes back into the labyrinth of life and time.

The burial of a child at Qafzeh Cave, Israel, about 100,000 years ago is the oldest for which we have good evidence of grave goods being added. Part of the skull of a fallow deer was clasped in the arms of the child's skeleton, and we can therefore infer that the head of a deer was placed on the child's body. The Qafzeh people were primitive modern humans – their skeletons show a modern body form, but the skulls have some archaic features including strong brow ridges, big faces, and large teeth.

THE PRIMATES' PROGRESS

Peter Andrews and Christopher Stringer

The pioneer of the modern system for classifying plants and animals was the Swedish botanist Karl Linné, better remembered as Linnaeus. In the key tenth edition of his classic study, *Systema Naturae*, published in 1758, he listed the various orders of mammals that make up the class Mammalia. To a naturalist of the eighteenth century it was obvious that the group containing human beings must be classed above the rest, so he named their order the Primates, from a medieval Latin word meaning "first in rank." It was Linnaeus who named the human species *Homo sapiens*, "wise man" – a classification that has come to seem more like a challenge than a scientific description.

An account of the Primates order has to begin with a physical description that links surviving species to the fossil record, but it has to end among a set of human qualities much harder to define in the present, let alone to trace back into the past. Memory, intelligence, language, and speech contribute toward making a social species. Somewhere along the line, they interacted to give rise to self-awareness, culture, and eventually history. It goes beyond the scope of science, a human invention, to account for all the elements that make up the measurable body and unmeasurable mind of humankind. Nevertheless, we know a lot more about the beginnings of the story than we did even a century ago, and about the order of some key events. New outlines are appearing, and if some of them seem to offer contradictory patterns, we should not be surprised. Few facts need a more skeptical, or more imaginative, scrutiny than those that bear upon the origins and infancy of the human family.

Those origins are an extraordinary episode in the longer, broader story of the primates. Its beginnings are remote, but because the story extends into modern times it is possible to describe the primate pattern hammered out since the end of the Mesozoic era by 65 My of

THE POSSIBLE CONNECTIONS OF THE HOMINOID PRIMATES

The relationships and evolution of the hominoid primates are shown on this time chart. On the left is the time scale in millions of years (My), from 34 My ago until the present. This period spans the Oligocene, Miocene, Pliocene, and Pleistocene periods. The living species of apes and humans are shown at the top, with their scientific names in plain type and their common names in bold, and their family groupings are shown above the top of the figure. Three families are presently recognized in the Hominoidea. During the first part of the Miocene, there is an extinct family, Proconsulidae, which is known only from East Africa. Secondly there are the gibbons, family Hylobatidae, for which there are no known fossil ancestors, although they must have diverged from the third family, Hominidae, at least 17 My ago.

The family Hominidae is used here to include all the living great apes and humans and includes also a number of fossil lineages, some of which cannot be related directly to any living descendant, for example the tribe Kenyapithecini known from both Kenya and Turkey. Two subfamilies are included in the Hominidae: Ponginae and Homininae. The former includes the living orangutan and *Sivapithecus*, a middle Miocene ancestor of the orang; the latter includes the African apes and humans, but unfortunately there are no known fossil ancestors for this group. The ape/human split was probably in the region of 4 to 6 My ago, for the earliest members of the human lineage are known from just under 4 My ago. Reconstructions of most of the fossil skulls are shown on the right, numbered to indicate their place in the phylogeny.

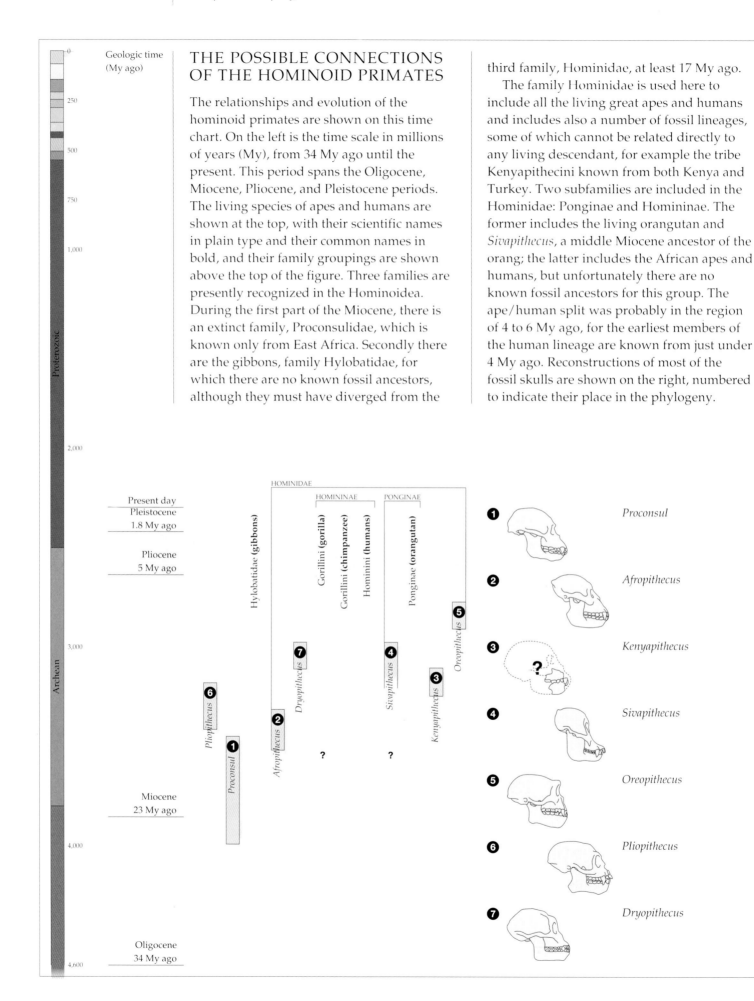

THE HOMINOID PRIMATES – THE IMPOSSIBILITY OF A FAMILY TREE

The diagram on the page opposite is a classic example of how an increase in knowledge over the last few years has served only to demonstrate the depth of earlier ignorance and the extent to which patterns of descent that were proposed in the past were little more than guesswork. Until a few years ago, paleoanthropologists, faced with the task of showing the relationships between the very early hominoid primates, would have produced a very different sort of diagram. Lines would have been drawn confidently across large stretches of geological time, ignoring the awkward fact that there was almost no evidence for the connections to be made.

The many people working in the field have now become far more cautious. Tantalizing gaps open up at precisely the points when evidence would be most welcome. For 9 My in the Oligocene and and again for 7 My in the late Miocene there is quite simply no evidence to go on. Those question marks floating in the middle of the diagram represent the highest state of modern knowledge about our own origins. Some recent authors have pointed out that there are more paleoanthropologists than there are fossils to work on, and that itself is a recipe for guesswork. Only one modern species, the orangutan, can confidently be linked to an ape from the middle Miocene, *Sivapithecus*, a chimpanzee-sized fruit-eater that resembles the modern animal in its jaw and palate.

Some wider trends can, however, be discerned. The pattern as you now see it in the diagram is intended to convey a picture of hominoid evolution in the Miocene which could be summed up as a punctuated series of adaptive radiations. In evolutionary theory, these bursts of diversification occur within a population very rapidly, making use of new environmental conditions. Because the radiation occurs – in geological terms – at speed, and because the fossil evidence is so patchy – the warm forests in which many of these creatures mostly lived are notoriously poor environments for the preservation of fossils – the paleontologist's task of disentangling the sequence of events is made very much harder. Rather than the classic picture of a slowly branching tree, one is left with these genetic "islands," animals whose precise relationships are almost impossible to determine. It is unlikely that to make the bridges between them will ever be more than guesswork.

An additional habit of mind, so often associated with the traditional "family-tree" diagram, must also be abandoned. These ancient animals were not in any way "primitive" precursors of existing species. They were highly efficient animals in their own right.

Proconsul africanus, for example, has been described, perhaps lightheartedly, as an "anatomical confection." Its shoulders and elbows are like those of a modern ape but the arm and hand bones are more monkey-like. In the rear legs, the affinities are reversed; the hip is not very apelike but the lower leg and the bones of the foot are close to those you would find in an ape today. This body-mosaic probably reflects the way in which *Proconsul* lived – a combination of running on all fours and climbing in the branches of the trees. As the diagram shows, *Proconsul* was certainly in existence for many millions of years, a success, at least in terms of longevity, which many other hominoids, including ourselves, have yet to emulate.

Finally, an absence should be explained. Anyone looking at this diagram who is familiar with its equivalent from the 1960s and 1970s will notice that *Ramapithecus*, a Eurasian ape of the late Miocene, is not here. It was first suggested in 1934 that *Ramapithecus* was the first hominid, pushing back discernible human origins to about 15 My ago. Its huge jaw and large, thickly enameled teeth were considered to be unique specializations, setting the hominid line apart from the other apes.

This theory is now known to be quite untrue. Thick tooth enamel was not unique to hominids, but was acquired by several different lineages. Evidence from molecular biology pointed to a more recent date than 15 My ago for the divergence of the hominids from the other apes. And *Ramapithecus* has anyway now been absorbed within the genus of *Sivapithecus*, the ancestor not of modern African apes but of the orangutan. From its unique position as our oldest ancestor, *Ramapithecus*, as a separate genus, has now been demoted out of existence.

evolution, and use it as a guide to the past. The guide is selective, of course; it omits discarded species, as well as features once important but made obsolete by time.

Primate bodies are less specialized than the bodies of other mammals, and they lack the detailed inventory of unique features that makes their fellow mammals easier to describe. We have the advantage of knowing a lot about the soft tissues of our own and other primate species, and about their habits and behavior; but soft tissues and behavior are unlikely to fossilize, though they may leave traces that science can learn to interpret. It is possible to make direct comparisons between the bones of ancient and modern species, but the earliest fossil species do not include all of the physical features now identified with primates.

The first primates developed in the tropics and subtropics, where all but a few species still live. Mostly they are tree-dwellers, with the versatile features that this habitat requires. The five digits on the end of each limb are specialized for grasping and handling, with sensitive pads backed by nails instead of claws, opposable thumbs for grasping branches, and usually opposable big toes. These five digits, along with the collarbone, are features of the earliest mammals that the primates have kept while some other mammalian orders were losing or altering theirs. The opposable thumb is not exclusive to primates, but it provides an ability to grasp and manipulate all sorts of objects. This primate skill predates their specialization into intelligence, although it offers the means for intelligence to express and improve itself.

In order to travel, to locate food or prey, and to detect predators in the hazardous three-dimensional world of the trees, most primates have developed sharp eyes in preference to keen noses. Instead of facing sideways, the eyes face forward, with overlapping fields of view that the brain has to coordinate so as to provide depth perception. They are set inside bars or rings of bone, and sometimes in a bony cup that holds them separate from the brain.

The brain itself is larger in primates than in other mammals of similar size, and more elaborate in design. It is serviced by a complex system of nervous pathways that carry information and instructions, and the plumbing of its fuel lines has to be particularly efficient because its functions can be quickly and even fatally damaged by quite brief interruptions to the blood supply. In human beings, the brain weighs about one-fiftieth of the total body weight, but uses about one-fifth of the energy budget. It must always have been an expensive organ to develop and maintain, so at every stage of its growth among the primates it must have delivered a valuable return on the energy invested.

Primates tend to bear one offspring at a time, and to carry it for months before giving birth. The offspring grow slowly, and are unusually dependent on parental care – features that appear to be closely connected to the development in many primate species of complex forms of social organization, whose history must be much longer than the fossil record can show. One of the main tasks of these primate societies is to protect and bring up their vulnerable young. This major investment of time and energy may have begun because the young had more skills to learn with their improving brains, and may have continued partly because social life itself turned out to have other advantages.

Paleontologists have described more than two hundred genera of primates, but only about two hundred species inhabit the modern world. In the chronicle of life their history is quite short, but it has also been eventful, and gaps of only a few million years at the wrong time and place conceal vital episodes. Because the order has survived into the present, the full resources of the life sciences – biochemistry, genetics, ecology, ethology, and a legion of others, backed by an army of technologies – have been focused on its study. Yet our own immersion in the subject, as both observers and observed, can make it hard to be objective. If we feel uneasy about the close relationship of humankind with apes and monkeys, we may be tempted to scale down the connection by pushing human origins further into the past. Racial or regional biases may cause scientists to claim or disclaim some vital primate event for their own country or continent. A wealth of information does not necessarily settle basic arguments, when human ingenuity is interpreting the facts.

Linnaeus classified the Chiroptera – the bats and flying foxes – among the order Primates. The order is hard to define, because its sources are mysterious. No undisputed primate fossils date back to the Cretaceous, and there is doubt about the status of the "plesiadapiform" mammals that are sometimes identified as early primates. Even their name is a record of scientific puzzlement. It means "close to an adapid in form." Adapids were later, and certain, primates that take their name – "toward Apis" (the sacred bull of Egypt) – from a description made in 1821 by Georges Cuvier, who thought they might be related to the even-toed hoofed mammals when he examined their fossils. So the name itself is a fossil from a bygone age of paleontology.

Several families of plesiadapiforms scattered at least twenty species through the Paleocene of Europe and North America, as part of the early expeditionary forces of small insect-eating mammals that took to the trees after the dinosaurs' departure, and developed capabilities to include fruit, nuts, and seeds in their diet. Forests and jungles covered most of the land, and were probably much more dense than they had been when massive dinosaurs had browsed and trampled through them. Gums, flowers, and nectars, and the insects that fed on them, were other new resources provided by modern flowering plants. Plesiadapiforms had broad skulls and long snouts, and looked more like tree shrews or large rats than like later "euprimates" (true primates). The pattern of their teeth reflects their range of diets, and links them to the primates, but they still had claws instead of nails, their eyes did not face forward, and their brains were not developed.

Are these the ancestors we long to trace? They were succeeded by recognizable primate groups so advanced that they must have sprung from simpler precursors. Perhaps we would not be so ready to co-opt them into the primate order if the record offered any more obvious candidates. Certainly the plesiadapiforms often included among the "archaic" primates had already developed too many specialized features, not found in later lineages, to be direct ancestors of the so-called "primates of modern aspect," including ourselves, that do not contain these features.

There is no argument about the primate identity of the waves of "omomyid" adapiform euprimates that flowed across Europe and North America during the Eocene, when

those regions were still connected by a land route through Iceland and Greenland. The omomyids were smaller than the adapids, ranging in weight from 2 oz to $2^1/_2$ lb (60 to 2,500 g), compared with the $2^1/_2$ oz to 22 lb (70 to 10,000 g) of the adapids. Both families had nails, forward-facing eyes, reduced snouts, and relatively larger brains than other contemporary mammals of the same size. The omomyids had resemblances, but no certain relationship, to modern tarsiers, which are rat-sized vertical clingers and leapers, preying on anything from insects to birds and snakes. Their faces are dominated by enormous eyes, night-sight adaptations not required by their Eocene ancestors, which had a much broader range of diets, and may have been outcompeted in the daytime by the "higher," apelike primates that poured onstage in the late Eocene and early Oligocene.

Adapiforms have been considered as possible ancestors of the modern lorises and bush babies of Asia and sub-Saharan Africa, and the various lemurs of Madagascar. Without more evidence, we cannot be certain whether this connection is direct, or whether these ancient and modern families derive their common features from unknown ancestors in the Paleocene. Rodents were a widespread and successful group in the Paleocene, and it may have been their presence on the ground that kept the Eocene primates to roles as smallish tree-dwellers.

Another puzzle is the origin of the primate inhabitants of present-day Central and South America, the platyrrhine ("flat-nosed") monkeys – marmosets, capuchins, spider, woolly, and howler monkeys, sakis, and uakaris. They differ from the Old World monkeys and hominoids because their nostrils are widely spaced and mainly sideways-facing, instead of close together and forward- or downward-facing. Most of them have long tails, often used as a grasping fifth limb, and none have ever taken to the ground. Two schools of thought trace their arrival, sometime in the late Eocene or the Oligocene, to accidental rafting and island-hopping either from North America or more likely from Africa, when falling sea levels may have exposed peaks along the mid-Atlantic ridge now long submerged.

Whatever their early history, the development of the New World primates

THE SPREAD OF LIFE ON LAND

Life on land dates back to the Silurian, about 420 My ago. This represents only a small proportion of the total time duration of life on Earth, when for thousands of millions of years life was confined to the seas. The first larger land animals were amphibians, followed by reptiles during the Permian and the rise of the dinosaurs and birds during the Mesozoic. Mammals are first known from the Triassic, about 230 My ago, but they did not become widespread until the end of the Mesozoic, 65 My ago. At this time there were several groups of mammals very similar to living primates, but the first definite primates are not known until the early Eocene, 55 to 60 My ago. The hominoid primates evolved during the Oligocene, 25 to 30 My ago.

The bar chart shows the time scale for terrestrial life. The scale at the top shows the age range from the Carboniferous, 360 My ago, up to the present. Below this, on a different scale, is the time distribution of the higher primates, first known from the late Eocene about 34 My ago. The hominoids appear in deposits 25 My old and are abundant throughout the Miocene. In more recent times hominoids have become rare, with the one exception of humans and their ancestors.

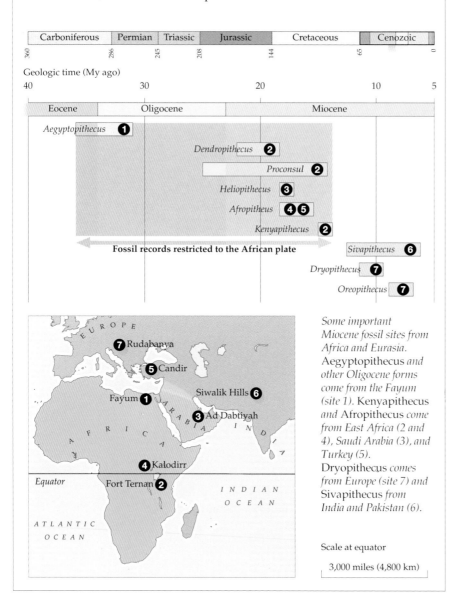

Some important Miocene fossil sites from Africa and Eurasia. Aegyptopithecus *and other Oligocene forms come from the Fayum (site 1).* Kenyapithecus *and* Afropithecus *come from East Africa (2 and 4), Saudi Arabia (3), and Turkey (5).* Dryopithecus *comes from Europe (site 7) and* Sivapithecus *from India and Pakistan (6).*

Scale at equator

3,000 miles (4,800 km)

offers a cautionary example to theorists of hominoid evolution. They need to explain what special causes may have drawn or driven the biped venture of the Old World, but tied those New World cousins to the trees, even when they had an entire continent to explore and be changed by, and more than 35 My to evolve in.

This chapter will center on the hominoid primates that developed in the Old World – the lesser apes, great apes, and humans. Here too, some tantalizing gaps erode our knowledge. The African continent is plainly vital in primate history, but the record peters out for some 9 My from halfway through the Oligocene into the early Miocene, and for another 7 My in the late Miocene. During the first interval (31 to 22 My ago), the gorillas, chimpanzees, and humankind launched their careers, but for most of this period there are hardly any fossils known from Africa. Writing about the key fossils from Europe, the Far East, and Africa that signpost human evolution, John Reader quipped in the 1980s that "even today the significant specimens could all be accommodated on a billiard table." A single new skull in an unexpected time or place could still rewrite the primate story. It has happened before.

The hominoids: from apes to humans

The hominoid primates consist of the lesser apes, great apes, and humans. The lesser apes (average weight about 15 lb, 7 kg) are the gibbons and siamangs of southeast Asia, a varied group, but all of them classified in the single genus *Hylobates*, because of the set of features that they share with no other primates. The most conspicuous and exhilarating of these is their "brachiating" style of travel, an acrobatic ability to swing from branch to branch using only their very long arms, whose bones and musculature have a range of adaptations for the purpose. A wide vocabulary of loud, complex howls and chatter allows them to keep contact with each other in the dense rainforest they inhabit. In their monogamous kin system, each family holds and defends its own patch of forest, usually by means of its rowdy duets and choruses.

The great apes used to be grouped in a

single family, the Pongidae, because they look superficially alike. Nowadays we recognize that their many resemblances often stem from their large size, which has caused a number of ancestral species to invent much the same "convergent" design solutions to similar problems with managing heavy bodies. One effect of this discovery has been the decision to class the orangutan of Borneo and Sumatra in its own subfamily, the Ponginae, together with various fossil species but separate from the African apes, which are grouped with humans in the Homininae. The evidence for this shift comes not from fossils but from breakthroughs made by molecular biology, which can analyze sequences of proteins, amino acids, and DNA taken from different species of living animals (including ourselves), and compare the findings in close detail.

Two broad insights provided by these techniques have revolutionized our view of the history and family relationships of hominoid primates. The first comes from measurements of the genetic distances between the various primate groups of today. Using the "molecular clock" countback system described overleaf, which estimates the rates of random DNA mutation to trace back to the point when species may have diverted from a common source, one survey has concluded that the gibbons parted from the great apes and humans line about 12 My ago, the orangutans about 10 My ago, and humans from the African apes about 5 My ago. This countback method is young and controversial, and it may not be safe to assume that rates of change have been constant over tens of millions of years. Different studies have timed the divergence of the Old World monkeys from the line leading to humans at 34 and 25 My ago, which is a wide margin. But the order of the main events is clear, and so is their general timing, which has proved to be far more recent than was once believed.

Molecular biology has shown that the orangutan differs from the ancestral hominoid state, and lacks some DNA sequences that humans have in common with the African apes. It is hard to find clinching evidence for this more distant relationship in the anatomy or physiology of the orangutan, which makes it reasonable to assume that there may be similar differences between fossil species that look physically alike.

The second clear insight from molecular biology is that it shows how small a genetic distance divides humankind from the African apes. These consist of two species of chimpanzees (the common chimpanzee and the bonobo, once called the pygmy chimpanzee) and three subspecies of gorillas. How did this group arise? Are chimpanzees more closely related to the other African ape, the gorilla, or to humans? Nothing in their morphology – their form and structure – offers a decisive answer, and the molecular evidence points several ways. Most of it suggests, but does not prove, that the closer kinship is between chimpanzees and humans. (One study has concluded that we share 98.4 percent of our DNA with chimpanzees, 97.7 percent with gorillas.) This would not mean that chimpanzees, or anything very like them, were ancestral to humans, but that their common ancestors were the last to diverge along the primate line.

This issue is plainly not settled, but it matters much less to human beings than the dramatic certainty that all the differences in appearance, physiology, intelligence, and behavior between the African apes and ourselves are contained in the roughly 2 percent of our DNA that differs from theirs. No huge and immensely long-established gulf distances human beings from the rest of the Hominidae group, and if we focus too closely on the differences that make us what we are, it may be at the expense of overlooking the sweep of formative features that we have in common.

Not that the African apes are all of a piece. Chimpanzees live in extended family groups, fluid enough to divide into small subgroups that vary from day to day. Paint one ear of a chimpanzee and of a gorilla the same (washable) color and stand them in front of a mirror, and the chimpanzee will reach for that ear, while the gorilla will ignore it. Does the chimpanzee have the more developed sense of self? The gorilla has developed a male-dominated social structure in which a leading male is shared by several females along with their newborn and adolescent young. Gorillas have outgrown the point of no return for a life in the trees, and the large adults spend much of their time on the ground. They eat more vegetation than the other apes, which are all proficient climbers, able to eat all sorts of fruit. When on the ground, both chimpanzees

MOLECULAR CLOCK

In recent years it has been possible to measure genetic differences between living species by identifying changes in proteins, in amino acids, or directly in the DNA. Most of these changes do not produce any change in actual morphology. Because they may occur at a regular rate, the length of time since the divergence between two species can be estimated by measuring the genetic difference.

Percentage differences are shown in the figure below based on the degree of DNA differences between the four living hominid species. Humans appear to be most closely related to chimpanzees. One estimate has found only a 1.2 percent difference between the DNA of humans and of chimpanzees, compared with a 1.4 percent difference between humans and gorillas. This would mean that chimpanzees and humans share 98.8 percent of their DNA.

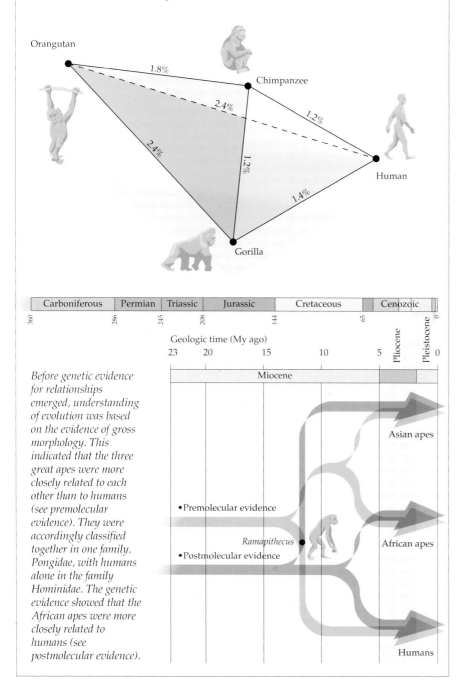

Before genetic evidence for relationships emerged, understanding of evolution was based on the evidence of gross morphology. This indicated that the three great apes were more closely related to each other than to humans (see premolecular evidence). They were accordingly classified together in one family, Pongidae, with humans alone in the family Hominidae. The genetic evidence showed that the African apes were more closely related to humans (see postmolecular evidence).

and gorillas walk on all fours, supporting their upper bodies on the extended knuckles of their hands.

This knuckle-walking gait is an adaptation just as individual as the brachiation of gibbons, the four-limbed climbing of orangutans, and the upright stance of humans. If any of these designs can be detected in the fossil record they will mark a possible ancestor of the user group, and in fact the first evidence for fossil humans comes from features identified as modifications for walking upright.

More clues to the affinities of early hominoids are their body size, the diet suggested by their dental layout, and the habitat recorded by the associated fossil finds and ancient environment. Paleontologists also have to make informed guesses about hominoid life histories and social structures on the basis of circumstantial evidence (if that is all there is), or of chains of reasoning that link the smallest fossil details. Modern techniques of excavation are rigorous in recording the exact location of every tooth, bone, stone, and pebble, and its relation to the others at the site. Tall structures have to be built on these foundations.

Hominoid origins

All primates that do not belong to the New World group, the Platyrrhini, belong to the much more varied Old World group, the Catarrhini ("hanging noses"). These consist of two superfamilies, the Hominoidea (hominoids) already described, and the Cercopithecoidea ("apes with tails") – the monkeys of Africa and Asia. In order to date the point when these two groups divided we have to identify clear-cut specimens of one or the other in the fossil record.

The very earliest example of a true cercopithecoid or hominoid obviously gives a date for the split that must lie sometime before the specimen. But although the two superfamilies are quite distinct from each other, they are not so distinct from their catarrhine ancestors before the split. So is fossil X, with its obviously advanced design, one of the last generations of original pre-split catarrhines, or an early species of one of the two new superfamilies? Only a full range of fossil primates that straddled the age when these two paths divided could settle that question. The evidence we do have is spasmodic, in both space and time.

A classic case is the varied group of primates preserved among the rich fossil flora and fauna collected during the last thirty years by the American paleontologist Elwyn Simons from the Fayum depression, not far from Cairo. They date from the early Oligocene of Egypt, about 35 to 31 My ago, when the Fayum was a region of low-lying swamp, bordering a broad river flowing into a shallow nearby sea. Water birds, crocodiles, turtles, and sea snakes teemed among mangrove swamps that lay side by side with dense tropical forest – hardwoods and palm trees, overgrown with lianas and infested by underground termites. Under a monsoon type of climate lived an assortment of mammals from rodents to elephants, hyraxes to the rhino-like *Arsinoitherium*.

Simons has retrieved at least ten species of fossil primates from the Fayum, of which four are certainly catarrhines, and the rest more primitive. *Propliopithecus* had teeth and jaws with a definite hominoid look, designed for eating fruit. *Parapithecus* had teeth with special cutting ridges designed for eating leaves, which are a monkey specialization. Many of the other, smaller primates were insect-eaters. The biggest animal in the group was a catarrhine that probably weighed no more than 20 lb (10 kg), but all of them had sturdy limb bones built more for strength than for agility.

The Fayum primates provide a useful island of knowledge. They are not closely connected to the earlier Eocene primates of the northern continents whose northward drift had devastated their primate populations. Nor do they contain evidence of the vital split between monkeys and hominoids that must have happened sometime later during the vast interval of 9 My, from 31 to 22 My ago, when we know practically nothing about primate history in Africa. It seems reasonable to suggest that the Fayum habitat was a typical tropical scene, and typical of Africa's coastal regions. All the Fayum primates lived in the upper forest canopy, like most of their present-day successors. Their career on the ground had not yet begun.

The earliest hominoids

In 1948 an expedition from the University of California found part of a fossil ape jaw at Lothidok in northern Kenya. It lay with other material between deposits of volcanic dust that have since been dated by the potassium-argon ("K-Ar") method, which measures the radioactive decay of an unstable isotope of potassium as it changes – at a known and regular rate – into inert argon gas. They place the jaw between 27.5 and 24.8 My ago. This jaw would be hard to identify on its own, but there is good evidence to tie it to an earlier discovery made at a site at Koru, Kenya, and described by the British Museum paleontologist A. T. Hopwood in 1933. He located it in a new genus, *Proconsul*, which was a humorous reference to a chimpanzee named Consul, a captive in Birmingham zoo. By calling the genus "before Consul" he was tagging it as a definite fossil ape, the first one ever found, and its hominoid identity is generally accepted. The Koru site has since been dated by K-Ar to the early Miocene between 20 and 19 My ago, so the Lothidok species, *Proconsul hamiltoni*, possibly 6 My older, is the earliest known hominoid.

Hopwood's specimen was brought to his attention by the now famous paleontologist Louis Leakey, who later found his own Miocene locality on Rusinga Island in Lake Victoria, Kenya, and made major *Proconsul* discoveries in a series of collections during the 1930s to 1950s. John Napier described the most complete skeleton found on Rusinga in 1959, and in the 1970s Peter Andrews carried out a series of excavations on Rusinga and at Songhor, also in western Kenya, and he described these and several hundred of Leakey's new finds in 1978. More recently Alan Walker has led several digs on Rusinga, adding to the partial skeleton described by Napier and finding a remarkable series of associated skeletons at a new site.

All this material allows us to reconstruct the anatomy of *Proconsul* in greater detail than for any other fossil ape, as well as to glimpse the profusion of other ape species from a golden age of hominoid development. *Proconsul* itself gave rise to several species, the smallest about the size of a siamang, the largest as big as a female gorilla. Sometimes they are found alongside the fossils of a related genus, *Rangwapithecus*, or with smaller apes whose kinships are not clear.

A genus with so many species, and such a range of sizes, is certain to practice all kinds of lifestyles, and to develop the necessary adaptations to pursue them. *Proconsul* designs may have been as varied as they are in living hominoids, but we know a lot more about the cranial details than about the rest of the "post-cranial" skeleton, which is less durable. It takes more than the usual great good fortune to preserve any bones at all from the acid, corrosive forest soils.

Most details of *Proconsul's* skull and teeth, including the facial shape, lack of brow ridges, and thin dental enamel, are primitive features inherited from non-hominoid ancestors. But the brain is comparatively large, and the increased surface area of the molar teeth and broadening of the incisors probably mean that it had a more fruit-based diet. These are strategic signposts that point toward some future directions among the hominoids, for the changes reinforce each other – teeth better suited to crushing and processing fruit before ingestion, brain

functions having to adapt to a food resource not always on tap in time or space, because fruit has seasons, and the trees that bear it may be scattered in the forest.

The rest of the skeleton has the same blend of advanced and primitive features. The shoulder gave the arm as much all-round mobility as in living apes, but the arm could not straighten fully at the elbow. The hand had a fully opposable thumb (unlike a monkey's), and its proportions were like those of a human hand, without the reduction of the thumb and lengthening of the hand seen in modern apes. The hind legs preserve the primitive catarrhine design at the hip and knee joints, but are more apelike in ankles and feet. *Proconsul's* style of four-legged movement differs from the monkeys' only because the shoulder and ankle are more mobile and the elbow joint more stable.

Yet the opposable thumb and "human" hand proportions would have uses impossible for living apes and only feasible now for humans. All or most of the living hominoids are tool-users, or even toolmakers in the case of chimpanzees, which will strip a twig

ADAPTIVE RADIATIONS

Human evolution has seen a gradual extension of adaptations to new habitats and a diversification of diets. It is assumed that hominids began in forested environments but progressively adapted to more open conditions, first in Africa (particularly with *Paranthropus* and *Homo habilis*) and subsequently in more temperate regions (*Homo erectus* and later forms). Early hominids would have been opportunistic feeders, and it was probably only the move into more seasonal environments in the temperate-cold regions of Europe and Asia that forced a greater reliance on meat, especially through the winter months when plant resources would have been scarce. Even in the later stages of human evolution, it is assumed that scavenging opportunities would still have been exploited.

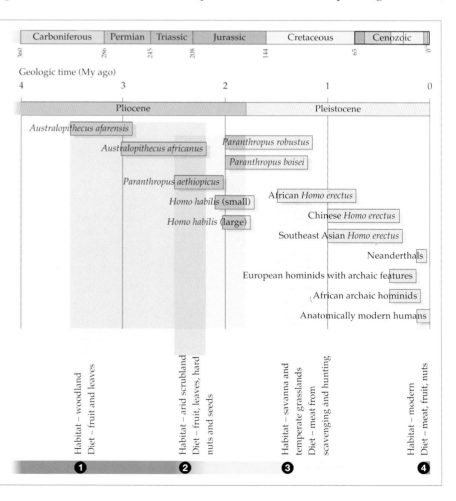

smooth so as to poke it into crevices and withdraw termites. So it is not too fanciful to speculate that *Proconsul* had the mental and manipulative power to use tools, although nothing could identify the very earliest tools for what they were, even if they somehow entered the fossil record. Tool use may come to be seen as a uniquely hominoid talent in primates, one of the non-human assets that interlocked with further physical changes as their potential was preserved and expanded by changing conditions.

Proconsul's tropical rainforest environment is preserved in all kinds of fossil seeds and fruits, including those of the West African mahogany *Entandophragma* and many species of liana. On Rusinga Island there is also evidence of less lush types of forest with a more seasonal climate. The preservation there is so good that one excavation revealed a forest floor litter that included buds, twigs, and leaves, mixed up with seeds and fruits. In general, *Proconsul* species are found alongside groups that are now rainforest specialists – large elephant shrews, the flying "squirrels" of the family Anomaluridae, and relations of living lorises – but it looks as if at least one species lived in drier, more seasonal conditions, either instead or as well. The genus would have looked much more like monkeys than like apes. That picture changes fast with the next key group.

Middle Miocene hominoids

Late in the early Miocene came a new group of hominoids that Peter Andrews proposes to recognize as a new subfamily, the Afropithecinae. This contains several genera successful mainly in the mid-Miocene: *Afropithecus* from northern Kenya, the similar but smaller *Heliopithecus* from Saudi Arabia, and *Kenyapithecus* from Kenya, together with the recently proposed new genus, *Otavipithecus*. This new subfamily may be linked to the later-occurring dryopithecines of the subfamily Dryopithecinae (see box).

Like all the other early hominoids so far known, and like the pre-hominoids of Fayum, the afropithecines are an African group. They lived on a continental plate that was separated from the Eurasian continent

throughout the Paleocene and Eocene, all the way through the Oligocene, and into the early Miocene, when Africa touched Eurasia again sometime between 18 and 15 My ago. All this points strongly toward an African origin for the higher primates, which did not appear in Eurasia till around the time the land bridge made travel possible from Africa.

The afropithecines

Both *Afropithecus* from northern Kenya and the closely related genus *Heliopithecus* from Saudi Arabia come from deposits about 17 to 18 My old. They share their extremely enlarged premolar teeth (located between the canines and the molars), and very strong canines, in a long-faced, robust skull that resembles the skull of the living great apes. This fossil group is obviously more closely related to the great apes and humans than are the later gibbons, or of course the earlier *Proconsul* species.

The molars in the early afropithecines show the beginning of one of the most important adaptations in later fossil and recent hominoids, the thickening of the molar tooth enamel. This was a key change because it would enable its makers to live in less friendly conditions. The function of enamel is to armor the teeth against the wear caused by chewing. A mammal's life expectancy can often be measured in terms of the length of time its teeth stay functional. Some animals with coarse or abrasive diets may starve to death when their teeth wear out and they can no longer process food. A whole series of changes in later hominoids would not have been possible without this key survival factor, and the longer life span it allowed.

Thickening the enamel may seem to be an obvious adaptation, but it takes radical improvements in physiology to absorb and deliver the minerals it requires, which are needed early in life, while the teeth are forming. Most mammals have relatively thin enamel.

These changes in the teeth of afropithecines, as well as their stronger skulls (buttressed against great pressure from the jaw muscles), and other adaptations, seem likely to indicate a change of diet, away from soft fruit and toward much harder items. Few details are known of the ecology they occupied, but there

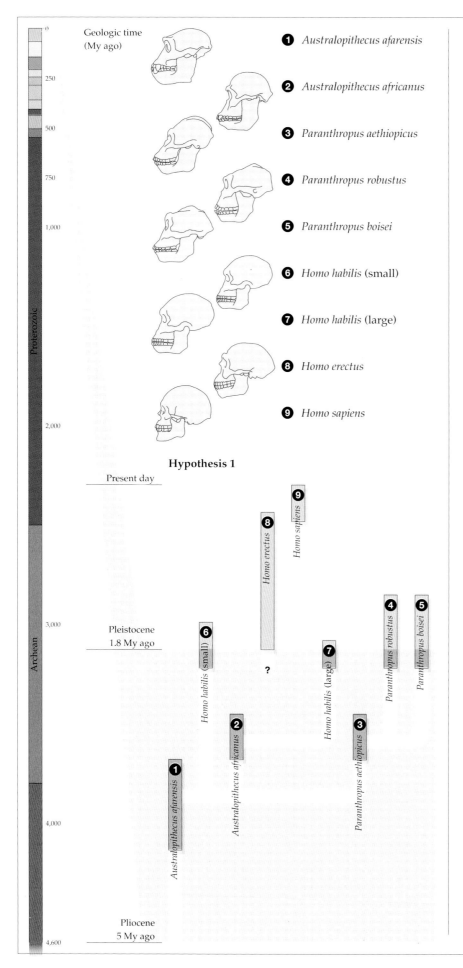

Geologic time (My ago)

1 *Australopithecus afarensis*
2 *Australopithecus africanus*
3 *Paranthropus aethiopicus*
4 *Paranthropus robustus*
5 *Paranthropus boisei*
6 *Homo habilis* (small)
7 *Homo habilis* (large)
8 *Homo erectus*
9 *Homo sapiens*

Hypothesis 1

Present day

Pleistocene
1.8 My ago

Pliocene
5 My ago

HOMINIDS

At least nine hominid species have probably existed over the last 4 My and there have clearly been overlaps between them, both in time and in space. The simple succession of *Australopithecus africanus* to *Homo habilis* to *Homo erectus* and *Homo sapiens* which was accepted by many scientists twenty years ago has been replaced by a much more complex pattern and much less certainty about ancestor–descendant relationships. There are very many ways in which the different species could be related to each other, and the gradual realization that even the species *Homo habilis* might consist of at least two distinct forms or species has meant that the early stages of evolution of the genus *Homo* are also clouded in uncertainty. Three possible evolutionary trees are given below.

Hypothesis 1

This model of hominid evolution has an early development of the small form of *Homo habilis*, with the larger form evolving later from *Australopithecus africanus*. *Paranthropus aethiopicus* is seen as the common ancestor of the robust australopithecines of both South and East Africa. As in all three hypotheses shown here, the origin of *Homo erectus* is left as uncertain, since it is unclear which of the *Homo habilis* forms is the more likely ancestor. However, most specialists are agreed that *Homo sapiens* evolved from *Homo erectus*, although it is not agreed whether that process occurred in one area (Africa?) or across the whole of the inhabited Old World. If the two forms of *habilis* evolved independently from the two forms of *Australopithecus*, they could not both be called *Homo*.

Hypothesis 2

This hypothesis is based on similarities in facial and dental form between the large type of *Homo habilis* and the robust australopithecines. The small form of *habilis* is seen as evolving from *Australopithecus africanus*, while the large form is regarded as a member of the robust australopithecine clade, probably sharing a common ancestor with *Paranthropus aethiopicus* about 2.5 My ago. As in the first hypothesis, there would be no justification for regarding the two types of *habilis* as *Homo*, and the large *habilis* form might even be regarded as a large-brained variety of *Paranthropus*.

Hypothesis 3

This hypothesis is close to the one favored by Don Johanson after the controversial creation of the species *Australopithecus afarensis*. He and Tim White argued that *afarensis* was the common ancestor of all later hominids and that *Australopithecus africanus*, contrary to common belief, was not ancestral to *Homo* but instead gave rise to the robust *Paranthropus* forms. *Homo habilis* had evolved directly from *afarensis*. However, if two species of *habilis* exist, the scheme becomes more complicated. In this example we have shown separate South African and East African lineages of *Paranthropus*, with the large *habilis* form related to the South African robusts. If this scheme were accurate, the two lineages of robusts could not both be called *Paranthropus*.

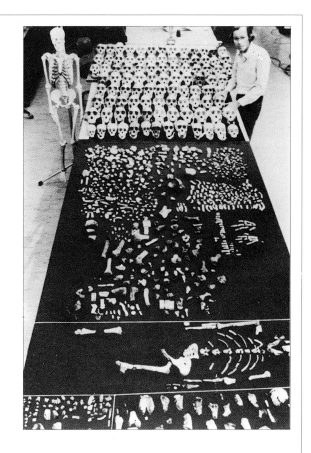

The Hadar fossil collection spread out in the Cleveland laboratory. Near the foreground is Lucy. Next is the entire First Family, arranged according to skeletal part. Tim White is in the background, standing next to some chimpanzee skulls.

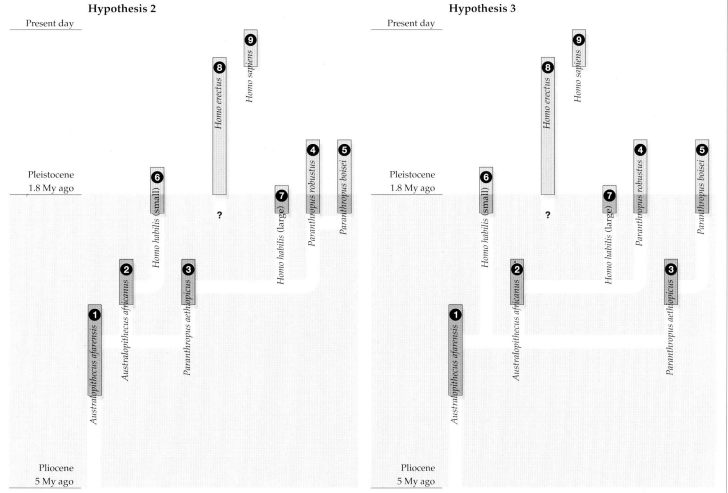

THE GENETIC STORY

Two models dominate current debate about modern human origins. These are the "monogenesis" or "out of Africa" model (top) and the "multiregional" model (bottom). Using the same fossil evidence, two contrasting pictures of human development have been constructed. The multiregional model envisages that the early human species *Homo erectus* spread around much of the Old World by about 1 My ago (e.g. Africa, Europe, China, and Australasia), and then began to develop local (regional or "racial") characteristics. These have continued through time up to the present day, but gene flow between the regions ensured that modern *Homo sapiens* developed in concert across the whole evolving population through the sharing of genes and characters.

In contrast, the "out of Africa" model suggests that, while *Homo erectus* did develop into subsequent populations across the Old World, there was only one continent where modern *Homo sapiens* evolved, and that was Africa. From this perspective, modern characteristics evolved first, and were then carried from Africa by migration. Pre-existing archaic populations outside Africa were replaced, and only then were regional ("racial") characteristics superimposed on the shared early modern pattern.

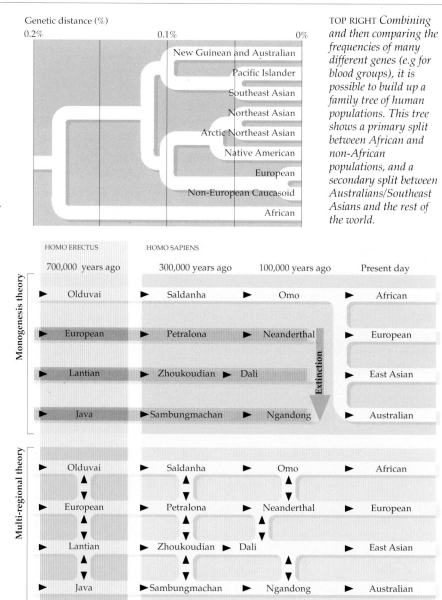

TOP RIGHT *Combining and then comparing the frequencies of many different genes (e.g for blood groups), it is possible to build up a family tree of human populations. This tree shows a primary split between African and non-African populations, and a secondary split between Australians/Southeast Asians and the rest of the world.*

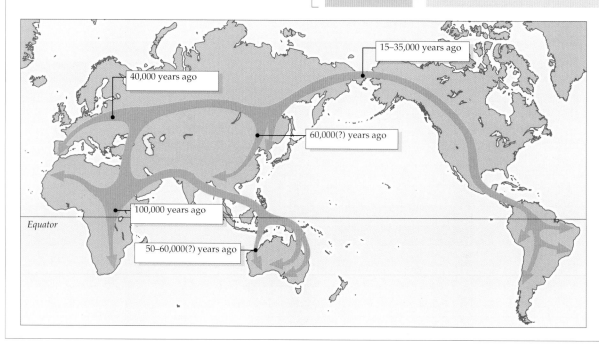

This map combines data from genetic relationships of present-day populations, and archaeological and fossil data for the first colonizations of different regions by Homo sapiens. Fossil evidence suggests that modern humans evolved in Africa by 100,000 years ago and soon spread from there to West Asia. They may have reached East Asia and Australia by 60,000 years ago, but probably only arrived in Europe about 40,000 years ago. Their migration into the Americas was probably more recent still.

are signs of a tougher environment, reflecting a long-term cooling and drying trend, with a more seasonal climate and therefore a less regular food supply. Without an all-year diet of soft fruit, flowers, and insects, animals need other sources of food. Hard fruits like nuts and acorns last longer, and may even be set aside as stores by species with good enough memories.

Later species to be classed among the afropithecines include the East African genus *Kenyapithecus* and perhaps the Namibian hominoid *Otavipithecus*, recently described by Glenn Conroy (who does not accept that the resemblance between the teeth and jaws of these geographically separated apes is proof of their relationship). At least two species of the small- to medium-sized *Kenyapithecus* seem to have been successful in Kenya during the middle Miocene, when they appear in deposits about 15 My old on Maboko Island in Lake Victoria, and on the nearby mainland, and in slightly younger sediments at Fort Ternan.

The dryopithecines

Dryopithecines are a mainly European group, the descendants of African immigrants, although one species of *Dryopithecus* has recently been described from China. They lived from the later middle to the late Miocene, and *Dryopithecus fontani* (chosen to characterize the species) from Saint Gaudens in France was the first fossil ape to be discovered, three years before Darwin published his *Origin of Species* in 1859. Larger collections have since been made from Spain and Hungary.

A word of warning here. If Gulliver's travels had ever taken him to the Hominoid Archipelago, he might have found it swarming with dedicated diggers, all of them looking for fossil clues to the roots of humankind, and many of them polarized into two great factions, the Lumpers and the Splitters. Lumpers simplify; they look for broad patterns, and tend to squeeze new finds into a handful of general categories. Splitters particularize; they see what is special about a new find, and tend to devise new categories so as to preserve the differences they see. Sometimes Splitters will give a new name to a newly discovered primate merely because they do not know the collection that houses a similar fossil. Sometimes the Lumpers will

miss key differences and stress some minor likeness rather than acknowledge some new complication. Some scientists have lumped an assortment of species and genera into a single genus, *Dryopithecus*, which in their view contains the subgenera *Proconsul* and *Sivapithecus* and so unites the entire diversity of fossil primates from early to late Miocene. Here these are seen as three distinct groups.

Two sets of contrasting features make it hard to pinpoint the role of *Dryopithecus* on the hominoid line. With its lightly built jaws and thinly enameled teeth, it can be counted more primitive than *Kenyapithecus*, whose thicker enamel appears to be an advanced development among later Miocene hominoids. On the other hand, its post-cranial bones contain improvements not found in afropithecines or proconsulids but shared with living hominoids. The even balance of the evidence makes it impossible as yet to clarify which of the two groups, dryopithecines or afropithecines, is the more advanced, and so the closer relation of living great apes and humans. Meanwhile, it is fair to conclude that, like the afropithecines, the dryopithecines

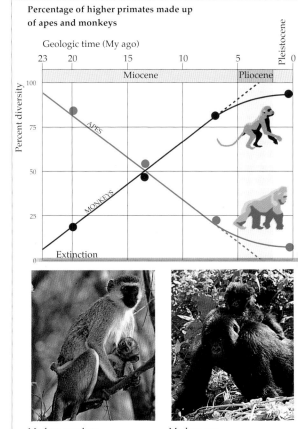

Percentage of higher primates made up of apes and monkeys

Modern monkey – vervet monkey

Modern ape – gorilla

Compared with the monkeys, the apes are in long-term decline. Twenty My ago, 80 percent of these higher primate families consisted of apes, against only 20 percent of monkeys. Nowadays, the position is precisely the opposite. Why should the apes have been evolutionary failures? It now seems clear that the critical distinction between apes and monkeys is that monkeys can tolerate many defensive chemicals used by plants. Apes have to avoid the unripe fruit which monkeys are able to digest. This may have been the critical selective advantage which set the monkeys on the road to success. If the trend in the past had continued, the apes would probably have become extinct about 3 My ago.

occupied a drier, more seasonal, and less complex woodland world than the one that sheltered both the early Miocene apes and most of the living hominoid species.

African apes and humans: the Homininae

All of the New World primates are tree-dwelling monkeys, none of them heavier than the 33 lb (15 kg) of the full-grown woolly spider monkey of Brazil, which is the world's most threatened primate species. None of the South American primates ever took to the ground, and no hominoids appeared in the New World until modern times. In Eurasia, we have no conclusive fossil record for the gibbons, the "lesser apes." Drier conditions in Europe eliminated the early hominoid population by about 8 My ago. *Sivapithecus* in Asia gave rise to the orangutan, the only remaining Asian "great ape," and another endangered species.

In South America, consolidation; in Eurasia, expansion and experiment, not always successful. But the center of development, and the great theater of hominoid innovation, was Africa, and in particular the East African Rift Valley system, where a fracture in the African continental plate runs for 2,000 miles (3,200 km) from the Red Sea and Ethiopia in the north, down through Kenya, Uganda, Tanzania, and Malawi, and into Mozambique in the south. For 20 My, tectonic activity along the length of this fracture has built great volcanoes, raised highlands, and collapsed the lowlands in between to form valleys that channeled the waters of East Africa into some of that continent's largest lakes. Here, the original tropical forest gave way to a patchwork environment grading through more open woodland to savanna, while the new uplands altered rainfall patterns and raised geographical barriers.

Other forces came to bear on this area. The mammal population came under pressure from the new immigrants from Eurasia. A long-term drying and cooling trend and more seasonal climate gave the original inhabitants the usual three choices: move, change, or die. Hominoid teeth adapted to a harder, scarcer diet. Some monkeys evolved the precious ability to digest cellulose, and leaves became a staple resource for them. Others could stomach green fruit not yet palatable for the hominoids. All kinds of pressures were focusing on the hominoids, in a "mosaic" of conditions which may have acted as a laboratory of evolution, offering species a range of options for change.

In the midst of this crisis for the hominoids of Africa, the fossil record runs out, and when it surfaces again the great variety of Miocene apes has disappeared, and a set of newcomers contains the ancestors of modern humans. We have hardly any evidence of what was happening between about 12 and 5 My ago. Some researchers have suggested that a fossil found in deposits in Greece some 10 My old is an African ape and human ancestor, but the evidence is weak. A single upper jaw from the Samburu Hills in northern Kenya, dated between 10.5 and 6.7 My ago, looks like a gorilla's, but is too isolated and scrappy a specimen to support a theory. Perhaps around 10 My ago the ancestors of the gorillas, chimpanzees, and humans split from a common stock, quite possibly a late form of African dryopithecine, perhaps a descendant of *Kenyapithecus*.

Yet the gorillas and chimpanzees have hardly any fossil history. From their present distribution in the equatorial forests of western and central Africa it looks as if their ancestors retreated westward toward more familiar conditions. But so scanty is the record that it remains uncertain whether chimpanzees are more closely related to modern humans or to the gorilla. Molecular evidence points to a closer human connection, but anatomical details – for example, in the arm bones, and the thin enamel of the molar teeth – favor the gorilla linkage. The later Miocene contains not only the origins of the living great apes and humans, the Hominidae, but the key innovation for the earliest humans, which was upright walking. Before we speculate about the biped revolution, the record allows us to look at some of its first beneficiaries.

The australopithecines

The first good evidence that the evolutionary split between apes and humans has already occurred comes from fossils of the early hominids, now called "australopithecines" –

southern apes. These were small-brained but bipedal primates that lived in southern and eastern Africa at a period between 5 and 2 My ago, and showed similarities both to apes and to humans. Their discovery and reluctant recognition forced humankind to revise some of its most central images of itself.

The Taung skull

At the start of this century the search for the supposed "missing link" between apes and humans focused on the Far East and Europe, where finds such as *"Pithecanthropus erectus"* ("Java man") had been made. This is now identified as *Homo erectus*, some way along the human line. In 1925 the anatomist Raymond Dart dared to identify a new kind of ape-man on the basis of the isolated fossil skull of a young primate found during mining work at Taung, near Kimberley, South Africa, the previous year. He knew he was challenging a powerful, if poorly founded, assumption: that humans first evolved in Europe, and that their big brains developed before other human features.

According to Dart, the shape of the skull and probable shape of the brain, together with the design of the teeth and jaws and the upright angle of the skull where it would have topped the (vanished) spinal column, all announced a recruit that was more human than any known ape. Dart labored for decades before he and another researcher, the aging but unwavering Robert Broom, eventually began to establish a valid new genus with the help of more skulls, jaws, teeth, and post-cranial bones from additional cave sites. The pre-human evolutionary stage formed by the australopithecines is widely accepted today.

Australopithecus afarensis

The species found in southern Africa and recognized as *Australopithecus africanus* spans a period from 3 to 2.5 My ago. The earliest known member of its genus comes from that hominoid test-bed, the East African Rift Valley. *Australopithecus afarensis*, the southern

FOOTSTEPS AT LAETOLI

Volcanic ash deposits at Laetoli in Tanzania, dated to 3.7 My ago, bear the imprints of mammals and birds which crossed the freshly fallen ash soon after it had been spewed out of a nearby volcano. Because of the lime in the ash, the rain hardened it, preserving the footprints precisely. Subsequent ashfalls then protected the layer from erosion. This plan view of part of the 1980s excavation shows the tracks of elephant, rhinoceros, giraffe, and many bovids, and abundant tracks of an African hare. Hyena tracks are also seen at the far left, and guinea fowl close to the elephant.

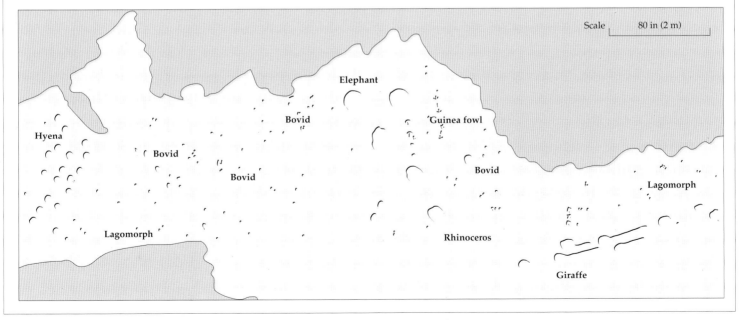

Two early hominids, identified as Australopithecus afarensis, *walk biped-fashion across an open ash field produced by an erupting volcano. The region as a whole is wooded, but here trees are absent and the volcanic ash, wetted by a light shower of rain, has formed a flat shallow layer in which the hominids' footprints are deeply implanted. The footprints filled up with yet more ash, and were thus preserved, to be uncovered over 3.5 My later in the 1980s, during the Laetoli (Tanzania) excavations directed by Mary Leakey. The footprints reveal that, even at this early stage of human evolution, our ancestors walked upright with a striding gait very similar to our own.*

ape of Afar, takes its species name from the Afar region of Ethiopia. Its remains have been found at Laetoli, near Olduvai Gorge, Tanzania, and at sites in Ethiopia such as Hadar and Omo. In 1984 a jaw fragment collected in Baringo, northern Kenya, appeared to extend its range. This find may date back to 5 My ago. Examples such as the famous "Lucy" skeleton from Hadar, the most complete known specimen, may be only 3 My old.

Australopithecus afarensis was still very apelike in many details of the skeleton, and probably in behavior, with a brain similar in size to those of modern apes. Whether or not the brain's structure had already begun to alter in a human direction is hotly disputed. The projecting jaws and face were large and apelike, and the jaws housed fairly large and thickly enameled back teeth, with canines and molars halfway, in some features, between an ape and a human form. The sheer variation between the largest and smallest individuals raises the possibility that more than one species could be represented by the fossils

unearthed, but most workers have concluded that the species was sexually dimorphic.

If so, the larger specimens were probably the males – about 4 ft 9 in (1.5 m) tall, and less heavy at about 100 lb (45 kg) than most males of today – and the smaller ones the females. "Lucy" was less than 3 ft 6 in (1.1 m) tall, and may have weighed less than 66 lb (30 kg). The males' skulls carried crests of bone like those in gorillas or large chimpanzees, which make more room to attach muscles reaching to the jaws and neck. Some differences in the skeleton of the supposed females could reflect behavioral factors; they may have spent more time in the trees to feed, rest, sleep, or escape.

What singles out *Australopithecus afarensis* from all known previous hominoids is that it walked upright. We know this not only through studies of hip and limb bones from Hadar, but thanks to one of the very rarest of time's gifts, the clinching evidence of a series of footprints preserved at Laetoli about 3.7 My ago when rain mixed with a fall of new volcanic ash to form a quick-drying cement. A team led by Mary Leakey excavated the trails left by three early hominids as they walked across ground where primitive elephants, three-toed horses, birds, and even insects and worms left their traces.

By the time we find good fossil evidence for upright walking, it was already an advanced technique, presumably developed by some late Miocene primate that has left no trace as yet. Why did it happen?

A number of early apes grew larger than any South American primate, so life in the trees was more precarious, and the ground was safer. But orangutans can weigh as much as 155 lb (70 kg), and they are arboreal; and an animal on the ground is more vulnerable to predators. Still, a heavier animal needs to eat more, and food became harder to obtain in East Africa while forest habitats were opening up or changing to savanna. Primates may have been forced on to the ground to look for food, and when they came down they brought with them considerable manual skills to pick, prod, and probe. None of the primates have ever specialized into fast four-legged running on the ground, and the ape that stood upright and learned to walk must have had fewer advantages to lose as a quadruped than it gained as a biped.

If that ape was having to forage further than before, it may have needed to carry some of the food it gathered back to a social group. Such a group might have been encouraged to evolve by the increased need for collective defense on the ground. Standing conferred other advantages: the ability to see further in exposed habitats, and to carry the young. Once the momentum had begun, the habitat may have selected for a division of labor. With less and less need for load-bearing, the arms and hands were available for handling simple tools, throwing, and possibly signaling; to carry more and more mass, the legs needed to be stronger, and therefore heavier, and that in turn would shift the body's center of gravity downward, and make an upright posture easier to maintain. What started as minor separate improvements may have begun to operate as a self-reinforcing system with a dynamic of its own.

Australopithecus afarensis, the walker at Laetoli, had a few hallmarks of humanity, but was probably apelike in its use of trees and in its diet – mainly plant materials, especially fruit and leaves, rather than meat. Its language and social systems are unknowable, but probably not remotely human. Studies of its teeth show that it probably grew up as apes do, with a shorter infancy and childhood than the long learning phase the human requires.

Australopithecus africanus

It was Dart who named the Taung species the southern ape of Africa. It too walked upright, but major questions remain to be settled about its lifestyle and its place in human evolution. *A. africanus* (the standard abbreviation) probably stood less than 4 ft 6 in (1.4 m) tall, and weighed around 65–90 lb (30–40 kg). The males would have been larger than the females, and the skull combined an ape-size brain with a large, flat, projecting face. Skull crests are rare, and the teeth may have been adapted for a hard abrasive diet of gritty roots, tubers, and seeds, as well as fruits, leaves, and some meat.

All sorts of bones, teeth, and horns were found alongside *A. africanus* fossils in cave deposits. The conditions in which the various fossils first entered these limestone caves and the later events that stirred and sometimes scrambled them, and set them in a matrix as solid as concrete, have created complex three-dimensional jigsaws. The evidence seems to show that carnivores such as hyenas and leopards may have dragged in the remains, and that the australopithecines were more likely the prey than the predators.

Little is known about the lifestyle of *A. africanus*. During its life span, the climate of southern Africa, originally more rainy than today's, shifted toward a drier scene with less bushland and more open savanna, probably less favorable for the diet of fruit and leaves that its fossil teeth suggest. There is no sign of tool use, or of any kind of living site.

Paranthropus robustus and Paranthropus boisei

Research has revealed that there were actually two types of man-ape from the South African caves. One was the lightly built ("gracile") form of Dart's *A. africanus*, the other a more solidly built biped with a larger skull, jaws, and teeth, which has been named as *Australopithecus robustus* or *Paranthropus robustus* ("robust near-man"). The gracile forms, dated between 3 and 2.5 My ago, existed before the robust ones (2 to 1.5 My ago), whose altered environment could account for evolutionary changes in the jaws and teeth of *Paranthropus* that fit them to grind hard food.

What little we know about the skeleton of *Paranthropus* suggests a body size on the scale of *A. africanus* and larger than the "Lucy" types of *A. afarensis*. A female *Paranthropus* probably weighed about 70 lb (32 kg) and stood some 3 ft 6 in (1.1 m) tall; a male about 90 lb (40 kg) and 4 ft 4 in (1.35 m) tall. The skulls of the larger males were equipped with bony crests at the top and back, anchors for powerful muscles. This design, and the size of the brain, gives some resemblance to a gorilla, but the peculiar flattened face, the tiny front teeth and large back teeth, and the rather human-looking base of the skull could not remotely be confused with a gorilla's.

These key features of the skull, jaws, and teeth were taken still further in the *Paranthropus* species of East Africa, famously typified by the skull discovered by Mary Leakey at Olduvai Gorge, Tanzania, in 1959, and first named as "*Zinjanthropus boisei*" ("Boise's East African man"), after a backer of the Leakeys, Charles Boise. The discovery of "Zinj" focused a beam of attention on the area, especially when dating techniques showed that the skull was probably almost 2 My old. This was the first decisive use of the K-Ar dating for a fossil hominid site, and gave an age so much greater than expected (though now quite commonplace for early hominids) that it has transformed the way we view their history.

The fossil itself was the skull of a near-adult individual that the media baptized "nutcracker man" in honor of its enormous back teeth, obviously high-powered grinders. Most experts now consider it, not a true human, but a form of *Paranthropus*. The Leakeys believed at first that they had found the maker of the primitive stone tools known as "Oldowan" and common at various Olduvai sites; the skull lay on what seemed to be a "living floor" – an occupied open space – next to both tools and bone remains. Soon afterward the more human form of *Homo habilis* turned up in the same levels, and it has

been identified as the true founder of the Oldowan industry, that straggling monument both to the first preserved technology and to the social networks that scattered it so widely.

A similar division of candidate toolmakers has emerged at Koobi Fora, Kenya, where *Paranthropus* and *Homo* both appear in levels rich in stone tools. *Paranthropus* may have used simple tools to unearth and process plant foods. This East African species was rather larger than the southern African form of *Paranthropus*; females measured about 4 ft (1.25 m) and weighed some 75 lb (34 kg), males 4 ft 6 in (1.4 m) and 110 lb (50 kg).

Human evolution: placing the australopithecines

Long after the discovery of the first australopithecines, most scientists still did not rank them as ancestors to humans. The "big brain" theory kept its grip on concepts of human evolution, and these southern apes seemed primitive with their small brains and apelike skull proportions. The new finds were hard to date, and the mistaken belief that they came from the period of *Homo erectus* fossils identified in other regions of the world made them appear to belong to a late surviving offshoot of some early ancestral stage, too late and far too backward to stand along the human line. When the advent of scientific dating methods showed in the 1960s that many australopithecine sites were over 2 My old, their occupants came to be seen as potential ancestors for humans. We know now that several of these sites are well over 3 My old. So which species is the most likely ancestor?

For a long time it was *A. africanus* that headed the list, but the coming of *A. afarensis* provided a more ancient contender, which in some aspects of its teeth and face seemed closer to *Homo habilis* and *Homo erectus*, true early humans, than was *A. africanus*. Many experts came to see *A. africanus* as a branch line leading only to the robust australopithecines, with possible new evolutionary trends that showed a growing adaptation to hard plant foods that required specialized grinding teeth.

This view is opposed by a minority that points out some human-like qualities of the robust australopithecines. These include larger brains than in *A. africanus* and *A. afarensis*, and details in the skull and teeth that could represent a shared evolutionary heritage. In this scenario, some recent common ancestor would have given rise both to the human lineage and to the robust australopithecines that specialized into extinction. *A. afarensis* could still be a more ancient ancestor, or could lie off the human path.

A find in 1985 has led to even more complications. The "Black Skull" – so called because of its dark fossilization – is a primitive robust variety from West Turkana, Kenya, dated to about 2.5 My ago, which is well before the other robust species so far known. Its small brain and very projecting face link it to *A. afarensis*, but details such as the shape of the face and upper jaw make it a possible ancestor of the later *Paranthropus robustus* and/or *P. boisei*. The "Black Skull" has been classed together with a more fragmentary fossil find made in Ethiopia in 1968 in the species *Australopithecus aethiopicus*, but if it can be shown that both specimens may be ancestral to later *Paranthropus* then they will be assigned to the same genus, and named as *P. aethiopicus*. Although the skull lacks the human-like features of a large brain, short cranial base, and reduced muzzle, some experts note a resemblance between its face and the face of some early *Homo* specimens.

In the growing gathering of resurrected hominids, it is impossible to identify a single line of development, and none of the various attempts to tie the best fossils on to a branching family tree is free of contradictions.

Evolution of the genus Homo: humankind

Footprints at Laetoli. Stone tools at Olduvai. Fire at Zhoukoudian. Female figurines carved in mammoth ivory or limestone, or shaped and baked in a mix of clay and bone, in the Ukraine, Austria, and elsewhere. Somewhere along the way, perhaps in a place where traces remain to be discovered, the genus *Homo*, humanity, appeared. We peer into the eyes of fossil skulls and ask ourselves whether they once brought sight to a mind that might have marveled at us, just as we do at their mere survival, and the mystery they hold.

To place the birth and growth of our own genus is no straightforward project either to define or to perform. The material facts alone are massively eroded by a haphazard fossil record. Much of what seems to distinguish our genus in essential, irrevocable ways from our closest primate relatives is embedded in behavior that is rarely if ever preserved in the fossil record. It has to be reconstructed out of our small but growing store of physical material, intensively scrutinized and described with the help of the most sensitive technology, and interpreted at the interface between a caustic skepticism and an open imagination.

Where among the relics of our remote prehistory can we identify signs of the transition to human society, with all its skills, customs, rules, and ceremonies to be passed on and modified as conditions demand and as our genetic programs permit? Language is plainly a key element in what happened, but it takes more than a keen ear to detect its presence from the surviving bones and stones turned up at ancient sites.

Because of these and a gaggle of other difficulties, scientists have in the past tried to define simple "Rubicons," turning-points that would signify that true humanity had evolved. One of these was toolmaking (not tool use: sea otters carry stone anvils to open shellfish), and it now seems likely that stone artifacts were being produced in Africa as early as 2.5 My ago. But it has been suggested recently that the non-human biped *Paranthropus* made and used simple tools, and since such behavior is also recorded among wild chimpanzees, toolmaking can no longer be counted as the exclusive ticket to being human. Another proposed Rubicon was the size of the brain, but while the gap between apes and *Homo sapiens* is wide – apes up to about 600 ml in brain volume, humans usually above 1,000 ml – if the earliest members of the genus *Homo* did evolve from some form of australopithecine their original brain size may well have fallen within the standard ape range.

More recently, scientists have veered away from relying on some single feature as the badge of genus *Homo*, and have opted instead for a weave of details that need not all be present in a single fossil. The physical traits include brain size, a smaller, less backward-sloping face, smaller teeth (especially molars and premolars), a prominent nose, and where the record preserves them, signs of a more human skeleton both in details of the hip region and in general body shape. There is no satisfactory periodic table of the non-physical elements – moral, psychological, cultural, and so on – that make a human being, but obviously they too make up a mosaic, not measurable by science but a part of the pattern.

Homo habilis

Vital discoveries made at Olduvai Gorge in 1959–61 led to recognition of the early human species *Homo habilis* ("handy man") so named because of its apparent connection with stone tools of the primitive Oldowan industry. (These were the first and simplest of worked tools, chipped to give a single rough cutting edge.) The finds were dated from associated volcanic rocks to between 2 and 1.5 My ago, and consisted of fragments of skull, jaw, tooth, and skeleton. Compared with

australopithecines, the estimated brain size was marginally larger, about 600–700 ml, while the premolar and molar teeth were narrow and set in what seemed to be more slender jaws. Hand, foot, and leg bones contained some human features but showed no obvious differences from an australopithecine's.

Louis Leakey, Phillip Tobias, and John Napier faced severe opposition when they named the new species in 1964, from anthropologists who doubted if the original fragments added up to a hominid radically different from specimens assigned to accepted species such as *Australopithecus africanus* and *Homo erectus*. The picture changed with further finds from Olduvai and from other African sites, especially East Turkana, in northern Kenya, which began to produce hominid fossils in 1967. Its tally of material linked to *H. habilis* includes more complete skulls than those from Olduvai, as well as hip and leg bones, and even partial skeletons.

The most famous East Turkana fossil is

The first real humans belonged to the species Homo habilis *and lived on the savanna plains and in the woodlands of South and East Africa. This group of males are feeding on small game and plants, watched by two robust australopithecines of the genus* Paranthropus. *Evidence suggests that the two lineages coexisted for several hundred thousand years, between about 1.5 and 2 My ago. While* Homo habilis *was known to be a regular toolmaker, it is also possible that* Paranthropus *used tools of stone, bone, or wood.*

about 1.9 My old, and is known by its catalog number, KNM ER-1470. It had a large flat face with a brain case unusually developed (about 750 ml) for such an early hominid, and must have had fairly big teeth, to judge from the sockets. By 1985 the fossil harvest unearthed in East Turkana and elsewhere had grown so lavish that its range was looking unnaturally elastic. Some scientists began to claim that all this variety could no longer be contained inside a single species, given the marked differences in their facial shape, teeth, and estimated brain size (from 500 to more than 750 ml).

The opinion is growing that *habilis* really consists of at least two different species: one large-bodied, with a bigger brain, face, and teeth; the other much smaller, with a more human face but a more primitive skeleton and smaller brain. No two species can share a name, so some of the fossils have been offered new ones: *Homo ergaster* ("working man"), or *Homo rudolfensis* – Lake Rudolf was the colonial name for Lake Turkana. The large species probably exceeded 5 ft (1.5 m) in average height, with a body weight of about 115 lb (52 kg). The small one may have been closer to 3 ft 3 in (1 m) tall, and about 70 lb (32 kg) in weight.

The larger species of "habilis" seems to be a convincing human, a toolmaker with a brain half as large again as an australopithecine's. Its presumed tools are found on sites that contain the bones of various species of antelope, pig, zebra, and occasional larger animals such as hippo, buffalo, and elephant. The sites occur by ancient lakes and rivers where these animals would naturally gather for water and for the shelter of the nearby trees and bushes. Both carnivores and humans would be drawn here too, so it is impossible to say whether it was humans or predators that killed a particular animal; perhaps humans found it after its death, or drove the killer away from its prey.

Modern hunter-gatherers seem to be successful foragers and scavengers rather than full-time killers, and *Homo habilis* may well have led a similar kind of life. Pre-humans and early humans shared their habitat with saber-tooth cats whose teeth could stab and slice meat, but could not crunch bones. They must have left plenty of meat and marrow available. The existence of so many sites where tools are plentiful strongly suggests that humans were starting to use temporary "living floors," and possibly home bases, though these cannot be proved.

Hunter or scavenger, it seems likely that the pre-human apes that stood up to explore a more and more open habitat were able to add some meat to their original diet. As meat is a concentrated and nutritious food, its consumption would reinforce the abilities that provided it; a smarter, more skillful hominid had greater access to whatever made it smarter and more skillful. The kind of brain that worked better to recall a likely food source, or to improve the group's performance in defense or attack, would also be capable of a lot more functions that only a more complex lifestyle would elicit.

One telltale feature of the human brain is its lopsided architecture, caused by a tendency to concentrate some important functions in the left-hand or right-hand lobe. It is this division of mental labor that makes most humans right-handed, and about one-fifth of us left-handed. No other primates specialize this way, but *Homo habilis* did, and left a record in the pressure patterns produced when making stone tools. A further suggestion of a vital human shift is a hint given by a faint impression left in some fossil skulls by the brain they protected. In *H. habilis*, the larger brain may have made room for the feature known in modern humans as "Broca's area," a region connected with speech production.

Speech is a skill with unlimited implications. It enables information to be passed on, and lies to be told. It encourages the invention of words, and therefore of a vocabulary to discuss what matters to the group – since words are social tools. Most of all, speech is the social glue that enables decisions to be made, intentions communicated, memories shared and passed on. *Homo habilis* cannot have developed the full range of sounds that a modern human brain and voice box can produce, but even a limited combination of vowels and consonants, reinforcing facial expressions and manual signals, must have offered practical advantages that paid off the higher running costs of a bigger brain with better food and improved survival chances. The tools made by *H. habilis* stayed much the same for more than half a million years, and show no sign of creative experiment. Humans had crossed another Rubicon, but then they rested.

Homo erectus

Until about 1.7 My ago, human ancestors seem to have been confined to their birthplace, Africa. Then came the outward drive, the phase when early humans explored the more habitable areas of the rest of the Old World. (In evolutionary time, Earth distances dwindle: traveling at 10 miles, 16 km, a year, it takes less than 2,500 years to walk around the globe.) A major change in the human skeleton had produced a species, *Homo erectus*, whose form was much more like our own. Its name derives from the one assigned in 1894 by Eugène Dubois to the fossil finds he made in Java, which included a quite modern-looking thighbone. He called the species "*Pithecanthropus erectus*" ("upright ape-man").

The Javanese fossils of *H. erectus* mostly date from 1 to 0.5 My ago, but some of the African specimens are far older. The species was living in East Turkana 1.7 My ago, overlapping in date, if not in habitat, with the last *H. habilis* populations, and producing a more versatile kit of stone tools. The relation between *habilis* and *erectus* is unclear. It is widely assumed that the first gave rise to the second, but since there seem to be at least two kinds of *habilis*, whose toolmaking skills could be independent of their successors', there is no obvious continuity.

The badge of *Homo erectus* is the solid construction of the skull, which is thicker-walled, more rugged, longer, and lower than in *Homo habilis*, and housed a brain whose volume may have grown over time from its earlier 850 to 900 ml. The brow ridge jutted further, but the face was smaller, with a more prominent nose and downward-facing nostrils. The teeth were like those of the smaller *habilis* type, and the lower jaw was still totally chinless.

For years we had only meager data on the body form of *H. erectus*, until the spectacular discovery in 1984 of a nearly complete skeleton in Nariokotome, West Turkana, known by its catalog number, KNM WT-15000. It belonged to a boy who was about eleven years old at the time of his death some 1.6 My ago. He appears to have died near a river that ferried his corpse to the site where it lay in sand and fossilized. KNM WT-15000 was already about 5 ft 3 in (1.6 m) tall and weighed about 106 lb (48 kg). Fully grown, he

might have reached a height of 5 ft 11 in (1.8 m) and a weight of 132 lb (60 kg). Tall, relatively narrow-hipped and long-legged, just like some present-day East Africans, he had an ideal physique for covering long distances in hot conditions.

As *Homo erectus* scattered across the Old World, the cooler climates in some regions of Europe and Asia must have led to a gradual adaptive change away from the ancestral tropical body shape toward a shorter, squatter form. In all versions of the species, the bones are adapted to take heavy muscular strains, and it looks as if *H. erectus* led an especially strenuous life. The species was present in Java and China by about 1 My ago, and a new discovery makes it possible that it had also reached western Asia and Europe. A lower jaw was recently unearthed at Dmanisi in Georgia, together with other animal bones and artifacts, and lying among volcanic rocks that suggest an age greater than 1 My.

The Chinese form of *erectus* is typified by the remains of "Peking man," so called when they were found in the 1930s because the huge cave site of Zhoukoudian lies near Beijing (formerly Peking). Chinese workers have since found fresh material. Zhoukoudian probably housed many different phases of human occupation between about 500 and 200 thousand years (Ky) ago.

Stone tools found at Zhoukoudian show little advance over the Oldowan industry of *Homo habilis*, but it is possible that *H. erectus* in Asia made greater use of perishable materials like bamboo. Further west, a remarkably standardized output of hand-ax tools occurs in a tradition that ranged widely in location (from southern Africa to England), materials (volcanic lava to flint), and dates (from over 1 My ago in Africa to less than 200 Ky ago in many areas). This industry is called the Acheulean, after St Acheul, a suburb of Amiens, in France, which has yielded many of these tools. Items in the Acheulean kit were designed to cut, pound, scrape, shred, and whittle. Wooden tools must also have been made, but only rarely survive. The earliest known wooden spear is a broken tip of yew excavated at Clacton, England, about 16 in (40 cm) long and 300 Ky old. No trace would be preserved of the simple shelters and windbreaks that a nomadic lifestyle constructs and abandons, nor of any traps or snares

The huge cave site of Zhoukoudian, near Beijing in China, was periodically occupied by Homo erectus *between about 500,000 and 250,000 years ago. The people used fire for warmth, cooking, and protection from carnivores and scavengers, such as the large striped hyenas shown here. Humans and hyenas were probably competitors for both the shelter of the cave and the carcasses of animals.*

made of woven grass, bark, or wood.

Eating more meat would help to fuel the growing brain size of *Homo habilis* and *Homo erectus*. Evidence comes from the bones of a possible female *H. erectus* from Koobi Fora, Kenya, which show signs of a fatal illness caused by overconsumption of vitamin A, a substance that concentrates in carnivore livers. In habitats rich in wildlife, natural deaths and the remains of carnivore kills, possibly located and hijacked in an organized way, would yield a safer meat supply than full-scale hunting. Such methods may have outcompeted *Paranthropus robustus* and *Homo habilis*, which had disappeared by about 1.5 My ago.

The increasing brain size of humans had to mean that there was more to learn, and therefore a longer time for learning, and a greater investment in child rearing. If the signs of a developing capacity for speech imply that information was becoming a human specialty – whether as facts, wants, or intentions – then social interactions were growing more complex, along with the subsistence activities that they reinforced. Humankind was discovering the variety of the world, and probably making an impact on other species, both its fellow primates and

perhaps some dangerous cohabitants. It was during the lifetime of *H. erectus* that saber-tooth cats became extinct in Africa.

The descendants of Homo erectus

In only one region does it appear that *Homo erectus* survived and developed into the late Pleistocene. A series of skullcaps and two leg bones from the site of Ngandong on the Solo River in Java belong to a still robust but more evolved *erectus* type, with an average brain size of about 1,100 ml, compared with the earliest forms' 850 ml. Studies suggest that the Ngandong populations managed to live on in Java for much of the later Pleistocene owing to its periodic isolation as an island.

Elsewhere, *Homo erectus* made room through evolutionary change or replacement for new populations which have often been graded under the label of "archaic" *Homo sapiens* (our own species). An alternative is to view them as a new species, and to name them after the locale where a lower jaw was found, at Mauer, near Heidelberg, Germany, in 1907, and assigned to *Homo heidelbergensis*. The new species kept much of the solid build of *H. erectus*, but its brain case was higher and more spacious, with an average capacity of about 1,250 ml, not far below the modern figure. The face sloped much less forward than in *erectus*, and the teeth were smaller – a trend that had started with *Homo habilis*. The body shape probably continued to vary according to local environments and climates.

At a possible 0.5 My old, the Mauer specimen is the oldest known human in Europe. Other fossils come from sites such as Petralona (Greece), Bilzingsleben (Germany), and Arago (France). In England, archaeological sites at Boxgrove (West Sussex) and High Lodge (Suffolk) have yielded evidence of occupation, but no fossils. One of the most massive human skulls known comes from Bodo (Ethiopia), and it is nearly matched by two crushed skulls from Yunxian (China). These specimens and sites probably grade in age from about 450 to 250 Ky ago, but younger examples have emerged from Broken Hill (Zambia), and Dali and Jinniu Shan (China). Jinniu Shan has provided the only reasonably complete skeleton of *heidelbergensis*, but except for its rugged construction, few details are yet available. The skull's advanced features include a brain size of perhaps 1,300 ml, and thin cranial walls, which contrast with the much thicker skulls found both in *erectus* and in some other *heidelbergensis* fossils, say from Bodo or Petralona.

It is assumed that *Homo heidelbergensis* evolved from *Homo erectus*, but the evidence does not tell whether this transition happened in one region (Europe has been suggested), or more widely. One striking possibility is that there was an overlap in China between *H. erectus* (dated from sites like Hexian and Zhoukoudian as late as 250 Ky ago) and more advanced types found at Yunxian and Jinniu Shan.

In spite of apparent advances in features such as brain size, *Homo heidelbergensis* has left us no record of great behavioral change or cultural progress. The colonization of what in those days were remote and marginal areas such as the British Isles and northeast China may indicate a gradual rise in human adaptability. Yet these were unsafe footholds, impossible to cling to when the Earth's climate switched into one of its regular glacial cycles. These long phases of colder, drier conditions would have forced early human populations either to quit more northerly or mountainous territory, or else to face local extinction.

The Neanderthals

It was the Neanderthal people who seem to have first adapted to life on the edge of an Ice Age world. They spanned an era from at least 200 to 35 Ky ago, and lived across much of Europe and western Asia from Wales in the northwest to Gibraltar in the southwest, from near Moscow in the north to Uzbekistan in the east. We also meet them in Israel and Iraq, cousins who test our concepts of humanity.

Although Neanderthals are named after the Neander Valley skeleton found near Düsseldorf, Germany, in 1856, there had been earlier finds in Belgium and Gibraltar in 1830 and 1848. At first the fossils tended to be viewed as abnormal forms of modern humans, diseased or deformed. By the early twentieth century, further finds, particularly from Belgium and France, had revealed them to be a distinct variety of early human, and perhaps a different species from our own: *Homo neanderthalensis*, as William King proposed in 1864. Even so, the word "Neanderthal" has picked up undertones of backwardness not justified by anything known about the species.

Although the Neanderthals must have evolved out of *Homo heidelbergensis*, the boundary between them is not sharp, and there is no general agreement that specifies the aspects of mid-Pleistocene fossils that make them conclusively Neanderthal or non-Neanderthal. The Petralona cranium from Greece is a controversial specimen often classed as Neanderthal owing to its prominent brow ridge and some other facial features, but it is not till about 300–200 Ky ago, with fossils from Swanscombe (England), Atapuerca (Spain), and Ehringsdorf (Germany), that a new pattern begins to be established in specimens that seem to combine features of both *H. heidelbergensis* and *H. neanderthalensis*.

The Atapuerca group is outstanding. It contains upward of 220 bones from at least twenty-three individuals, and includes two well-preserved skulls in a fossil treasure that far outnumbers the entire world fossil hoard for the previous million years. Their presence deep in a cave system far from any entrance is hard to account for. They may have been carried there by a disaster such as a flash flood that swept some single social gathering to its death, or else collected bodies intentionally buried by other humans somewhere higher in the cave system, or deposited by the repeated activities of carnivores.

By 120 Ky ago, true Neanderthals had evolved, and by 70 Ky ago the standard "classic Neanderthals" were living in western Europe. Most of their fossils come from caves used either as living quarters or as burial sites. Usually they are found together with the bones of animals at home in cold conditions – reindeer, arctic fox, lemming, mammoth. Neanderthal people were likewise adapted for a glaciated Europe, with a bulky, squat physique, heavy muscles, and barrel chests in men, women, and children. Males were about 5 ft 6 in (1.7 m) tall and weighed about 155 lb (70 kg), females closer to 5 ft 2 in (1.6m) and about 120 lb (54 kg). The limb bones were strongly reinforced by thick walls, yet even so they show signs of routine strains and injuries. Some features of the hip region, together with the thickness of the leg bones, suggest that they did not walk exactly as we do, and may have had a more rolling gait.

Neanderthal brains were at least as large as our own, though differently shaped, but there is little clue here to their mental capabilities. Their skulls were long and low, like those of earlier humans, with a brown ridge molded as a twin-arched unit jutting above a massive nose accentuated by swept-back cheekbones. The external area of the nasal opening, as well as its internal volume, was larger than in any humans before or since. One theory links this generous organ to the high energy levels implied by the Neanderthal physique, and suggests that it worked as a cooling device. Or it may have helped to warm and humidify the ice age air on its way to the lungs.

As toolmakers, the Neanderthals were generally linked to the Mousterian stone-working industry named after the French site of Le Moustier, from the middle Paleolithic (middle Old Stone Age). Products included borers, scrapers, points, knives, and hand-axes, and designs changed little if at all over tens of thousands of years. Although they used wood, they seem never to have perceived the possibilities of bone, antler, and ivory, and there is no hard evidence of Neanderthal ornaments or decoration until the very end of their career in Europe – one of the sharpest contrasts with the people who followed them.

Neanderthals had the use of fire, both

The Neanderthals, the "cave people" of popular legend, were nomadic hunter-gatherers who formed large family groups. The group seen here is in southern Europe about 50,000 years ago during the last ice age. They used caves as convenient and comfortable resting places providing shelter from the freezing conditions beyond the mouth of the cave. The Neanderthals' resemblance to the modern human form, Homo sapiens, *is striking – theirs was a species so closely related to our own that some scientists believe that interbreeding between the two took place. They disappeared from Europe, and parts of Asia, some 35,000 years ago, soon after the spread of* Homo sapiens.

FOSSIL SITES OF THE WORLD

Our earliest direct ancestors were probably restricted to Africa until about 1.5 My ago, and a series of rich fossil sites in southern and eastern Africa document this (below). Thus the earliest human species are not certainly known from Eurasia. However, *Homo erectus* did spread from Africa soon afterward, as indicated by sites in Georgia (Dmanisi), China (Lantian), and Java (Sangiran).

For the latest stages of human evolution, Europe becomes much more important for the richness of its sites (left), but scattered finds from elsewhere show that similar developments were occurring everywhere that modern humans migrated. Recent finds show that regions such as Australia were not backwaters in the story of modern human evolution, and that the earliest colonizers had apparently reached Australia by boat long before the Cro-Magnons arrived in Europe to replace the Neanderthals.

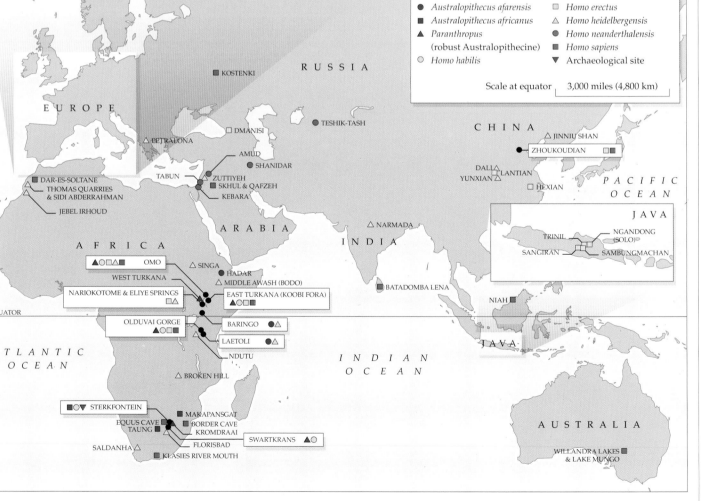

Legend:

- ● *Australopithecus afarensis*
- ■ *Australopithecus africanus*
- ▲ *Paranthropus* (robust Australopithecine)
- ○ *Homo habilis*
- □ *Homo erectus*
- △ *Homo heidelbergensis*
- ● *Homo neanderthalensis*
- ■ *Homo sapiens*
- ▼ Archaeological site

Scale at equator   3,000 miles (4,800 km)

for warmth and perhaps to defrost meat scavenged from snowy countryside or recovered from underground storage. The most obvious use for the stone scrapers they made would be to work hides for clothing and shelter. They buried their dead – the first humans known to have done so, and surely a significant event, though it tells us nothing for certain about the meaning that death may have had in Neanderthal culture, or the role of remembrance or respect. Some objects found associated with burials have been interpreted as grave goods, but no pattern of deliberate placing can be shown as yet, or any other clue to the possibility of ceremonies. Some skeletons show the marks of obvious injuries or illnesses suffered sometime before death, evidence that there must have been social care to support certain ailing or disabled individuals.

Whether the Neanderthal species had a language capacity like our own is a topic of dispute. Anatomical reconstructions have suggested that their voice box (larynx) was differently positioned, and so limited in the sounds it could produce. On the other hand, a Neanderthal hyoid bone (from near the larynx) found in Israel looks modern, and that could indicate a throat shape like our own. The modern human vocal equipment can produce an enormously broader range of sounds than, say, a chimpanzee's, and can articulate tens of thousands of words. At the same time, a more limited spoken language might be supplemented by intonation, gesture, facial expression, and body language. How much thought is possible without a store of words to "think with" is a question that only leads to more questions. It may be that the long-lived technical conservatism of Neanderthal people expresses limitations counterbalanced by capabilities that we no longer understand, or undervalue. *Homo "sapiens,"* the "wise" human, had better be cautious not to dismiss the possibility that there are other ways of being "human."

Homo sapiens

The Neanderthals lived through the earlier part of the last ice age in Europe, but had disappeared by about 30 Ky ago. Since 40 Ky ago, early modern people had also been present in Europe, and the extinction of the Neanderthals seems to be linked to their arrival. It is likely that the newcomers, sometimes known as Cro-Magnons after a site in the Dordogne department of France, but clearly an immigrant species, outcompeted and gradually replaced the original Europeans. The Cro-Magnons produced much more sophisticated tools, but some of the last Neanderthal tools show signs of change, and it is not impossible that there was contact or even interbreeding during the coexistence phase, before the Neanderthals died out. There is no evidence of warfare between the two species, or within either species, but recent history has shown that an active, expanding culture does not have to attack a stable, more passive one in order to destabilize it through disease and "culture shock."

The Cro-Magnons were tall, long-legged, and narrow-hipped. Although they too were quite heavily muscled, they lacked the thick bones of the Neanderthals, and with a height and weight of about 5 ft 6 in (1.7 m) and 120 lb (55 kg) in females, 6 ft (1.8 m) and 155 lb (70 kg) in males, their physique was much more like the tropical African pattern of the Nariokotome boy of 1.6 My ago. Their capabilities invite comparison with those of modern hunter-gatherers. Clearly they had complex societies, language, symbolism, and ceremonies. They produced art: not a scatter of experiment but embedded in their lives – engravings, sculptures, clay models, and, most famously, cave paintings. If the source of these lies outside Europe, it has not yet appeared. They also made ornaments such as bracelets and necklaces, some of them intricately designed, with dozens of carved pieces. Probably they painted their bodies; there is widespread evidence of their use of pigments such as red ocher, which they sprinkled over bodies at burial and left to outline handprints on subterranean walls.

Cro-Magnon tools included long blades of flint that they could modify to produce more specialized items such as engraving kits and all sorts of scrapers and smaller blades. New crafts explored the material and artistic potential of bone, antler, and ivory to make statuettes, beads, rings, buttons, needles, spear points, and many other objects, including flutes and whistles, although some students of early humans believe that music must have originated long before this period.

These "early modern" people were contemporaries of the last Neanderthals, not their descendants. So where did they come from? Their innovative tool industry seems to appear sooner in eastern than in western Europe, which suggests an influx from that direction, but there are no obvious close ancestors there, and no significant settlements, except for some obscure or isolated finds from areas such as the Crimea and the Lebanon. Further east still, there is evidence from skeletons found in the Upper Cave at Zhoukoudian of early modern people, looking rather like the Cro-Magnons, first appearing in China about 30 Ky ago, so their origin must be elsewhere.

People seem to have reached Australia earlier still, perhaps more than 50 Ky ago. We do not know what the first arrivals looked like, but they must have traveled by raft or boat, because Australia and New Guinea have been cut off by sea from southeast Asia for the whole of human history. By 30 Ky ago, two sets of very different humans were living in Australia, some of them looking like Cro-Magnons, the others much more strongly built. The roots of these differences remain to be traced. One theory argues that they

to modern Europeans. In both cases, evolution did not cease with the advent of modern humans: comparisons of ancient and modern skeletons show sizable changes, for example in tooth size and body size, which both decreased.

The pre-European history of Cro-Magnon people belongs to the broader theme of tracing the source of the modern populations throughout the world. In Europe, as in China and Australia, there seems to be little evidence of any close relationship between the regional descendants of *Homo erectus* or *Homo heidelbergensis*, on the one hand, and the early modern forms, on the other.

There is now good evidence that early modern people lived in the Middle East and Africa about 100 Ky ago, which is long before their first appearance in any other region. Their burials have been excavated from the caves of Skhul and Qafzeh in Israel, and less complete remains come from the Border and Klasies River Mouth caves in South Africa, Omo-Kibish in Ethiopia, and Guomde in Kenya. The dates and locations of these earliest modern humans, together with the contrast between similar but much later modern humans in other parts of the world, and the human populations that preceded them, point to one likely mainstream sequence. It looks as if an evolutionary shift from pre-modern to modern humans took place uniquely in some part of Africa as yet unidentified between 150 and 100 Ky ago, and then later these African moderns spread to other continents, changing along the way according to climate and conditions. (One of the more trivial symptoms of these changes is our present-day spectrum of skin colors, estimated to be controlled by a possible five to seven genes, out of a total of about 300,000.)

Further information comes from the genetic makeup of modern humans. Several studies have shown that African peoples are the most genetically diverse of all continental populations. Further, there is a greater genetic difference between African peoples and the peoples of the other continents than there is between the "gene pools" of any other two continents. The most straightforward explanation of these facts is that the African genetic assortment was the first to appear, and therefore has had more time than any other to develop its broad range of variations. A subgroup left Africa, and branched out to

The early modern people of Europe (often called Cro-Magnons after a site in the Dordogne, France) were not only skilled hunters of game and foragers for plant foods, they were also among the first people to fish in a systematic way. Some were probably equipped with boats, nets, and fish traps; but in this depiction, a man simply wields a barbed spear-point to take salmon from a fast-flowing river. Some Cro-Magnon cave sites have huge accumulations of fish bones, while others are adorned with engravings of fish or contain the remains of fishing equipment, such as harpoons or fish hooks made of bone or antler.

embody changes caused by contrasting environments inside the continent acting on a single original population; another believes that two distinct groups must have arrived in separate waves. The second theory assumes that a wave from the east was more lightly built, while western arrivals could have met and interbred with, or even evolved from, "Solo man," the late-surviving stock of *Homo erectus* that seems to have lived on the island of Java less than 100 Ky ago.

It is assumed that these early Australians are the ancestors of the present-day Aborigine population, just as the Cro-Magnons gave rise

ANATOMICAL HERITAGE

Nine characters of modern humans are illustrated with data to show when they evolved. For example, post-orbital closure is an ancient characteristic which evolved at least 40 My ago and is present in all higher primates, whereas a human knee joint is present only in living humans and human ancestors.

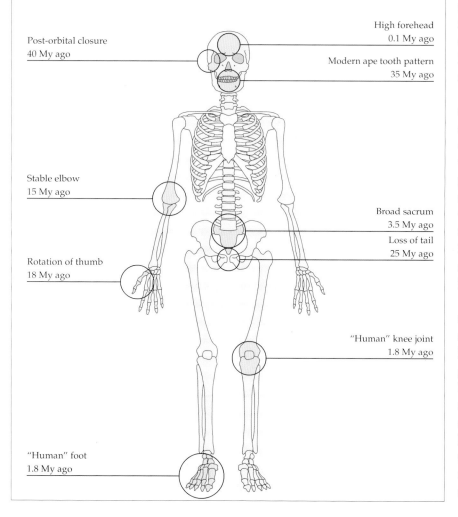

Post-orbital closure
40 My ago

High forehead
0.1 My ago

Modern ape tooth pattern
35 My ago

Stable elbow
15 My ago

Broad sacrum
3.5 My ago

Loss of tail
25 My ago

Rotation of thumb
18 My ago

"Human" knee joint
1.8 My ago

"Human" foot
1.8 My ago

give rise to the rest of modern humanity, which displays fewer genetic variations because it has had less time to produce them. There is less difference between the peoples of any two other continents because fewer generations divide their founders from each other, than from their African ancestors.

The alternative would be for each continent to have developed its own form of modern humanity, probably out of the local *H. erectus* people. This would contradict the genetic evidence, because it ought to give rise to much the same amount of difference between any two continents. Nor would it account for the obvious differences between ancient and early modern populations – between Neanderthals and Cro-Magnons, for example.

It would also require extremely high levels of gene-flow right from the start, communicated by habitual interbreeding across continental boundaries, to prevent each of these separate human subspecies from developing into separate species, no longer able to interbreed as all modern humans can. And it would require each of these human subsets, arising in their different regions, to develop a series of skeletal features, universal among modern humans but not among ancient populations, more or less independently of each other.

This would write the human story as an elaborate plot, bristling with improbabilities and driven by coincidence. The weight of the evidence yields a more logical narrative whose first chapter is set in Africa, and which gains momentum as the characters move across Europe and Asia. Eventually they cross water to reach Australia, and they walk over the land bridge where the Bering Strait now lies, to enter Alaska and the Americas. The date of that first colonization is controversial, but it probably happened less than 30 Ky ago.

The driving force for this worldwide spread of modern *Homo sapiens* was probably a gradual increase in population numbers, leading to an outward press of hunter-gatherer territories, and a trickle of advance by a species whose adaptability was no longer measured solely by its potential for genetic and physical change. Where the climate was hostile, they could make clothes and build shelter. What their unaided bodies could not manage, their artifacts, tools, weapons, and collective action could achieve.

Modern *H. sapiens* seems to have gained powers of invention and creativity that might derive from improvements in the brain, real neurological changes. But it is also possible that these were qualities generated by humankind out of its own history and experience. Each stage of development, whether of social organization, tools, or language, forced humans to learn to manage more and more complex lives. Tools, language, memory, social organization – all these became extensions of humanity, permanent attachments added to the original design. If some of these stepping stones were also thresholds of understanding, we may have pushed ourselves into the mental dimension where experiment, creativity, art, and imagination were necessary skills. At the point where people became able to enter culture, society, and history, this story closes.

It closes, but it is not, and cannot be, complete. Like the rest of the book of life, the human chapter has lost words, paragraphs, handfuls of pages. Vast tracts of time and space are scientific deserts. We know something is happening there, but we don't know what it is, and we can only hope that a few more lost treasures will turn out to be merely buried, not obliterated.

Yet paleontology can do more than wait and hope, and chance is not its chief resource. Aerial reconnaissance uses exotic new photographic techniques to identify worthwhile sites. With laser disk technology and computer algorithms, huge data bases can be analyzed for patterns and resemblances among analog images, in order to identify fresh finds, or examine the possible mechanics of a fossil modeled in three-dimensional detail. Scanning electron microscopy achieves brilliantly sharp images at magnifications of x 1,000 and more. For dating, we can call on techniques like fission-track analysis, electron-spin resonance, thermoluminescence, and the whole new discipline of geochronometry. It has become possible to extract DNA from fossil remains, and so to study past species at a molecular level. A Stone Age ax maker damages a finger and throws away the flint, and a thousand generations later that smear of blood may one day illustrate its producer's relationship to the modern human family.

Humankind has given evolution a new field of action. Our powers no longer die when we do, but survive and adapt in our arts, sciences, and technologies. Natural selection cannot reduce our knowledge by eliminating its possessors. Lethal or life-enhancing, the best and worst of what we know is indelibly printed, recorded, encoded. We have shrunk the world to a few hours' flight, annexed its lands, tamed or destroyed its species, and altered its atmosphere and oceans. Our plans and decisions now outreach us, and the greater their scale, the harder to repair. Our contradictions define us. *Homo sapiens* calls its best qualities "humane," "humanitarian." At the same time, compared with *Homo sapiens*, even *Tyrannosaurus rex* looks limited. But *T. rex* was locked biologically into its role. The story does not tell us whether the same is true of humankind.

TIBIA / FEMUR LENGTH RATIO

The ratio of the length of the tibia (the shin bone) and the femur (the thigh bone) is called the crural index. In modern humans it is related to the climate in which the population currently lives or previously lived. Populations derived from hot regions (e.g. Central African Pygmies, Afro-Americans) have a relatively long tibia, while those derived from cooler conditions (e.g. South African whites, Inuit) have lower values. The crural index reflects the tendency of human bodies to be relatively taller, longer-legged, and narrower in hot climates and the opposite in cold ones. These differences may have evolved to lose heat in hot conditions and conserve it in the cold. Fossil skeletons appear to follow the same general rules; the Neanderthals of the European ice age have a crural index like modern Lapps, while the Kenyan *Homo erectus* boy is similar to modern Africans. The earliest modern skeletons in Europe and Israel are unlike the Neanderthals and look as though they came from a hot climate, possibly African.

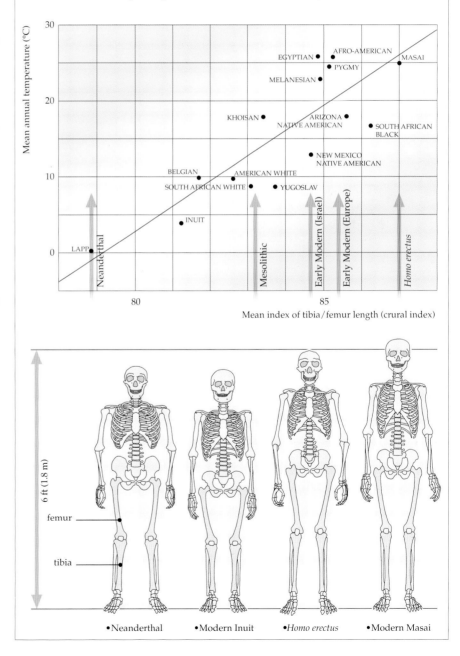

•Neanderthal   •Modern Inuit   •*Homo erectus*   •Modern Masai

INDEX

Page numbers in *italics* indicate captions to illustrations

PICTURE CREDITS

The publishers further acknowledge with thanks the permission of the following to reproduce or adapt copyright illustrations.

Courtesy Department Library Services, American Museum of Natural History 11 (2417), 18 bottom (1775), 19 (2425), 131 (328401); Ardea 125/Pat Morris 147 top right/Adrian Warren 233 bottom; British Library, London 14-15, 18 top; Jean-Loup Charmet 6, 179; Mary Evans Picture Library 69 top left, 155 bottom; S.J. Gould 7; © 1986 Mark Hallett. All Rights Reserved 12-13; Eva Hochmanova 20; Hulton Deutsch Collection 131 bottom/Bettmann Archive 131 top right; Institute of Human Origins, Berkeley, CA, USA 231; The Natural History Museum, London 10, 71, 74, 83, 139, 153; Gregory S. Paul 21; © 1966, 1975, 1985, 1989 Peabody Museum of Natural History, Yale University, CT, USA 8-9; Scottish National Portrait Gallery 69 centre right; N.H. Trewin 67, 69 bottom left, top & bottom right, 77.

All outline paleomaps based on work by Cambridge Paleomap Services Ltd., Cambridge, England; (pp.24/25 Shark) "Relationships of fossil and living elasmobranchs," B. Schaffer and M.

Williams, *American Zool*, 1977; (Dolphin) *A review of the Archaeoceti,* R.M. Kellogg, Carnegie Institution Washington 1936; (p.52 Cichlids) after Greenwood, 1974; (p.72 Drepanaspis) *Paleozoic Fishes* (2nd edn.), J.A. Moy-Thomas and R.S. Miles, Chapman & Hall, 1971; (p.72 Pteraspis) (E.I. White) *Phil. Trans. R. Soc. Lond.* B225 381 (1935); (p.76 main Scotland diagram) "Environmental controls on fish faunas of the Middle Devonian Orcadian Basin," R.F.M. Hamilton and N.H. Trewin, "Devonian of the World," *Memoirs of the Canadian Society of Petroleum Geology,* Calgary 1988; (figure 1) "Palaeoecology and sedimentology of the Ancharras fish bed of the Middle Old Red Sandstone, Scotland," N.H. Trewin, *Transactions of the Royal Society of Edinburgh: Earth Sciences,* vol. 77 (1986) pp. 21-46; (p.81 Eusthenopteron skeleton) S.M. Andrews and T.S. Westholl "The postcranial skeleton of *Eusthenopteron foordi* Whiteaves," *Transactions of the Royal Society of Edinburgh: Earth Sciences,* vol. 68(9) (1968-9) pp. 207-329; (p.81 Ichthyostega skeleton) after E. Jarvik; (p.96 *Thrinaxodon*) after Parrington 1946; (p.96 *Morganucodon* skull) "The skull of *Morganucodon,*" K.A. Kermack, F. Mussett and H.W. Rigney, *Zoological Journal of the Linnean Society* 1981; (p.117 Proterosuchus) after Greg Paul in Parrish 1972; (p.117 Euparkeria) (R.F. Ewer) *Phil. Trans. R. Soc. Lond.* B248 379 (1965); (p.117 *Saurouchus*) after J.F.

Bonaparte 1981; (p.117 *Ornithosuchus*) (A.D. Walker) *Phil. Trans. R. Soc. Lond.* B248 53 (1964); (p.117 Posture tree) *Vertebrate Palaeontology,* M. J. Benton, Harper Collins Academic 1990; (p.138 Cephalopod adaptation) "The Ammonoidea," M.R. House and J.R. Senior, Academic Press London 1981; (p. 140 Leg circumference/body mass graph) "Mechanics of posture and gait of some large dinosaurs," R. McN. Alexander, *Zoological Journal of the Linnean Society* 1985; (p.140/1 *Brachiosaurus*) after W. Janesch; (p.146 *Dromaeosaurus*) "The small Cretaceous dinosaur *Dromaeosaurus,*" E.S. Colbert and D.A. Russell, American Museum Novitiates 1969; (p.146/7 *Archaeopteryx*) after Yalden 1984; (p. 163 Hadrosaurs) "The evolution of cranial display structures in hadrosaurian dinosaurs," J.A. Hopson, *Paleobiology* 1975; (p.166 K-T diagrams) adapted from *Scientific American,* Oct. 1990 pp.45-56; (p.175 *Thrinaxodon*) *Mammal-like Reptiles and the Origin of Mammals,* T.S. Kemp, Academic Press London 1982; (p. 199 Skull, feet and tooth detail) *Mammal Evolution: An Illustrated Guide,* R.J.G. Savage and M.R. Long, Natural History Museum, London; (p.226 Molecular distance pyramid) *Human Evolution – an Illustrated Introduction,* R. Lewin, Blackwell Scientific Publications 1984 p.20; (p.226 Pre/post molecular distance chart) *Human Evolution – an Illustrated Introduction,* R. Lewin, Blackwell Scientific Publications

Second Edition p.37; (p.230 *Proconsul*) "Human Evolution – an illustrated introduction," R. Lewin, Blackwell Scientific Publications 1984 p.40; (p.230 *Pliopithecus*) after H. Zapfe 1960; (p.230 *Oreopithecus*) after H. Schafer 1960; (p.230 *Sivapithecus*) "Maxillofacial morphology of Miocene hominoids from Africa and Indo-Pakistan," S.C. Ward and D.R. Pilbeam, part of *New Interpretations of Ape and Human Ancestry* ed. R.L. Ciochon and R.S. Corruccini, Plenum (New York) 1983; (p.232 Genetic distance tree) adapted from *Scientific American* Nov. 1991, p. 75; (p.232 Mono-multi-regional theories) adapted from *Scientific American* Dec. 1990, p.100; (p.232 Migration map) adapted from "Genes, Peoples and Languages," by Luigi Luca Cavalli-Sforza, *Scientific American* Nov. 1991 p.75; (p.235 Laetoli site map) from Mary Leakey and John Harris: *Laetoli* (1988). Reprinted by permission of Oxford University Press; (p.251 Crural index graph) *Hominid Evolution and Community Ecology,* ed. R. Foley, Academic Press London 1984.